ELECTRONIC INSTRUMENTATION AND MEASUREMENTS

For Engineering Students

By

J.B. Gupta

B.Sc. (Elec. Engg.); C. Eng. (I);
F.I.E. (India); M.I.S.T.E.

S.K. KATARIA & SONS®

Publisher of Engineering Books

4885/109, Prakash Mahal, Dr. Subhash Bhargav Lane,
Opposite Delhi Medical Association, Daryaganj, New Delhi–110002 (INDIA)
Phone: +91-11-43551243 Mobile: +91-9871775858, 7678425209
e-mail: katariabook@gmail.com, katariabooks@yahoo.com
Website: www.skkatariaandsons.com
(e-books, subscriptions and bulk buying also available on our website)

Published by:

S.K. Kataria & Sons®
4885/109, Prakash Mahal, Dr. Subhash Bhargav Lane,
Opposite Delhi Medical Association, Daryaganj, New Delhi-110002 (INDIA)
Phone: +91-11-43551243 Mobile: +91-9871775858, 7678425209
e-mail: katariabook@gmail.com, katariabooks@yahoo.com
Website: www.skkatariaandsons.com

Edited By: **Rajeev Manglik**
B.E. (Elec.) Pb. University
and
Rohit Manglik
B.E. (Elec.) University of Roorkee

First Edition: 2005-06, Second Edition: 2006-07, Third Edition: 2009-10, Reprint: 2024

Price : ₹ **465/-**

Laser Typesetting by:
SARA ASSIGNMENTS

PREFACE TO THE THIRD EDITION

The author feels great pleasure in presenting the third edition of the book with marked improvements over the previous one. The book has been thoroughly revised and rewritten keeping in view the latest revised syllabus (effective from session 2009-10) prescribed by U.P. Technical University. The matter has been reorganised accordingly so as to facilitate the students as well as teachers. The subject matter has been presented in an easily comprehensible manner and logical sequence using simple language. A large number of worked out examples, theoretical questions, short answer type questions with answers and numerical problems included in the revised text are from latest question papers of U.P. Technical University.

The book has been divided into 5 units according to the revised syllabus (w.e.f. Session 2009-10) prescribed by U.P. Technical University and consists of 10 chapters in all.

The solution to illustrative examples covering almost all topics/subtopics of different chapters makes the entire presentation easy to follow. Selected exercises and numerical problems with answers given at the end of each chapter help the students in evaluating their understanding of the subject. Highlights given at the end of almost each chapter help the students in refreshing their knowledge while short answer type questions with answers help them in preparing for viva-voce and practical examinations.

It is earnestly hoped that with the extensive additions, revisions, modifications and improvement, the present edition will facilitate the students in preparing well for their regular examinations and prove more useful to the students than the earlier edition.

Every care has been taken to eliminate misprints, omissions and errors but it is too much to expect that no inaccuracy/error/misprint has crept in and the author would be grateful to the readers for bringing to his notice any of such errors/misprints/omissions they may come across while going through the book.

The author shall feel satisfied if the book meets the needs of the students for whom it is meant. Suggestions and criticism for the further improvement of the book shall be welcomed.

The author will like to appreciate the sincere efforts and contribution made by his sons Rajeev Manglik and Rohit Manglik in presentation of the new edition in this form.

The author will like to thank all the scientists, engineers, authors and publishers whose works and text have been the source of enlightenment, inspiration and guidance in presenting this new edition.

Lastly, but not the least, the author is thankful to his publisher, Sanjeev Kumar Kataria for bringing out the book in such a short time and for pricing it moderately despite heavy cost of paper and printing.

January 9, 2010 **J.B. Gupta**

PREFACE TO THE FIRST EDITION

This book is written to serve as a text, for students, those who are studying in Electronics, Electrical, Instrumentation Engineering and other similar disciplines.

It is not possible to name all the many people to whom I am indebted for suggestions, comments and help directly or indirectly. But I express my special thanks to **Mr. K.M. Soni,** Assistant Professor, Electronics and Instrumentation Department of Galgotia's College of Engg. and Technology, Greater Noida (U.P.) for putting major efforts in revising and editing this book. Keeping in mind the difficulties faced by the students through the lack of easily accessible reference and back-ground material and especially of the need for a handy single volume which would provide the complete syllabus. This book attempts to fill the gap.

In addition, I wish to thank **Mr. Sanjeev Kataria** of S.K. Kataria and Sons, Delhi for giving cooperation in the process of publishing and bringing out this book in a very short span. The members of Sara Assignments deserves my special thanks for his efficient and careful typing and drawing work.

Though every effort has been made to ensure that all the facts, block-diagrams and figures presented in this book are accurate, there may still be some short coming. I request the readers to point these out so that improvements can be made in the next edition.

August 10, 2005 **J.B. Gupta**

OTHER USEFUL BOOKS

J.B. Gupta	A Course in Electrical Technology
J.B. Gupta	Basic Electrical Engineering [A Course in Electrical Technology Vol. 1]
J.B. Gupta	Electrical Machines [A Course in Electrical Technology Vol. 2]
J.B. Gupta	Electronic Devices and Circuits [A Course in Electrical Technology Vol. 3]
J.B. Gupta	Theory and Performance of Electrical Machines [DC Machines, AC Machines and Polyphase Circuits]
J.B. Gupta	Electrical Machines-I [DC Machines]
J.B. Gupta	Electrical Machines-II [AC Machines]
J.B. Gupta	An Integrated Course in Electronics Engineering
J.B. Gupta	Fundamentals of Electrical and Electronics Engineering
J.B. Gupta	A Course in Electrical Power [Generation & Economic Considerations; Transmission & Distribution; Switchgear & Protection; Utilization of Electrical Power & Electric Traction]
J.B. Gupta	Generation and Economic Considerations [Electrical Power Systems Vol. 1]
J.B. Gupta	Transmission and Distribution [Electrical Power Systems Vol. 2]
J.B. Gupta	Switchgear and Protection [Electrical Power Systems Vol. 3]
J.B. Gupta	Electrical Power Systems [Vol. 1 of A Course in Electrical Power]
J.B. Gupta	Utilization of Electrical Power and Electric Traction [Vol. 2 of A Course in Electrical Power]
J.B. Gupta	A Course in Electrical and Electronic Measurements and Instrumentation
J.B. Gupta	Question Bank in Electrical Engineering
J.B. Gupta	An Integrated Course in Electrical Engineering [Objective Type Questions With Answers & Notes]
J.B. Gupta	Electrical Installation Estimating & Costing
J.B. Gupta	Elements of Electrical Engineering & Electronics
J.B. Gupta	Electrical and Electronics Engineering [Rajasthan Technical Univ.]
J.B. Gupta	Electrical Engineering [U.P. Technical Univ.]
J.B. Gupta	Electrical Science [G.G.S.I.P. Univ.]
J.B. Gupta	Basic Electrical and Electronics Engineering [Pb. Technical University]
J.B. Gupta	Basic Electrical Engineering [Rajiv Gandhi Technical Univ., MP, CVTU Durg]
J.B. Gupta	Electrical Technology [M.D. Univ. and Kurukshetra Univ.]
J.B. Gupta	Electrical Science [AMITY Univ.]
J.B. Gupta	Electronics Engineering [U.P. Technical Univ.]
J.B. Gupta	Electrical Engineering Materials and Semiconductor Devices [M.D. Univ.]
J.B. Gupta	Electrical Measurements and Measuring Instruments [M.D.U., P.T.U. and U.P.T.U.]
J.B. Gupta	Electronic Measurements and Instrumentation [3rd Semester Pb. Technical Univ.]
J.B. Gupta	Electronic Instrumentation and Measurements [4th Semester U.P. Technical Univ.]
J.B. Gupta	Basic Electronics [M.D. Univ.]
J.B. Gupta	Electronic Devices and Circuits [3rd Semester Pb Technical Univ.]
J.B. Gupta	Electrical Machines [4th Semester U.P. Technical Univ.]
J.B. Gupta	Electrical and Electronics Engineering Materials [4th Semester U.P. Technical Univ.]
J.B. Gupta	Semiconductor Materials and Devices
J.B. Gupta	Electromechanical Energy Conversion-I [4th Semester U.P. Technical Univ.]
J.B. Gupta	Energy Conversion
J.B. Gupta	Analog Electronics [4th Semester Pb Technical Univ.]
J.B. Gupta	Analog Integrated Circuits [5th Semester U.P. Technical Univ.]
J.B. Gupta	Pulse and Digital Switching Circuits [5th Semester Pb Technical Univ.]
J.B. Gupta	Linear Integrated Circuits [5th Semester Pb Technical Univ.]
J.B. Gupta	Electromechanical Energy Conversion-II [5th Semester U.P. Technical Univ.]
J.B. Gupta	Electronics Measurements and Instrumentation [3rd Semester M.D. Univ.]
J.B. Gupta	Electromechanical Energy Conversion [5th Semester M.D.U.]
J.B. Gupta	Basic Electronics [R.G.T.U. M.P.]
J.B. Gupta	Basic Electricity
J.B. Gupta	Analog Electronics [4th Semester M.D. Univ.]
J.B. Gupta	Analog Electronic Circuits [M.D. Univ.]
J.B. Gupta	Fundamentals of Electronic Devices [3rd Semester U.P. Technical Univ.]
J.B. Gupta	Elements of Electrical Engineering [Gujarat Technical Univ.]

CONTENTS

Chapters *Pages*

UNIT-I
UNITS, DIMENSIONS, AND STANDARDS, MEASUREMENT ERRORS AND PMMC INSTRUMENTS 1—78

1. Units, Dimensions and Standards 3–21

1.1 Nature of Units 3
1.2 Fundamental and Derived Quantities 3
1.3 SI Mechanical Units 4
1.4 Scientific Notation and Metric Prefixes 5
1.5 Electric Current 6
1.6 SI Temperature Scales 8
1.7 Other Unit Systems 9
1.8 Dimensions 11
1.9 Standards 16
1.10 International and Absolute Units 17
Exercises 18
Short Answer Type Questions With Answers 19
Problems 20

2. Measurement Errors 22–49

2.1 Introduction 22
2.2 Absolute Error, Relative Error and Limiting or Guarantee Errors 22
2.3 Resolution and Sensitivity 26
2.4 Accuracy and Precision 27
2.5 Significant Figures 28
2.6 Types of Errors 29
2.7 Measurement Error Combination 32
2.8 Statistical Analysis 38
2.9 Characteristics of Experimental Data 40
2.10 Normal Law of Error 41
2.11 Probable Error 43
2.12 Measurement Data Specification 44
Exercises 46
Short Answer Type Questions With Answers 47
Problems 48

3. PMMC Instruments 50–78

3.1 Introduction 50
3.2 Permanent Magnet Moving Coil (PMMC) Instrument 50
3.3 Galvanometer 58
3.4 DC Ammeter 63
3.5 DC Voltmeter 69
3.6 Series Ohmmeter 73
Exercises 75
Short Answer Type Questions With Answers 76
Problems 77

UNIT-II
ANALOG ELECTRONIC METERS AND DIGITAL METERS 79—150

4. Analog Electronic Meters 81–110

4.1 Introduction 81
4.2 Transistor Voltmeter Circuits 82
4.3 Operational Amplifier Voltmeter Circuits 91
4.4 AC Electronic Voltmeters 93
4.5 Current Measurement With Electronic Instruments 100
4.6 Analog Electronic Multimeter 101
4.7 Multimeter Probes 103
Exercises 106
Short Answer Type Questions With Answers 107
Problems 110

5. Digital Meters 111–150

5.1 Introduction 111
5.2 Digital-to-Analog and Analog-to-digital Conversion 112
5.3 Digital-to-Analog (D/A) Conversion 112
5.4 Analog-to-Digital (A/D) Conversion 118
5.5 ADC Resolution and Accuracy 123
5.6 Seven Segment Displays 125
5.7 Digital Voltmeters (DVMs) 126
5.8 Range Selection 133

5.9 Characteristics of Digital Meters 133
5.10 Digital Multimeters 135
5.11 Frequency Meter (Counter) 141

Exercises 147
Short Answer Type Questions With Answers 148
Problems 150

UNIT-III
RESISTANCE MEASUREMENTS AND INDUCTANCE AND CAPACITANCE MEASUREMENTS 151—220

6. Resistance Measurements 153–175

6.1 Introduction 153
6.2 Voltmeter and Ammeter Methods 154
6.3 Wheatstone Bridge Method 157
6.4 Low Resistance Measurement 168
6.5 Low-Resistance Measuring Instruments 172

Exercises 173
Short Answer Type Questions With Answers 174
Problems 175

7. Inductance and Capacitance Measurements 176–220

7.1 Introduction 176
7.2 AC Bridge Theory 176
7.3 Capacitance Bridges 180
7.4 Inductance Bridges 194
7.5 Q Meter 206

Exercises 215
Short Answer Type Questions With Answers 216
Problems 218

UNIT-IV
CATHODE RAY OSCILLOSCOPES 221—279

8. Cathode Ray Oscilloscopes 223–267

8.1 Introduction 223
8.2 Cathode Ray Tube 225
8.3 Focusing Device 228
8.4 Post Deflection Acceleration 233
8.5 Basic Controls 235
8.6 Oscilloscope Time Base 239
8.7 Multiple Trace Oscilloscope 244
8.8 Measurement of Voltage, Frequency and Phase 248
8.9 Oscilloscope Probes 258
8.10 Oscilloscope Specifications and Performance 261

Exercises 263
Short Answer Type Questions With Answers 264
Problems 267

9. Special Oscilloscopes 268–279

9.1 Introduction 268
9.2 Delayed Time-Base Oscilloscopes 268
9.3 Sampling Oscilloscopes 271
9.4 Digital Storage Oscilloscopes 273
9.5 Digital Storage Oscilloscope Applications 274

Exercises 278
Short Answer Type Questions With Answers 278

UNIT-V
INSTRUMENT CALIBRATION AND RECORDERS 281—300

10. Instrument Calibration and Recorders 283–300

10.1 Instrument Calibration 283
10.2 Comparison Methods 284
10.3 Digital Multimeters as Standard Instruments 287
10.4 Calibration Instruments 288
10.5 Recorders 290
10.6 Potentiometric Strip Chart Recorder 291
10.7 X-Y Recorder 292
10.8 Plotters 296

Exercises 298
Short Answer Type Questions With Answers 298

Model Test Papers 301–304
Examination Question Paper 305–309
Index 310–312

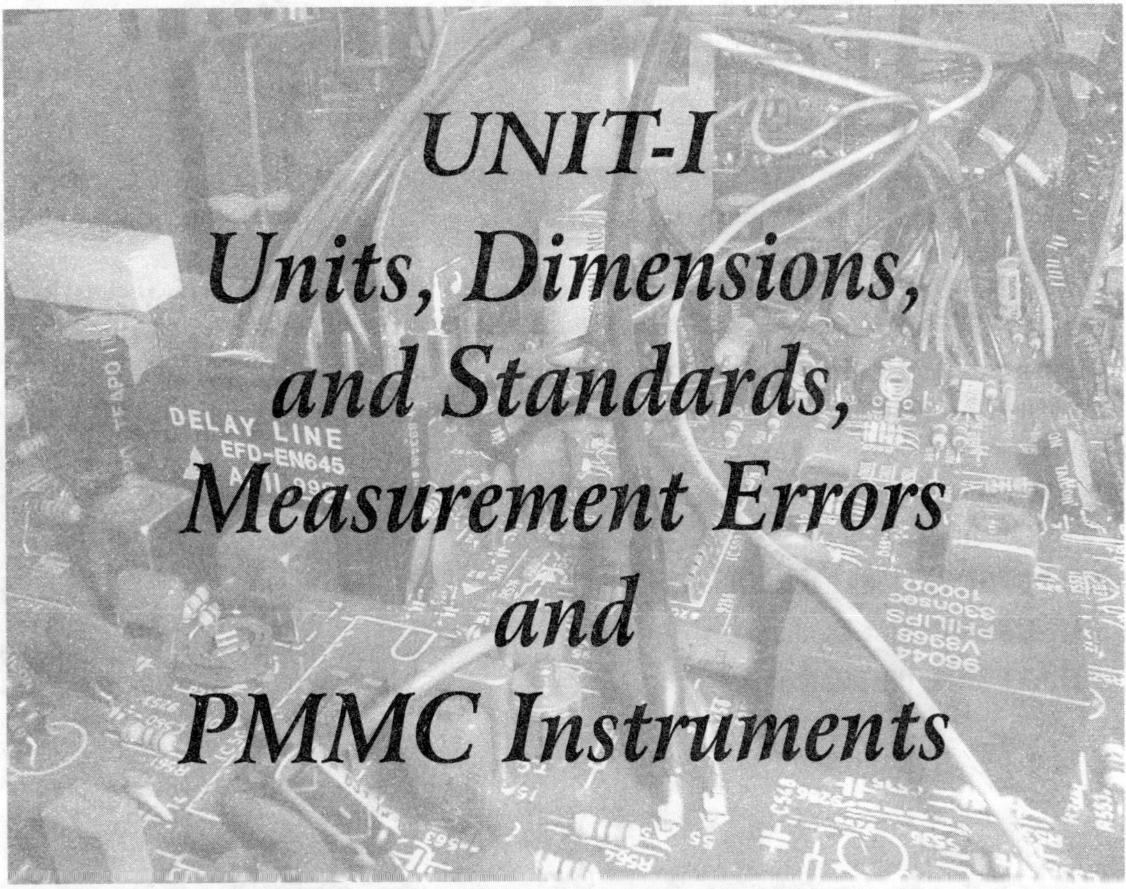

UNIT-I

Units, Dimensions, and Standards, Measurement Errors and PMMC Instruments

CONTENTS IN THIS UNIT

Units, Dimensions and Standards: Scientific notations and metric prefixes, SI electrical units, SI temperature scales, Other unit systems, Dimensions, Standards.

Measurement Errors: Gross errors, Systematic errors, Absolute errors, Accuracy, Precision, Resolution and Significant figures, Measurement error combinations, Basics of statistic analysis.

PMMC Instruments: PMMC instruments, Galvanometer, DC ammeter, DC voltmeter, Series ohmmeter.

CHAPTER 1: Units, Dimensions and Standards.

CHAPTER 2: Measurement Errors.

CHAPTER 3: PMMC Instruments.

1

Units, Dimensions and Standards

1.1 NATURE OF UNITS

In engineering, quantities of different kinds are involved, including physical, chemical, mechanical, thermal, electrical and physiological ones. In order to record or to compare magnitudes of quantities, some one magnitude of each kind must be taken as basis or unit. Each unit, either must be represented by a physical standard of some kind, natural or artificial or must be derived from a combination of other units represented by such standards.

So the number of times the unit occurs in any given amount of some quantity is the number of measure. For example when it is said that length A is 5 metres, it indicates that the metre is the unit of length and that the number of units of length is five. So the physical quantity, length is defined by the unit metre. Without unit the number of measure has no physical meaning.

In brief it can be said that

$$\text{Quantity or Measure} = \text{a number} \times \text{unit size}$$

1.2 FUNDAMENTAL AND DERIVED QUANTITIES

Experience shows that three basic ideas or units are sufficient to describe quantitatively all the phenomena encountered in mechanical science. These are length, mass and time and are called *fundamental units*. Since apart from mechanics these units are fundamental to most other physical quantities, they are called *primary fundamental units*. Measure of certain physical quantities in the thermal, electrical and illumination science are also represented by fundamental units and these are used only when these particular physical quantities are involved and so these units are called *auxiliary fundamental units*.

All other units are expressed in terms of the fundamental units and are, therefore, called *derived units*. Each derived unit originates from some physical law which defines that unit. For example speed of any body is proportional to the distance travelled and inversely proportional to the time interval during which the distance has been travelled, so speed = l/t. If metre is taken as the unit of length and second is taken as unit of time then the speed of the body covering distance of 4 metres in 2 second time interval will be equal to 4/2 *i.e.* 2 m/s. Now here the number of measures as well as the units are divided. So the derived unit of speed is metre/second (or m/s).

1.3 SI MECHANICAL UNITS

As already mentioned above, the units of length, mass and time are called the fundamental units. In SI system, the three fundamental units are: Metre (unit of length), kilogram (unit of mass) and second (unit of time), which are defined below:

1. Metre (m). The metre is the length equal to 1,650,763.73 wavelengths of the orange line in the spectrum of an internationally — specified krypton discharge lamp.

2. Kilogram (kg). The kilogram is the mass of a platinum-iridium cylinder preserved at the International Bureau of Weights and Measures at Sevres near Paris.

3. Second (s). The second is the interval occupied by 9,192,631,770 cycles of the radiation corresponding to the transition of the caesium-133 atom.

4. Force: The force is an external effort in the form of a push or pull which either changes or tends to change the state of rest or of uniform motion of a body along a straight line.

A force applied to a body equals the mass of the body multiplied by the acceleration produced in the body in the direction of force *i.e.*

$$F = ma \qquad \text{...(1.1)}$$

There are two types of unit of force: absolute unit and gravitational unit.

In SI system of units, *newton* (N) is the absolute unit of force and is defined as the force which produces an acceleration of 1 m/s^2 in a body of mass of 1 kg.

In SI system of unit, the gravitational unit of force is *kilogram weight (kg-wt)* or *kilogram force (kg-f)* which is defined as that force which produces an acceleration of 9.81 m/s^2 in a mass of 1 kg. Thus

$$1 \text{ kg-wt (or 1 kg-f)} = 1 \text{ kg} \times 9.81 \text{ m/s}^2 = 9.81 \text{ N}$$

5. Work is said to be done by or against a force, when its point of application moves in or opposite to the direction of the force and is measured by the product of the force and the displacement of the point of application in the direction of force.

i.e. Work done, W = Force [F] × distance [*d*]

The SI or MKS unit of work is the joule, which is defined as the work done when a force of one newton acts through a distance of one metre in the direction of the force. Hence if a force F acts through distance *d* in its own direction

$$W = F \text{ [newtons]} \times d \text{ [metres]} = Fd \text{ joules}$$

6. Power is defined as the rate of doing work or the amount of work done in unit time.

The MKS or SI unit of power is the joule/second or watt. In practice, the watt is often found to be inconveniently small and so a bigger unit, the *kilowatt* is frequently used.

1 kilowatt = 1,000 watts

The bigger unit of power, most commonly used in engineering practice (not at all in SI system) is horse power defined as below :

Metric Horse Power : It is the practical unit of power in MKS system (not in SI system) which according to ISI specifications is equal to 75 kgf-m of work done per second.

7. Energy is defined as the capacity of doing work. Its units are same as those of work, mentioned above.

If a body having mass *m*, in kg, is moving with velocity v, in metres/second,

$$\text{Kinetic energy} = \frac{1}{2} mv^2 \text{ joules}$$

If a body having mass m, in kg, is lifted vertically through height h, in metres, and if g is the gravitational acceleration, in metres/second^2 in that region, potential energy acquired by the body

$$= \text{Work done in lifting the body}$$
$$= mgh \text{ joules} = 9.81\ mh \text{ joules}$$

As already stated, in SI system the unit of energy of all forms is joule. Bigger unit of energy is mega joules (MJ) where 1 MJ = 10^6 J.

The thermal units of energy, calorie (gm-calorie) and kilo-calorie (kilogram-calorie), are defined below :

Calorie : It is the amount of heat required to raise the temperature of one gram of water through 1ºC.

$$1 \text{ calorie} = 4.18 \text{ J} \simeq 4.2 \text{ J}$$

Kilo-Calorie : It is the amount of heat required to raise the temperature of 1 kg of water through 1ºC.

$$1 \text{ k. calorie} = 1{,}000 \text{ calories} = 4{,}180 \text{ joules} \simeq 4{,}200 \text{ J}$$

1.4 SCIENTIFIC NOTATION AND METRIC PREFIXES

Scientific Notation: Multiples of 10 and fractional number of 10 may be written by raising plus and minus powers respectively on 10. For example

$1 = 10^0$	$1 = 10^0$
$10 = 10^1$	$0.1 = 10^{-1}$
$100 = 10^2$	$0.01 = 10^{-2}$
$1000 = 10^3$	$0.001 = 10^{-3}$
$10000 = 10^4$	$0.0001 = 10^{-4}$
$100000 = 10^5$	$0.00001 = 10^{-5}$
$1000000 = 10^6$	$0.000001 = 10^{-6}$

On this basis, we can write any number in terms of power of 10 in the following manner:

$$500 = 5 \times 100 = 5 \times 10^2$$
$$22000 = 2.2 \times 10000 = 2.2 \times 10^4$$
$$6250000 = 6.25 \times 1000000 = 6.25 \times 10^6$$
$$0.00525 = 5.25 \times 0.001 = 5.25 \times 10^{-3}$$
$$0.00000415 = 4.15 \times 0.000001 = 4.15 \times 10^{-6}$$

Thus, any number, large or small, can be expressed in terms of plus or minus powers of 10 and multiplied by a number which is greater than 1 but less than 10 (in above examples these numbers are 5, 2.2, 6.25, 5.25, 4.15).

In SI system of units, spaces are used instead of commas when writing large numbers. Four-numeral numbers are an exception. One thousand is written as 1000, while ten thousand is 10 000.

Metric Prefixes. The magnitudes of very large and very small quantities are written compactly by using prefixes for powers of 10. Thus, 1000 Ω can be written as 1 *kilohm*. Here *kilo* is the prefix that represents 1000, and *k* is the symbol for *kilo*. Similarly, 1×10^{-3} A can be written as 1 milliampere, or 1 mA.

Recommended prefixes for formation of multiples and submultiples of units are given below:

Multiple	*Prefix*	*Symbol*	*Fraction*	*Prefix*	*Symbol*
10	deca	*da**	0.1	deci	*d**
100	hecto	*h**	0.01	centi	*c**
1000	kilo	k	0.001	milli	*m*
1000 000	mega	M	0.000 001	micro	μ
1000 000 000	giga	G	0.000 000 001	nano	*n*
1000 000 000 000	tera	T	0.000 000 000 001	pico	*p*
			0.000 000 000 000 001	fento	*f*
			0.000 000 000 000 000 001	atto	*a*

* Should be restricted as much as possible.

Engineering Notation: As already mentioned, 1 kΩ = 1×10^3 Ω and 1 mA = 1×10^{-3} A. Similarly, 1×10^6 Ω may be expressed as 1 MΩ, and 1×10^{-6} A can be written as 1 μA. These quantities, and most of the metric prefixes in Table 1.1 involve multiples of 10^3 or 10^{-3}. Quantities using 10^3 or 10^{-3} are said to be written in *engineering notations.* The terms used in engineering notations are giga (G) meaning 10^9, mega (M) meaning 10^6, kilo (k) meaning 10^3, milli (m) meaning 10^{-3}, micro (μ) meaning 10^{-6}, nano (n) meaning 10^{-9} and pico (p) meaning 10^{-12}. For electrical calculations, engineering notation is more convenient than ordinary scientific notation.

1.5 ELECTRIC CURRENT

Electric current may be defined as the time rate of net motion of an electric charge across a cross-sectional boundary (Fig. 1.1). A random motion of electrons in a metal does not constitute a current unless there is a net transfer of charge with time

i.e. electric current, i = Rate of transfer of electric charge

$$= \frac{\text{Quantity of electric charge transferred during a given time duration}}{\text{Time duration}} = \frac{dQ}{dt} \quad ...(1.2)$$

Coulomb is the practical as well as SI unit for measurement of electric charge. One coulomb is approximately equal to 624×10^{16} electrons.

Since current is the rate of flow of electric charge through a conductor and coulomb is the unit of electric charge, the current may be specified in coulombs per second.

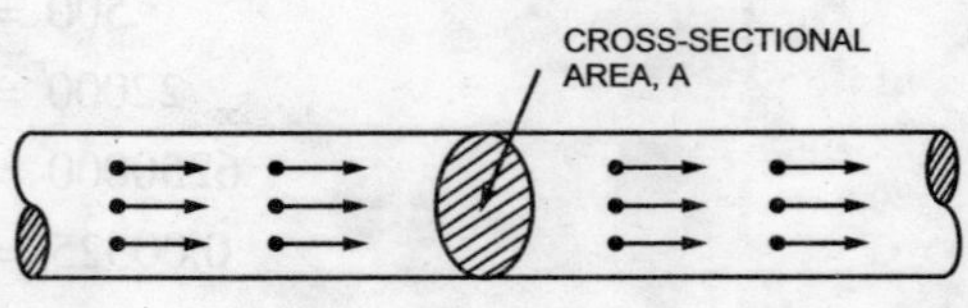

Fig. 1.1

In practice the term coulomb per second is seldom used, a shorter term, *ampere* is used instead.

The coulomb (C) was originally selected as the fundamental electrical unit and all other electrical units were derived in terms of unit of charge. However, as it is much easier to measure current accurately than it is to measure charge, so the unit of current is taken as the fundamental electrical unit in SI system of units. Thus, the coulomb is a derived unit and is defined in terms of the unit of electric current.

Ampere, the SI unit of current, is defined as that current which, if maintained in two straight parallel conductors of infinite length, of negligible circular cross-section and placed one metre apart in vacuum, would produce between these conductors a force of 2×10^{-7} newtons per metre length.

The **coulomb** is defined as that charge which passes a given point in a conductor each second, when a current of 1 A flows.

From above definitions, the coulomb could be termed as *ampere-second*. Conversely, the ampere can be expressed as *coulomb per second*.

$$i.e.\ \text{Amperes} = \frac{\text{Coulombs}}{\text{Seconds}}$$

As already mentioned, 1 coulomb is equal to the total charge carried by 6.24×10^{18} electrons. Therefore, the charge carried by one electron is

$$Q = \frac{1}{6.24 \times 10^{18}}\ C = 1.602 \times 10^{-19}\ C$$

EMF, Potential Difference and Volt: *Electromotive force* (emf) is the force that causes an electric current to flow in an electric circuit while the *potential difference* (pd) between two points in an electric circuit is that difference in their electrical state which tends to cause flow of electric current between them.

Volt is a unit of electromotive force as well as potential difference in practical as well as in SI system of units.

The volt is defined as that potential difference between two points of a conductor carrying a current of one ampere when the power dissipated between these points is equal to one watt.

For a discussion of energies involved in electronic devices the joule is too large a unit. A unit of work or energy, called the *electron volt* (eV) is given as

$$1\ eV = 1.602 \times 10^{-19}\ J$$

The name *electron volt* arises from the fact that, if an electron falls through a potential of 1 V, its kinetic energy will increase by the decrease in potential energy, or by one electron volt.

$$i.e.\ 1\ eV = 1.602 \times 10^{-19}\ C \times 1\ V = 1.602 \times 10^{-19}\ J$$

The abbreviations MeV and BeV are used to designate million and billion electron volts respectively.

Resistance and Conductance. Resistance may be defined as that property of a substance which opposes (or restricts) the flow of an electric current (or electrons) through it.

The practical as well as mks (or SI) unit of resistance is ohm (Ω), which is defined as that resistance between two points of a conductor when a potential difference of one volt, applied between these points, produces in this conductor a current of one ampere, the conductor not being a source of any emf.

The reciprocal of resistance *i.e.* 1/R is called the *conductance* and is denoted by English letter G. It is defined as the inducement offered by the conductor to the flow of current and is measured in siemens (S). Earlier, the unit of conductance was mho (℧).

Magnetic Flux and Flux Density. Magnetic flux is the total number of lines of force comprising the magnetic field. In SI unit system unit north pole *i.e.* north pole of strength of one weber is supposed to radiate out the flux of one weber. It is represented by symbol ϕ and is measured in webers.

∴ Flux emanated from north pole of strength m webers = m Wb.

The **weber (Wb)** is the SI unit of magnetic flux and is defined as the magnetic flux which, linking a single turn-coil, develops an emf of 1 V when the flux is reduced to zero at a constant rate in 1 second.

Magnetic flux density is defined as the magnetic flux passing per unit area through any material through a plane at right angles to the direction of flux. This is also known as *magnetic induction.* It is represented by English letter B and is measured in Tesla (weber/m^2)

$$i.e.\ B = \frac{\phi}{a} \quad ...(1.3)$$

Magnetic flux density is a scalar quantity.

The tesla (T), the SI unit of magnetic flux density, is defined as the flux density in a magnetic field when 1 weber of flux occurs in a plane of 1 m^2, that is, the tesla can be expressed as 1 Wb/m^2.

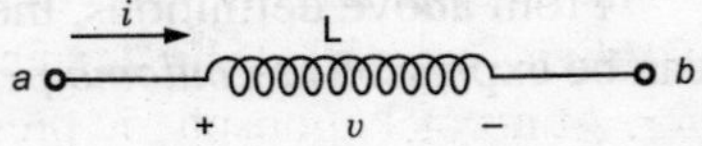

Fig. 1.2 *Schematic Representation of Inductance*

Inductance. This circuit element is used to represent the energy stored in a magnetic field and is defined by the relation

$$v = L\frac{di}{dt}$$

The above equation describes a situation in which the voltage across the element is proportional to the time rate of change of current through it. The constant of proportionality L is the *self inductance* or simply the *inductance* of the element and is measured in henrys (abbreviated H).

The inductance of a circuit is 1 **henry**, when an emf of one volt is induced by the current changing at the rate of 1 ampere per second.

Capacitance. Capacitor is a two-terminal element that has the capability of charge storage and, consequently, energy storage. The energy stored can be fully retrieved.

When a charge of Q coulombs is given to a conductor and if its potential rises from 0 to V volts then the capacitance of the conductor will be given by $C = \frac{Q}{V}$.

In SI system the capacitance is measured in farads.

If one coulomb of charge is required to raise the potential of the conductor through one volt, then the conductor is said to be of one farad capacity.

By definition one farad is equal to one coulomb per volt.

Farad, the unit of capacitance is very large, so microfarad or micro-microfarad is usually used.

$$1\ \mu F = 10^{-6}\ F \text{ and } 1\ \mu\mu F = 10^{-12}\ F$$

Power and Energy. The unit of work done and of energy expended is joule. It is equal to the energy expended in passing 1 coulomb of charge through a resistance of 1 ohm *i.e.* the energy expended in passing one ampere current for 1 second through a resistance of one ohm is taken as one joule. It may also be expressed as 1 watt-second *i.e.* one watt of power consumed for one second.

i.e. 1 joule = 1 watt-second

The unit of energy, joule or watt-second is too small for practical purposes, so a bigger unit Mega joule (MJ) or kilowatt-hour (kWh) is used in electrical engineering.

$$1 \text{ kWh} = 1{,}000 \text{ watt-hours} = 1{,}000 \times 3{,}600 \text{ watt-seconds or joules} = 3.6 \text{ MJ}$$

Watt **:** It is defined as the power expended when there is an unvarying current of one ampere between two points having a potential difference of one volt. As already stated the bigger unit of power is kW or Mega watt.

$$1 \text{ kW} = 1{,}000 \text{ watts}$$
$$1 \text{ MW} = 1{,}000 \text{ kW} = 1 \times 10^6 \text{ watts}$$

1.6 SI TEMPERATURE SCALES

Kelvin is the basic unit of temperature in SI system of units and is defined as the fraction $\frac{1}{273.15}$ of the thermodynamic temperature of the tripple point of water.

The **Celsius** and the **Kelvin** scales are the two SI temperature scales. The Celsius scale has 100 equal divisions, called the degrees, between the freezing temperature and the boiling temperature of water. At normal atmospheric pressure, water freezes at zero degree Celsius (0ºC) and boils at 100ºC.

The Kelvin temperature scale is also called the *absolute scale* and commences at absolute zero of temperature corresponding to – 273.15ºC. Thus 0ºC corresponds to 273.15 K, and 100ºC corresponds to 373.15 K. A temperature difference of 1 K is the same as a temperature difference of 1ºC.

1.7 OTHER UNIT SYSTEMS

Besides SI system of units, there are four general systems of mechanical units used in engineering, Foot-pound second system, the English system, Centimetre-gram-second system and Metre-kilogram-second system. In FPS system the units of fundamental quantities, length, mass and time are foot, pound and second respectively. In English system the corresponding units are foot, slug and second. In CGS system the units of fundamental quantities length, mass and time are centimetre, gram and second respectively. In MKS system the respective fundamental units of length, mass and time are metre, kilogram and second, defined below.

The Metre. The metre is a distance at 15ºC between two marks on a platinum bar kept at International Bureau of Weights and Measures at Sevres in France.

The Kilogram: The kilogram is the mass of a block of platinum kept at Sevres.

The Second: The second is the mean solar second *i.e.* $\frac{1}{86,400}$ of a mean solar day.

The relation between some of the principal mechanical units in MKS, CGS and FPS systems are given below.

Quantity	*MKS Units*	*CGS Equivalent Units*	*FPS Equivalent Units*
Length	1 metre	100 cm	3.281 ft
Mass	1 kg	1,000 gm	2.205 lb
Velocity	1 metre/second	100 cm/second	3.281 ft/second
Acceleration	1 metre/second^2	100 cm/second^2	3.281 ft/second^2
Force	1 Newton	100,000 dynes	7.23 poundals
Energy or work	Joule or N-m	10^7 ergs	23.76 ft-poundals
Power	1 Joule/second	10^7 ergs/second	23.76 ft-poundals/second

Whereas, three basic concepts and three fundamental units are sufficient for description and measurement in mechanical science, experience shows that, in electrical science, four concepts or dimensions and four arbitrarily defined fundamental units are necessary to obtain a complete system of dimensions and units. At least one of these four units must be electrical in character.

The various electrical unit systems used are:

1. *The CGS Electrostatic System of Units (CGS ESU).* It is an absolute system based on the centimetre, gram and second as the fundamental mechanical units and permittivity (ε) of the media as fourth fundamental unit. The unit of permittivity (ε) is of such a size that the measure number of permittivity for free space is unity, *i.e.* $\varepsilon_0 = 1$. These units are commonly designated by using the prefix *stat* with the name of corresponding practical units *e.g. stat* volt.

2. *The CGS Electromagnetic System of Units (CGS EMU).* It is another absolute system based on the centimetre, gram and second as the fundamental mechanical units and permeability (μ) of the media as fourth fundamental unit. The unit of permeability is of such a size that the measure number of permeability for free space is unity *i.e.* $\mu_0 = 1$. These units are commonly designated by using *ab* with the name of the corresponding practical units *e.g. ab* volts.

The electromagnetic system is more convenient from the point of view of most electrical measurements, and is, therefore, much more generally used than the electrostatic system.

3. *The Practical System of Units.* Since neither CGS electrostatic units nor CGS electromagnetic units were of convenient size for the purpose of practical work, a practical system of units in which unit current (ampere) = $\frac{1}{10}$ cgs emu of current and unit of potential difference (volt) = 10^8 cgs emu of potential difference was employed. Though the practical system is sufficiently extensive to deal with the vast proportion of everyday calculations of electrical engineering, but it has got following deficiencies:

(*i*) The absence of recognised units for magnetic and electrostatic relations, therefore, some or all the magnetic units used to be expressed in cgs emu system.

(*ii*) The practical units are related by conversion factors to those systems of units (cgs esu and cgs emu) in which theoretical relations of the science have been commonly developed in the past.

4. *The MKS System of Units.* It is an absolute system based on the metre, kg and second as fundamental mechanical units and permeability of the medium as fourth unit. The unit of permeability is of such a size that measure number of permeability for free space is 10^{-7}.

Its great advantage is that its units are identical with those of practical system and are the same whether built up from the electromagnetic or electrostatic theory.

The value of permittivity for free space, ε_o in mks (unrationalised) system can be determined from the relation

$$\frac{1}{\sqrt{\mu_0 \, \varepsilon_0}} = \text{C, velocity of light}$$

$$\text{or } \varepsilon_0 = \frac{1}{\mu_0 C^2} = \frac{1}{10^{-7} \times (2.998 \times 10^8)^2} = 1.113 \times 10^{-10}$$

5. *Rationalised MKS System of Units.* The above unrationalised system has been further rationalised by assigning different values to μ_0 and ε_0. Rationalisation means the removal of quantity π from the places, where its appearance is unnatural, irrational and geometrically unjustified to places where it properly belongs. By doing so, most of the equations and formulae in electrical engineering have become simple and logical. In RMKS system, the permeability of free space, $\mu_0 = 4\,\pi \times 10^{-7}$ H/metre and permittivity of free space $\varepsilon_0 = 8.854 \times 10^{-12}$ F/metre.

The change in the values of μ_0 and ε_0 does not change the values of electrical units but does change the units of mmf, magnetizing force etc.

It was Prof. G. Giorgi, who first of all, suggested this system in 1901. After the name of Prof. Giorgi, it is also known as *Giorgi system of units.*

The important characteristics of the system are given below:

1. A single set of units covers all electrical and magnetic quantities and is applicable to both electromagnetic and electrostatic effects.
2. The values of units used in circuit theory are identical with the units of 'practical' system described above.
3. A unit magnetic pole emits a unit magnetic flux and not 4π units of flux.

4. In cgs systems described above, the numerical values of magnetising force H and flux density B are equal in vacuum, but in RMKS system the values of B and H in vacuum are not equal but are linked through a constant μ_0.

$$B = \mu_0 H$$

where μ_0 is named as permeability of free space and its value has been apparently chosen as $4\pi \times 10^{-7}$.

In magnetic materials $B = \mu_0 \mu_r H$, where μ_r is the relative permeability of the material and is equal to permeability value in cgs system.

5. Similarly in a dielectric flux density, $D = \varepsilon_0 \varepsilon_r E$ (field strength) where ε_0 and ε_r are the permittivity of free space and relative permittivity of the medium.
6. The relation $\left(1/\sqrt{\mu_0 \varepsilon_0}\right)$ = Velocity of light holds good numerically as well as dimensionally.

6. The Rationalized MKSA System of Units. Since these properties (permeability and permittivity of free space) although eminently suitable as physical reference standards, are less easily comprehended than such concept as ampere, charge, potential difference or resistance, it has been agreed to make, ampere defined from its magnetic effect as fourth fundamental quantity and the system developed so *i.e.* employing metre, kilogram, second and ampere as fundamental units is called as RMKSA system.

1.8 DIMENSIONS

Disregarding the problem of measurement and all concepts of magnitude, it is obvious that every quantity has a quality which distinguishes it from all other quantities. This unique quantity is called *dimension.* The dimension is written in a characteristic notation, as for example [L] for length, [M] for mass etc.

A derived unit is recognised by its dimensions and can be defined as the complete algebraic formula for the derived unit. The dimensional symbols for fundamental units of length, mass and time are L, M and T respectively. The various powers of the fundamental units represent the dimensions of any derived unit. For example the dimensional symbol for the derived unit of speed is LT^{-1} and that for acceleration is LT^{-2}. These dimensional formulae of the derived units are very useful for converting units from one system to another.

For convenience, some derived units have been given new names such as the derived unit of force in the SI system is called the newton (N) instead of dimensionally correct name kg-m/s^2.

All other mechanical quantities can be expressed in terms of three fundamental quantities *i.e.* length, mass and time. A few of them are given below:

$$\text{Thus velocity, } [V] = \frac{[\text{Length}]}{[\text{Time}]} = \frac{[L]}{[T]} = [LT^{-1}]$$

$$\text{Acceleration, } [\alpha] = \frac{[\text{Velocity}]}{[\text{Time}]} = \frac{\left[LT^{-1}\right]}{[T]} = [LT^{-2}]$$

$$\text{Force, } [F] = [\text{Mass}] \times [\text{Acceleration}] = [M][LT^{-2}] = [MLT^{-2}]$$

$$\text{Work, } [W] = [\text{Force}] \times [\text{Length}] = [MLT^{-2}][L] = [ML^2T^{-2}]$$

$$\text{Power, } [P] = \frac{[\text{Work}]}{[\text{Time}]} = \frac{\left[ML^2T^{-2}\right]}{[T]} = [ML^2T^{-3}]$$

$$[\text{Surface Tension}] = \frac{[\text{Force}]}{[\text{Length}]} = \frac{[MLT^{-2}]}{[L]} = [MT^{-2}]$$

Such expressions indicate the nature of derived quantities in terms of fundamental dimensions [L], [M] and [T] and are known as *dimensional equations.*

The dimensional equations help (*i*) in conversion from one system of units to another one (*ii*) in derivation of equations for physical quantities and (*iii*) in checking the accuracy of an equation.

1.8.1. Dimensions of Electrical and Magnetic Quantities. As already stated, all mechanical quantities can be expressed in terms of the three quantities mass, length and time. But for expressing electrical quantities dimensionally, the choice of fourth dimensional concept is necessary. In establishing our system of definitions, Coulomb's law of force between charges made use of space permittivity as connecting link between the mechanical and electrical quantities with which we were dealing. Similarly, in Ampere's law of force between currents, space permeability was used as the connecting link. Although such a connecting link between electrical and mechanical quantities is necessary, we are by no means restricted to either space permittivity or permeability as fourth dimensional concept. A number of other choices are possible, and some are certainly more convenient for the dimensional checking of physical equations than ε_0 or μ_0 since either of these choice requires the use of fractional exponents. If for example, charge is chosen as fourth dimensional unit with mass, length, and time, the use of fractional exponents is avoided altogether in expressing the dimensions of any electrical quantity. The system of units employing mass, length, time and charge as four fundamental concepts is termed as L, M, T, Q system of parameters. The system of units employing mass, length, time and current as four fundamental concepts is termed as SI system as already mentioned.

The dimensions of various electrical and magnetic quantities can be derived from the known relationship between them, as given below:

(*i*) Current [I]

(*ii*) Charge

Since quantity of electricity = Current × Time, [Q] = [TI]

(*iii*) Potential Difference

$$\text{Since Potential Difference} = \frac{\text{Work done}}{\text{Quantity of electricity}},$$

$$[V] = \frac{[E]}{[Q]} = \frac{[ML^2T^{-2}]}{[TI]} = [ML^2T^{-3}I^{-1}]$$

(*iv*) Resistance Since by Ohm's law

$$\text{Resistance} = \frac{\text{Potential difference}}{\text{Current}}$$

$$[R] = \frac{[V]}{[I]} = \frac{[ML^2T^{-3}I^{-1}]}{[I]} = [ML^2T^{-3}I^{-2}]$$

(*v*) Magnetic flux

Since emf = Rate of change of flux

$$[V] = \frac{[\phi]}{[T]}$$

$$\text{or } [\phi] = [V][T] = [ML^2T^{-3}I^{-1}][T] = [ML^2T^{-2}I^{-1}]$$

***(vi)* Flux density**

$$\text{Since flux density} = \frac{\text{Magnetic flux}}{\text{Area}}$$

$$[B] = \frac{[\phi]}{[A]} = \frac{[ML^2T^{-2}I^{-1}]}{[L^2]} = [MT^{-2}I^{-1}]$$

***(vii)* Magnetomotive Force** Dimensions of mmf are identical with those of current since mmf can be defined by current-turns

$$[MMF] = [I]$$

***(viii)* Magnetising Force**

$$\text{Since magnetising force} = \frac{\text{MMF}}{\text{Length}}$$

$$[H] = \frac{[MMF]}{[L]} = [IL^{-1}]$$

***(ix)* Reluctance**

$$\text{Since reluctance} = \frac{\text{MMF}}{\text{flux}}$$

$$[\text{Reluctance}] = \frac{[I]}{[ML^2T^{-2}I^{-1}]} = [M^{-1}L^{-2}T^2I^2]$$

***(x)* Inductance**

Since emf = Inductance × rate of change of current,

$$[\text{Inductance}] = \frac{[ML^2T^{-3}I^{-1}]}{[IT^{-1}]} = [ML^2T^{-2}I^{-2}]$$

***(xi)* Electric Flux** Since electric flux is equal to the charge producing it,

$$[\psi] = [Q] = [TI]$$

***(xii)* Electric Flux Density**

$$\text{Since electric flux density, } D = \frac{\text{Electric flux}}{\text{Area}}$$

$$[D] = \frac{[\psi]}{[A]} = \frac{[TI]}{[L^2]} = [L^{-2}TI]$$

***(xiii)* Electric Field Strength**

Since electric field strength = Potential gradient,

$$[E] = \frac{[V]}{[L]} = \frac{[ML^2T^{-3}I^{-1}]}{[L]} = [MLT^{-3}I^{-1}]$$

***(xiv)* Capacitance**

$$\text{Since Capacitance} = \frac{\text{Charge}}{\text{Potential}}$$

$$[C] = \frac{[Q]}{[V]} = \frac{[TI]}{[ML^2T^{-3}I^{-1}]} = [M^{-1}L^{-2}T^4I^2]$$

(*xv*) Energy Stored in Capacitor

$$\text{Since energy stored in capacitor} = \frac{1}{2}QV$$

$$[\text{Energy stored}] = [TI]\,[ML^2T^{-3}I^{-1}] = [ML^2T^{-2}]$$

(*xvi*) Permittivity

$$\text{Since by Coulomb's law}\quad F = \frac{Q_1Q_2}{\varepsilon d^2}$$

$$[\varepsilon] = \frac{[Q^2]}{[F][L^2]} = \frac{[T^2I^2]}{[MLT^{-2}][L^2]} = [M^{-1}L^{-3}T^4I^2]$$

(*xvii*) Permeability

$$\text{Since permeability, } \mu = \frac{\text{Flux density}}{\text{Magnetising force}}$$

$$[\mu] = \frac{[B]}{[H]} = \frac{[MT^{-2}I^{-1}]}{[L^{-1}I]} = [MLT^{-2}I^{-2}]$$

(*xviii*) Pole Strength

$$\text{Since by Coulomb's law } F = \frac{m_1m_2}{\mu d^2}$$

$$[m^2] = [F][\mu][d^2] = [MLT^{-2}][MLT^{-2}I^{-2}][L^2] = [M^2L^4T^{-4}I^{-2}]$$

$$\text{or } [m] = [ML^2T^{-2}I^{-1}]$$

Similarly dimensions of other quantities can be obtained.

Example 1.1. If we adopt new units of voltage and current such that one new unit of voltage = 100 volts and one new unit of current = 10 amperes; how would the units of resistance, inductance and capacitance be affected ? *[Pb. Univ. Elec. Measurements & Measuring Instruments I, April 1975]*

Solution: Let the new unit of resistance be R′, when the new unit of voltage is 100 times and that of current 10 times.

$$\text{Since resistance} = \frac{\text{Voltage}}{\text{Current}}$$

$$R' = \frac{V'}{I'} = \frac{100\,V}{10\,I} = 10\,R$$

i.e. New unit of resistance will be equal to 10 ohms **Ans.**

$$\textit{Similarly for capacitance, } C' = \frac{\text{Charge}}{\text{Voltage}} = \frac{TI'}{V'} = \frac{10\,I}{100\,V} \times T = \frac{IT}{10\,V} = \frac{C}{10}$$

i.e. New unit of capacitance will be 0.1 Farad **Ans.**

$$\textit{and for Inductance, } L' = \frac{V'}{\frac{dI'}{dt}} = \frac{100\,V}{10\frac{dI}{dt}} = 10\frac{V}{\frac{dI}{dt}} = 10\,L$$

i.e. New unit of inductance will be 10 Henrys **Ans.**

Example 1.2. Show that dimensions of product of voltage and current are dimensions of power.

Solution:

$$[V] = [ML^2T^{-3}I^{-1}]$$

$$[I] = [I]$$

$$[\text{Product of voltage and current}] = [V][I] = [ML^2T^{-3}I^{-1}][I] = [ML^2T^{-3}] = [\text{Power}]$$ **Proved.**

Example 1.3. Show that $\omega^2 LC$ is non-dimensional, ω being the angular frequency of the applied voltage.

Solution:

$$[\omega] = [T^{-1}]$$
$$[L] = [ML^2T^{-2}I^{-2}]$$
$$[C] = [M^{-1}L^{-2}T^4I^2]$$
$$[\omega^2 LC] = [\omega^2][L][C] = [T^{-2}][ML^2T^{-2}I^{-2}][M^{-1}L^{-2}T^4I^2] = [M^0L^0T^0I^0]$$

Hence $\omega^2 LC$ is non-dimensional quantity.

Example 1.4. Noting that the term $\frac{L}{R}$ is the time constant, derive the dimensions of R from those of L.

Solution: Since $\frac{L}{R} = T$,

$$[R] = \frac{[L]}{[T]} = \frac{[ML^2T^{-2}I^{-2}]}{[T]} = [ML^2T^{-3}I^{-2}]$$ **Ans.**

Example 1.5. Eddy current loss W in watts per unit length of a conductor of diameter d, resistivity ρ can be expressed as:

$$\mathbf{W} \propto f^a B_m^b d^c \rho^g$$

where f is the frequency in Hz, B_m is the maximum flux density, and a, b, c, g are constants. Use RMKSA system of dimensional analysis to find the values of a, b, c, g.

Solution: Let $W = K f^a B_m^b d^c \rho^g$ ...(*i*) where K is a dimensionless constant number

Dimensionally, we can write

$$[V] = \frac{[F][d]}{[Q]} = \frac{[MLT^{-2}][L]}{[TI]} = [ML^2T^{-3}I^{-1}]$$

$$[W] = \frac{[V][I]}{[L]} = \frac{[ML^2T^{-3}I^{-1}][I]}{[L]} = [MLT^{-3}]$$

$$[f] = [T^{-1}]$$

$$[B_m] = \frac{[\phi]}{[a]} = \frac{[V][T]}{[a]} = \frac{[ML^2T^{-3}I^{-1}][T]}{L^2} = [MT^{-2}I^{-1}]$$

$$[d] = [L]$$

$$[\rho] = \frac{[R][a]}{[L]} = \frac{\left[\frac{V}{I}\right][L^2]}{[L]} = \frac{[ML^2T^{-3}I^{-1}][L^2]}{[I][L]} = [ML^3T^{-3}I^{-2}]$$

Substituting the dimensions of quantities involved in the given equation we have

$$[MLT^{-3}] = [T^{-1}]^a\,[MT^{-2}I^{-1}]^b\,[L]^c\,[ML^3T^{-3}I^{-2}]^g$$

$$\text{or } [MLT^{-3}] = [M^{b+g}L^{c+3g}T^{-a-2b-3g}I^{-b-2g}]$$

Equating corresponding indices we have

For M $\quad b + g = 1$ ...(*ii*)

For L $\quad c + 3g = 1$ ...(*iii*)

For T $\quad -a - 2b - 3g = -3$ ...(*iv*)

and for I $\quad -b - 2g = 0$ ...(*v*)

Adding Eqs. (*ii*) and (*v*), we get

$$-g = 1$$

$$\text{or } g = -1 \textbf{ Ans.}$$

Substituting $g = -1$ in Eqs. (*ii*) and (*iii*) we get

$$b - 1 = 1$$

$$\text{or } b = 2 \textbf{ Ans.}$$

$$\text{and } c - 3 = 1$$

$$\text{or } c = 4 \textbf{ Ans.}$$

Substituting, $g = -1$ and $b = 2$ in Eq. (*iv*), we get

$$-a - 2 \times 2 - 3 \times (-1) = -3$$

$$\text{or } a = 2 \textbf{ Ans.}$$

$$\text{Hence} \quad W \propto f^2 B_m^2 d^4 \rho^{-1} \propto \frac{f^2 B_m^2 d^4}{\rho}$$

Example 1.6. The expression for mean torque T of an electrodynamic wattmeter is written in the form

$$T \propto M^p E^q Z^t$$

where M is mutual inductance between fixed and moving coils, E-applied voltage, Z-impedance of coil. From first principles determine the values of constants *p, q, t*. *[Pb. Univ. Elec. Measurements-I June 1993]*

Solution: Let $T = KM^pE^qZ^t$ ...(*i*)

where K is a dimensionless constant number.

Dimensionally we can write

$$[T] = [ML^2T^{-2}]$$

$$[M] = [ML^2T^{-2}I^{-2}]$$

$$[E] = [ML^2T^{-3}I^{-1}]$$

$$[Z] = [ML^2T^{-3}I^{-2}]$$

Substituting the dimensions of quantities involved in Eq. (*i*) we get

$$[ML^2T^{-2}] = [ML^2T^{-2}I^{-2}]^p [ML^2T^{-3}I^{-1}]^q [ML^2T^{-3}I^{-2}]^t$$

$$\text{or} \quad [ML^2T^{-2}] = [M^{p+q+t} L^{2p+2q+2t} T^{-2p-3q-3t} I^{-2p-q-2t}]$$

Equating the corresponding indices, we have

For M $\quad p+q+t = 1$ ...(*ii*)

For T $\quad -2p-3q-3t = -2$ ...(*iii*)

For I $\quad -2p-q-2t = 0$ ...(*iv*)

Solving Eqs. (*ii*), (*iii*) and (*iv*) we get

$$p = 1 \text{ ; } q = 2 \text{ and } t = -2 \textbf{ Ans.}$$

$$\text{Hence} \quad T \propto ME^2Z^{-2} \propto \frac{ME^2}{Z^2}$$

1.9 STANDARDS

A standard of measurement is a physical representation of a unit of measurement. A unit is realized by reference to an arbitrary material standard or to natural phenomena including physical and atomic constants. For example, the fundamental unit of mass in SI system is kilogram, defined as the mass of a cubic decimetre of water at its temperature of maximum density of 4°C. This unit of mass is represented by a material standard; the mass of the International prototype kilogram, consisting of a platinum-iridium alloy cylinder. The cylinder is preserved at the International Bureau of Weights and Measures at Sevres, near Paris, and is the *material representation* of kilogram. Similar standards have been developed for other units of measurement, including standards for the fundamental unit as well as for some of the derived units.

There are different types of *standards of measurement* on the basis of their function and applications such as:

(*i*) International Standards (*ii*) Primary Standards
(*iii*) Secondary Standards (*iv*) Working Standards

The *International Standards* are defined by international agreement and they represent certain units of measurement to the closest possible accuracy that production and measurement technology allow. These standards are periodically evaluated and checked by *absolute measurements* in terms of the fundamental units. Such standards are maintained at the International Bureau of Weights and Measures and are not available to the ordinary users of measuring instruments for the purpose of comparison or calibration.

The *primary standards* are maintained by National Standard Laboratories in different parts of the world. These standards represent the fundamental units and some of the derived mechanical and electrical units. They are calibrated independently by absolute measurements at each of the National Laboratories. The results of such measurements are compared against each other, leading to a world average figure for the primary standard. Primary standards are not available for use outside the national laboratories. Main function of primary standards is the verification and calibration of secondary standards.

Secondary standards are the basic reference standards used in industrial measurement laboratories. These standards are maintained by the particular involved industry and are checked locally against reference standards available in the area. Secondary standards available in the area are maintained as their calibration is checked by that particular industrial laboratory. Such standards are calibrated and compared against the primary standards in national standards laboratories on a periodic basis. These laboratories issue a certificate of their measured value in terms of the primary standards.

Working standards are the principal tools of a measurement laboratory. These are used to check and calibrate general laboratory instruments for accuracy and performance or to perform comparison measurements in industrial applications.

For example, manufacturers of electronic components such as resistors, capacitors etc make use of working standard for checking the component values being manufactured, *e.g.* a *standard resistor* (a working standard) for checking of value of resistance manufactured. Working standards are periodically checked against secondary standards.

1.10 INTERNATIONAL AND ABSOLUTE UNITS

Although the absolute system of units was adopted by the British Association Committee on Electrical Measurements in 1863 and was confirmed at an International Conference on Electrical Units in London in 1908. This conference decided to specify material standards to be calibrated in absolute units and thereafter set up or maintained as working standards. This decision was taken because of the difficulty of making accurate absolute measurements; but the fundamental standards determined on the electromagnetic system based on the centimetre, gramme and second—the standards for which were more permanent than electrical one remained.

The four units established by these specifications (defined below) were known as International Units. The Ohm was chosen as the primary standard.

The *International Ohm* is the resistance offered, by a column of mercury of mass 14.4521 g of uniform cross-section and of length 106.300 cm to the passage of an unvarying electric current, at the temperature of melting ice.

The International Ampere is the unvarying electric current which when passed through a solution of silver nitrate ($AgNO_3$) in water; as said in specifications, will deposit silver at the rate of 0.00111800 g/second.

The International Volt is the steady electric pressure which, when applied to a conductor of resistance of one International ohm, causes a current of one International ampere.

The International Watt is the electrical energy expended per second, when unvarying electric current of one International ampere flows through a resistance of one international ohm.

The centimetre, gramme, and second were selected as the units of length, mass, and time, by an International Electrical Congress at Paris in 1881, and were defined by them.

The absolute system of units could not be adopted owing to the difficulty of constructing standards which did not vary appreciably with time. Since then, however, wire resistance standards have been developed which are sufficiently permanent for their use to act as a better method of maintaining the international ohm. Thus it became the practice of the national standardising laboratories to maintain the international ohm by wire resistance standards.

As also, by 1930, it was clear that the absolute ohm and ampere could be determined as accurately as the international units, a decision was finally taken in 1946 to abandon the international units, reverting to the fundamental units defined in 1908. The date of the changeover was 1st January 1948.

Following a number of determinations of the absolute ohm and ampere by different national standards laboratories during the period 1934-42 the National Physical Laboratory has adopted the following conversion factors

$$1 \text{ international ohm} = 1.00049 \text{ absolute ohms.}$$
$$1 \text{ international ampere} = 0.99985 \text{ absolute ampere}$$

From the above we have

$$1 \text{ international volt} = 1.00034 \text{ absolute volts}$$
$$1 \text{ international watt} = 1.00019 \text{ absolute watts}$$
$$1 \text{ international henry} = 1.00049 \text{ absolute henrys}$$
$$1 \text{ international farad} = 0.99951 \text{ absolute farad.}$$

EXERCISES

1. What is necessity of units in the measurements ? What are three basic SI mechanical units ?
2. Define "fundamental units" and "derived units". Give examples of each.
3. State and define the SI units of force, work, power and energy.
4. List the names of the various metric prefixes and corresponding symbols.
5. How does engineering notation differ from scientific notation and what is the advantage of engineering notation ?
6. State and define the units of electric current, charge, electrical resistance, inductance and capacitance.
7. What are the two SI temperature scales ? Explain them.
8. Give the expressions of the derived units of velocity, acceleration and force in terms of fundamental units.
9. Derive the dimensions of (*i*) electric current (*ii*) magnetic flux (*iii*) force (*iv*) voltage.
[Pb. Univ. Elec. Measurement-I December 1992]
10. What do you understand by the dimension of a quantity ? Obtain the dimensions of resistance R, inductance L and capacitance C in terms of mass, length, time and current. Hence, check dimensionally if the equation $R^2C = L$ is balanced. If not indicate the missing dimension.
[U.P.S.C. I.E.S. Elec. Engineering-I, 1999]

11. Differentiate clearly between primary and secondary standards.

[U.P. Technical Univ. Elec. Measurements and Measuring Instruments 2002-03]

12. What are the differences between the primary and secondary standards? How the accuracy of the secondary standards is ensured ?

13. Classify different types of standards used in measurement and explain their scope of application.

14. Distinguish between international, primary, secondary and working standards.

[U.P.S.C. I.E.S. Elec. Engineering-I, 1998]

15. Give the salient features of primary and secondary standards. **[U.P.S.C. I.E.S. Elec. Engineering-I, 2000;**

16. What is the difference between primary and secondary standards ? How is the standard metre defined ?

[Bangalore Univ. 1980]

17. Describe briefly the International, primary, secondary and working standards with suitable examples.

[U.P. Technical Univ. Elec. Measurements and Measuring Instruments 2005-06]

18. Define the term 'standards' in measurement system. How are they classified ?

[U.P. Technical Univ. Electrical Measurements and Measuring Instruments, 2008-09]

SHORT ANSWER TYPE QUESTIONS WITH ANSWERS

Q. 1. What is meant by the term 'unit' in the measurements ?

Ans. The result of measurement of a physical quantity must be defined both in kind and magnitude. The standard measure of each kind of physical quantity is called a unit.

Q. 2. What is the difference between fundamental and derived units? **[U.P.S.C. I.E.S. Elec. Engineering-I, 2000]**

Ans. The size of the fundamental units is quite arbitrary and can be selected to fit a certain set of circumstances *i.e.*, mass, length, time etc. whereas the derived units can be expressed in terms of fundamental units with the help of physical equations *i.e.* area, volume, force etc.

Q. 3. Distinguish between absolute, primary and auxiliary fundamental and derived units.

[U.P.S.C. I.E.S. Elec. Engineering-I, 1998]

Ans. **Absolute Units.** An absolute system of units is defined as a system in which the various units are expressed in terms of a small number of fundamental units. Absolute measurements do not compare the measured quantity with arbitrary units of the same type but made in terms of fundamental unit.

Primary Fundamental Units. The size of primary fundamental units is quite arbitrary and can be selected to fit a certain set of circumstances *i.e.* mass, length, time etc.

Auxiliary Fundamental Units. These units are used only where particular disciplines are involved i.e. thermal, illumination etc.

Derived Units. The units which can be expressed in terms of fundamental units with the help of physical equations are called the derived units *i.e.* area, volume, density etc.

Q. 4. What are three basic units in the SI system of units ?

Ans. The three basic units in the SI system of units are:

Length — Metre (m)
Mass — Kilogram (kg)
Time — Second (s)

Q. 5. What are SI units of force, work, power and energy ?

Ans. The SI units of force, work, power and energy are Newton, joule, joule per second or watt and joule respectively.

Q. 6. What are SI units of current, charge, emf, pd, resistance, conductance, magnetic flux, flux density, inductance and capacitance ?

Ans. The SI units of different electrical and magnetic quantities are given below in tabular form ?

Quantity	*Unit*	*Quantity*	*Unit*
Current (I)	Ampere (A)	Conductance (G)	Siemens (S)
Charge (Q)	Coulomb (C)	Magnetic flux (ϕ)	Weber (Wb)
EMF, PD	Volt (V)	Flux density (B)	Tesla (T)
Resistance (R)	Ohm (Ω)	Inductance (L)	Henry (H)
		Capacitance (C)	Farad (F)

Q. 7. The normal human body temperature is given as 98.4°F. Give the equivalent Celsius and Kelvin scale temperatures.

Ans. Celsius temperature $= (°\text{F} - 32) \times \frac{5}{9} = \frac{98.4 - 32}{1.8} = 36.9°\text{C}$ **Ans.**

Kelvin temperature = Celsius temperature + 273.15 = 310.05 K **Ans.**

Q. 8. What do you understand by the dimension of a quantity ? **[U.P.S.C. I.E.S. Elec. Engineering-I, 1999]**

Ans. Every quantity has a quality which distinguishes it from all other quantities. This unique quantity is called dimension. The dimension is written in a characteristic notation, as for example [L] for length, [T] for time etc.

Q. 9. What is meant by 'dimensional equation' ?

Ans. The dimensional symbol for the derived unit of area is L^2 and that for volume, L^3. The dimensional symbol for the unit of force is LMT^{-2}, which follows from the defining equation for force. Expressions like $[a] = [L^2]$, $[v] = [L^3]$, $[F] = [MLT^{-2}]$, $[P] = [ML^2T^{-3}]$ indicate the nature of derived quantities in terms of the fundamental dimensions [L], [M] and [T] and are known as dimensional equations.

Q. 10. What is the difference between primary and secondary standards ?

Ans. Primary standards are absolute standards of such high accuracy that they can be used as the ultimate reference standards while the secondary standards are the basic reference standards used in industrial measurement laboratories. Secondary standards are calibrated and compared against the primary standards in national standard laboratories on a period basic.

Q. 11. How the accuracy of the secondary standards is ensured ?

Ans. To ensure the accuracy of secondary standards, these standards are checked locally against reference standards available in the area and are also normally sent periodically for calibration and comparison against the primary standards.

Q. 12. What is meant by absolute measurement of electrical quantities ?

Ans. Measurement of electrical quantities in terms of the units of mass, length and time are called absolute measurement.

PROBLEMS

1. If the practical units of voltage and current were made 100 times as large as they are at present, what would be the consequent alteration in the size of the units of capacitance and inductance.

   [Ans. No change]

2. The energy stored in a parallel plate capacitor per unit area of the dielectric is expressed as

$$W \propto \varepsilon^a V^b d^c$$

where ε = the absolute permittivity of the dielectric,

V = the potential difference across the plates,

and d = the distance between the plates.

Considering the dimensions of the quantities involved, determine the values of the indices a, b and c.

[Allahabad Univ. II BE (Computer Science) Measurements I 1980]

[Ans. $a = 1$; $b = 2$ and $c = -1$]

3. Check the following equation dimensionally

$$e = \mathrm{R}i + \mathrm{L}\frac{di}{dt} + \frac{1}{\mathrm{C}}\int i\,dt$$

4. The pull of an electromagnet is given by the equation

$$\mathrm{F} \propto \mathrm{B}^a \mathrm{A}^b \mu^c$$

where B = the flux density in the air gap

A = cross-sectional area of the gap

μ = permeability of the air in the gap.

Find the values of a, b and c by analysing the equation dimensionally. **[Ans.** $a = 2$; $b = 1$ and $c = -1$]

5. Determine a, b, c and g in the following equation for the eddy current loss W per cm length of the wire

$$\mathrm{W} \propto f^a \mathrm{B}_{max}^b d^c \rho^g$$

where f = frequency

B_{max} = maximum flux density

d = diameter of wire

r = specific resistance of the material.

[Gorakhpur Univ. Elec. Measurements & Measuring Instruments 1976]

[Ans. $a = 2$; $b = 2$, $c = 4$ and $g = 1$]

6. The mean torque of an electrodynamic wattmeter is expressed as $\mathrm{T} \propto \mathrm{M}^p\mathrm{E}^q\mathrm{Z}^n$

where M = Mutual inductance between two coils of the wattmeter,

E = Applied voltage, and

Z = Impedance of coil.

Use dimensional analysis technique to determine p, q and n. **[Ans.** $p = 1$, $q = 2$ and $n = -2$]

7. Check dimensionally if the equation $\mathrm{R}^2\mathrm{C} = \mathrm{L}$ is balanced. If not indicate the missing dimension.

[U.P.S.C. I.E..S. Elec. Engineering-I, 1999]

[Ans. Given equation is balanced and needs no correction]

8. Using dimensional analysis check whether the following equation is correct or not ?

$$\mathrm{R} = \frac{1 + \omega^2\mathrm{C}^2\mathrm{R}^2}{\omega^2\mathrm{CR}}$$

[Pb. Univ. Elec. Measurements & Measuring Instruments-I April 1975]

$$\left[\textbf{Ans.}\ \text{No, correct equation will be } \mathrm{R} = \frac{1 + \omega^2\mathrm{C}^2\mathrm{R}^2}{\omega^2\mathrm{C}^2\mathrm{R}}\right]$$

❑❑❑

2

Measurement Errors

2.1 INTRODUCTION

Measurement is the process of comparing an unknown quantity with an accepted standard quantity and a measuring instrument is a device used for comparing the unknown quantity with the standard quantity or a unit of measurement. No measurement can be made with complete accuracy, therefore, a study of errors is necessary in the study of measuring processes. The fact that we are studying the subject of errors does not mean that all measurements can be made with an extreme degree of accuracy. Measurements cannot be called good or bad merely on the basis of the degree of accuracy, but rather on their adequacy under the given conditions. In general, closer results can be achieved by time, care and expense; whether the improvement is justified in a particular case is, of course another question. In any event, we must know what the accuracy of our results really is, whether high or low, otherwise we are in the dark, and have figures which we do not know how to evaluate. We cannot study the matter of accuracy without knowing about different types of errors that may enter, and what we can do about them.

2.2 ABSOLUTE ERROR, RELATIVE ERROR AND LIMITING OR GUARANTEE ERRORS

The value of the unknown quantity obtained on making measurements with standards and measuring instruments is considered to be its *true value*, though practically it is never. There is always some difference, small or large, between the measured value and the true or exact value of the unknown quantity. The difference between the measured value A_m and the true value A of the unknown quantity is known as the *absolute error of measurement*, δA*

$$i.e., \quad \delta A = A_m - A \qquad \text{...(2.1)}$$

2.2.1. Relative Error and Percentage Error. The absolute value of error δA does not indicate precisely the accuracy of measurements. For example, an absolute error of one cm is infinitesimal in determining

* **Note:** δA is also called the reading correction or simply correction, being of the same magnitude as the error, but of opposite sign.

the circumference of earth, negligible in setting out a cricket pitch, but intolerable in a metre stick. The quality of measurement is, therefore, indicated preferably in terms of *relative error*. The relative error is the ratio of absolute error to the true value of the quantity to be measured.

$$\textit{i.e., } \text{Relative error, } \varepsilon_r = \frac{\delta A}{A} = \frac{\varepsilon_0}{A} = \frac{\text{Absolute error}}{\text{True value}} \quad \text{...(2.2)}$$

When the absolute error $\varepsilon_0 = \delta A$ is negligible, *i.e.*, when the difference between measured value A_m and true value A is negligible then relative error may be expressed as

$$\varepsilon_r = \frac{\delta A}{A_m} \quad \text{...(2.3)}$$

The relative error may be quoted as a fraction *e.g.*, 5 parts in 1,000 or may be expressed as a percentage.

$$\textit{i.e., } \text{Percentage error } = \varepsilon_r \times 100 = \frac{\varepsilon_0}{A} \times 100 \quad \text{...(2.4)}$$

The amount of relative error is often an indication of the class of use to which a piece of apparatus may be put. For example, a resistor which is within 5% of its nominal value may be quite satisfactory in a shunt field regulator or in a radio set but useless as a standard or in a decade resistance box. A resistor within 1% of its nominal value can be quite useful for general measurements but not for very precise measurements. In order to assure the purchaser of the quality of the circuit components or measuring instruments, the manufacturers guarantee a certain accuracy. Circuit components such as resistors, inductors and capacitors are guaranteed to be within a certain percentage of the rated value whereas in indicating instruments the accuracy is mostly guaranteed to be within a certain percentage of full-scale reading.

2.2.2. Limiting or Guarantee Errors. Manufacturers of equipment/apparatus give guarantee about the accuracy of the equipment/apparatus with some limiting deviations from the specified value in order to enable the purchaser to make proper selection according to his requirement. As mentioned above in Art. 2.2.1, in most of the indicating instruments the accuracy is guaranteed to be within a certain percentage of full-scale reading while circuit components such as resistors, inductors and capacitors are guaranteed to be within a certain percentage of the rated value. The limits of these deviations from the specified values are known as *limiting errors.*

For example, if a certain ammeter of 0–25 A range is guaranteed to be accurate within ± 2 percent of full-scale reading by the manufacturer, it means that the magnitude of limiting error will be 2 percent of 25 A *i.e.* $\frac{2 \times 25}{100} = 0.5$ A.

If the resistance of a resistor is given as 800 ± 5% the manufacturer guarantees that resistance falls between the limits of (800 ± 40) Ω *i.e.*, between 760 Ω and 840 Ω. Here the maker is not specifying a standard deviation* or a probable error* but assures that the error is not more than the limits set.

The magnitude of a given quantity having a measured value A_m and a maximum or a limiting error $\pm \delta A$ must have a magnitude between the limits

$$A_m - \delta A \text{ and } A_m + \delta A$$

$$\text{or } A = A_m \pm \delta A \quad \text{...(2.5)}$$

For example, the measured value of a capacitor is 100 μF with a limiting error of ± 5 μF. The true value of the capacitor will be between the limits 100 ± 5 *i.e.* between 95 and 105 μF.

---

* **Note:** The terms standard deviation and probable error will be explained later on in Arts. 2.8 and 2.11 respectively.

Example 2.1. The measured value of a resistance is 10.25 Ω, whereas its value is 10.22 Ω. Determine the absolute error of measurement.

Solution: Measured value, $A_m = 10.25\ \Omega$

True value, $A = 10.22\ \Omega$

Absolute error, $\delta A = A_m - A = 10.25 - 10.22 = 0.03\ \Omega$ **Ans.**

Example 2.2. An ammeter reads 6.7 A and the true value of current is 6.54 A. Determine the error and the correction for this instrument.

[U.P. Technical Univ. Electronic Measurements and Instrumentation 2008-09]

Solution: Measured value, $A_m = 6.7$ A

True value, $A = 6.54$ A

Absolute error or error, $\delta A = A_m - A = 6.7 - 6.54 = 0.16$ A **Ans.**

Correction $= -\delta A = -0.16$ A **Ans.**

Example 2.3. A wattmeter reads 25.34 watts. The absolute error in the measurement is – 0.11 watt. Determine the true value of power.

Solution: Measured value, $A_m = 25.34$ watts

Absolute error, $\delta A = -0.11$ watt

True value, A = Measured value – absolute error

$= A_m - \delta A = 25.34 - (-0.11) = 25.45$ watts **Ans.**

Example 2.4. The measured value of a capacitor is 205.3 μF, whereas its true value is 201.4 μF. Determine the relative error.

Solution: Measured value, $A_m = 205.3 \times 10^{-6}$ F

True value, $A = 201.4 \times 10^{-6}$ F

Absolute error, $\varepsilon_0 = A_m - A = 205.3 \times 10^{-6} - 201.4 \times 10^{-6} = 3.9 \times 10^{-6}$ F

$$\text{Relative error, } \varepsilon_r = \frac{\varepsilon_0}{A} = \frac{3.9 \times 10^{-6}}{201.4 \times 10^{-6}} = 0.0194 \text{ or } 1.94\% \textbf{ Ans.}$$

Example 2.5. The inductance of an inductor is specified as 20 H ± 5 percent by a manufacturer. Determine the limits of inductance between which it is guaranteed.

Solution:

$$\text{Relative error, } \varepsilon_r = \frac{\text{Percentage error}}{100} = \frac{5}{100} = 0.05$$

Limiting value of inductance, $A = A_m \pm \delta A = A_m \pm \varepsilon_r A_m = A_m(1 \pm \varepsilon_r) = 20\,(1 \pm 0.05) = 20 \pm 1$ H **Ans.**

Example 2.6. A 600 V voltmeter is specified to be accurate within ± 2.5% of FSD. Calculate the limiting error when the instrument is used to measure a voltage of 400 V.

[U.P. Technical Univ. Electronic Measurements and Instrumentation 2006-07]

Solution: The magnitude of the limiting error of the voltmeter,

$$\delta V = \varepsilon_r V = \pm 2.5\% \text{ of } 600 \text{ V} = \pm \frac{2.5}{100} \times 600 = \pm 15 \text{ V}$$

The magnitude of voltage under measurement is 400 V.

$$\text{The percentage limiting error at this voltage} = \pm \frac{15}{400} \times 100 = \pm 3.75\% \textbf{ Ans.}$$

Example 2.7. A wattmeter having a range of 500 W has an error of ± 1.5 percent of full scale deflection. If the true power is 500 W, what would be the range of the reading ?

[U.P. Technical Univ. Electronic Measurements and Instrumentation 2007-08]

Solution: Power under measurement is between the limits of

$$A = A_m (1 \pm \varepsilon_r) = 500\left(1 \pm \frac{1.5}{100}\right) = 500\ (1 \pm 0.015) = (500 \pm 7.5)\ \text{W}$$

So range of reading of wattmeter is (500 – 7.5) to (500 + 7.5) = 492.5 W to 507.5 W **Ans.**

Example 2.8. Define limiting errors.

A 0 –10 A ammeter has a guaranteed accuracy of 1.5 percent of full-scale reading. The current measured by the instrument is 2.5 A. Calculate the limiting values of current and the percentage limiting error. *[U.P. Technical Univ. Elec. Measurements and Measuring Instruments 2005-06]*

Solution: The magnitude of limiting error of the instrument, $\delta A = \varepsilon_r \times A = \frac{1.5}{100} \times 10 = 0.15$ A

The magnitude of current being measured is 2.5 A. The relative error at this current is

$$\varepsilon_r = \frac{\delta A}{A} = \frac{0.15}{2.5} = 0.06$$

Hence, the current under measurement is between the limits of

$$A = A_m(1 \pm \varepsilon_r) = 2.5\ (1 \pm 0.06) = (2.5 \pm 0.15)\ \text{A}$$

So limiting values of current under measurement are 2.35 A and 2.65 A **Ans.**

$$\text{Limiting error} = \frac{0.15}{2.5} \times 100 = 6\%\ \textbf{Ans.}$$

Example 2.9. What do you understand by 'limiting error' ? A wattmeter has a range of 1,000 watts with an error of ± 1% of full-scale deflection. If the true power passed through it is 100 watts, calculate the limiting error.

Solution: The magnitude of limiting error of the instrument,

$$\delta P = \varepsilon_r \times P = 0.01 \times 1{,}000 = 10\ \text{W}$$

The magnitude of the power being measured is 100 watts.

$$\text{The percentage limiting error at this power} = \frac{10}{100} \times 100 = 10\%\ \textbf{Ans.}$$

From the above examples it is worthnoting that meters are guaranteed for better accuracies for full-scale reading but when the meters are used for lower readings the limiting error increases (say, from 1.5% to 6% in case of Example 2.8; and from 1% to 10% in case of Example 2.9). If the quantity (current or voltage or any other one) under measurement is further reduced, the limiting error will further increase because the magnitude of the limiting error is a fixed quantity based on the full-scale deflection of the meter. The above examples also show the importance of taking measurements as close to full scale as possible.

Measurements or computations, combining guarantee errors, are usually made. Such a computation is illustrated in the following examples.

Example 2.10. Find the value of ($R_1 + R_2$). Considering the errors in their values as standard deviation when

$$\mathbf{R_1 = 100 \pm 2\% \quad \text{and} \quad R_2 = 200 \pm 2.5\%}$$

[U.P. Technical Univ. Electrical Measurements and Measuring Instruments, 2008-09]

Solution: $R_1 + R_2 = (100 + 2\%) + (200 \pm 2.5\%) = (100 \pm 2) + (200 \pm 5) = (300 \pm 7)\ \Omega$ **Ans.**

Example 2.11. In a Wheatstone bridge three decade resistance boxes are used which are guaranteed for ± 0.2%. An unknown resistor of R ohms is measured with this bridge. Determine the limits on resistor R imposed by the decade boxes.

Solution: Under balanced condition of Wheatstone bridge, unknown resistance R can be determined in terms of the resistances of three decade boxes, *i.e.* $R = \frac{R_1R_2}{R_3}$. Here R_1, R_2 and R_3 are the resistances of the decade boxes and each is guaranteed to ± 0.2%. In worst case, the two terms in numerator may both be + ve to the full limit of 0.2% and the denominator may be –ve to the full limit of 0.2% giving a resultant error of 0.6%.

So the guarantee error is obtained by taking the direct sum of all the possible errors, adopting the algebraic signs giving the worst possible combination.

Example 2.12. A certain resistor has a voltage drop of 110.2 V and a current of 5.3 A. The uncertainties in the measurements are ± 0.2 V and ± 0.6 A respectively. Calculate the power dissipated in the resistor and the uncertainty in power.

[*U.P. Technical Univ. Elec. Measurements and Measuring Instruments 2006-07*]

Solution: Percentage limiting error to voltage drop $= \pm\frac{0.2}{110.2}\times 100 = 0.18\%$.

Percentage limiting error to current $= \pm\frac{0.6}{5.3}\times 100 = \pm 11.32\%$

The power dissipated in the resistor, P = Voltage drop across resistor × current flowing through resistor

$= 110.2 \times 5.3 = 584.06$ W **Ans.**

In worst possible combination of errors, the limiting error in the power dissipation

$= \pm (0.18 + 11.32) = \pm 11.5\%$ **Ans.**

So power dissipation is given by the expression P = 584.06 W + 11.5 % of 584.06 W

= (584.06 ± 67.167) W **Ans.**

Example 2.13. The current passing through a resistor of 100 ± 0.2 Ω is 2.00 ± 0.01 A. Using the relationship, $P = I^2R$, calculate the limiting error in the computed value of power dissipation.

Solution: Percentage limiting error to resistance $= \pm \frac{0.2}{100} \times 100 = \pm 0.2\%$

Percentage limiting error to current $= \pm \frac{0.01}{2} \times 100 = \pm 0.5\%$

The power dissipated in resistance R due to flow of current I is given by the equation

$$P = I^2 R$$

In worst possible combination of errors the limiting error in the power dessipation is

$$2 \times 0.5 + 0.2 = 1.2\%$$

Power dissipated in resistor, $P = I^2 R = (2.0)^2 \times 100 = 400$ W

So power dissipation is given by the equation as

P = 400 W ± 1.2% of 400 watts or 400 ± 4.8 watts **Ans.**

2.3 RESOLUTION AND SENSITIVITY

If the input is slowly increased from some arbitrary value it will be noticed that the output does not change at all until the increment exceeds the certain value called the *resolution* or *discrimination* of the instrument. Thus the resolution or discrimination of any instrument is the smallest change in the input signal (quantity under measurement) which can be detected by the instrument. It may be expressed as an actual value or as a fraction or percentage of the full-scale value. In some texts this quantity is referred to as '*sensitivity*'.

The sensitivity gives the relationship between the input signal to an instrument or a part of an instrument system and the output. Thus the sensitivity is defined as the ratio of output signal or response of the instrument to a change of input signal or the quantity under measurement.

The sensitivity will, therefore, be a constant in a linear instrument or element *i.e.* where equal changes in the input signal cause equal changes in the output.

Sensitivity is usually needed to be high, and an instrument should, therefore, not have a range largely exceeding the quantity under measurement, although some margin of excess over the expected values should be allowed inorder to prevent accidental overload.

Example 2.14. A moving coil voltmeter has a uniform scale with 100 divisions and gives full-scale reading of 200 V. The instrument can read upto $\frac{1}{5}$th of a scale division with a fair degree of certainty. Determine the resolution of the instrument in volt.

Solution:

$$\text{Full-scale reading} = 200 \text{ V}$$

$$\text{Number of divisions of scale} = 100$$

$$1 \text{ scale division} = \frac{200}{100} = 2 \text{ V}$$

$$\text{Resolution} = \frac{1}{5}\text{th of a scale-division} = \frac{2}{5} \text{ V} = 0.4 \text{ V} \quad \textbf{Ans.}$$

2.4 ACCURACY AND PRECISION

Accuracy is a closeness with which the instrument reading approaches the true value of the variable under measurement while precision is a measure of the reproducibility of the measurement *i.e.,* precision is a measure of the degree to which successive measurements differ from one another. In brief, accuracy can be defined as *conforming to truth* and precision can be defined as *sharply* or *closely defined.*

For illustration of distinction between accuracy and precision, we may compare two decade resistance boxes, each with four dials, giving resistance increments of 1,10,100 and 1,000 Ω per step. Suppose one box is cheaper one of low grade quality guaranteed to be within 1% while another one is high grade quality guaranteed to be within 0.1%. Both boxes may be set to the same precision *i.e.,* both boxes can be set to any value upto 10,000 Ω by one ohm step but the accuracies, however, are very different.

Difference between accuracy and precision can be illustrated by another example in which two voltmeters of the same make and model are compared. Both voltmeters have carefully calibrated scales and knife-edged pointers with mirror backed scales to avoid parallax. So both instruments may be read to same precision. If the value of the series resistance in one meter changes considerably, its reading may be in error by fairly large amount. Hence accuracy of the two meters may be quite different. For determination of the meter in error a comparison measurement with a standard meter should be made.

Here it should be clearly understood that accuracy of a reading is not necessarily guaranteed by its precision and good measurement technique demands continuous skepticism as to the accuracy of the results.

Precision is composed of two characteristics; *conformity* and the number of *significant* figures to which a measurement may be made. Consider an example in which a voltage reading of true value 1.952 V is measured by a voltmeter which consistently and repeatedly indicates 2 V. This is as close to the true value as an observer can read the scale by estimation. Although there are no deviations from the observed value, the error caused by the limitation of the scale reading is a precision error. So, conformity is a necessary, but not sufficient condition, for precision, because of lack of significant figures obtained. Similarly precision is a necessary, but not sufficient condition, for accuracy as explained below.

If a magnitude is to be determined with accuracy to a required number of digits, it is necessary that the measuring instrument have precision of this order. Thus precision is an essential condition for achievement of the accuracy. Accuracy is a matter of careful measurement in terms of an accurately known standard.

If the results of measurement of the same variable agree among themselves then that set of readings shows precision but it does not guarantee the accuracy as there may be some systematic disturbing effects which introduce error in the readings. For this reason, in important measurements, an independent set of observations by second method is also taken to avoid the effects of some systematic errors, where this is not possible to apply second method then method under use should be studied carefully to find out and eliminate any systematic disturbing error.

2.5 SIGNIFICANT FIGURES

An indication of the precision of the measurement is obtained from the number of significant figures in which result is expressed. Significant figures convey actual information regarding the magnitude and the measurement precision of a quantity. More the significant figures, greater the precision of measurement.

For example if the value of resistance is specified as 138 Ω then its resistance is closer to 138 Ω than to 137 Ω or 139 Ω and there are three significant figures. If the resistance is specified as 138.0 Ω then its resistance is closer to 138.0 Ω than to 137.9 Ω or 138.1 Ω. In this case there are four significant figures. The latter, with more significant figures, expresses a measurement of greater precision than the former. A resistance of 138 Ω might also be written as 0.000138 MΩ. Here there are also three significant figures and one should not count zeros between the decimal point and 138 as significant figures because these zeros are present on account of the size of the unit used but give no greater precision to the result.

In some situations, the total number of digits may not represent the measurement precision. Frequently larger number of zeros before a decimal point are used for approximate population or amounts of money. For example, the population of a city is reported in six figures as 150,000. This may imply that the true value of the population lies between 149,999 and 150,001, which is six significant figures. What is meant, however, is that the population is closer to 150,000 than to 140,000 or 160,000. So this information has only two significant numbers, and a more technically correct notation uses power of ten as 15×10^4 or 1.5×10^5. This method indicates that the population figure is only accurate to two significant figures.

In measurement works, all those digits are recorded of which we are sure nearest to the true value. For example in reading an ammeter a current of 5.7 A simply indicates that the current read by the observer to best estimation, is closer to 5.7 A than to 5.6 A or 5.8 A. Another way of expressing this result indicates the range of possible error *i.e.,* the current may be expressed as 5.7 ± 0.05 amperes, indicating that the value of the current lies between 5.65 A and 5.75 A.

The number of significant figures in a quantity is the one measure of precision, though not definite as a percentage statement. For example 100 A covers values between 99.5 and 100.5, which is ± 0.5% or a total range of doubt of 1%. The value of 999 A also has three significant figures and a range of doubt of 1 A, but as a percentage, the range is not 0.1%. So three significant figures are indefinite as a measure of precision as it covers total range of doubt from 1 to 0.1%. It is, however, a convenient approximate index, and is used to a considerable extent.

When two or more measurements with different degrees of accuracy are added, the result is only as accurate as the least accurate measurement. Suppose two quantities 15.3 and 6.245 are added, the sum of these two quantities will be 21.545. Because one of the quantity is accurate only to three significant figures, so the result will also be reduced to three significant figures *i.e.*, 21.5. Here one point is to be added that in such cases extra digits accumulated in the answer should be discarded or rounded off. In usual practice, if the digit in the first place to be discarded is less than 5 then this digit and the following digits are discarded from the answer.

But if the first digit to be discarded is 5 or higher, the previous digit is increased by 1. For three digit precision 5.4546 should be rounded to 5.45 and 5.4556 to 5.46.

In multiplication, significant figures increase rapidly, but again only the approximate figures are retained in the answer and extra digits accumulated in the answer are discarded or rounded off.

In addition of figures with a range of doubt, doubtful parts are also added, since the ± sign means that one number may be high and the other low. The worst possible combination of range of doubt should be taken in the answer. For example if two quantities A = 430 ± 5 and B = 225 ± 3 are added then answer will be represented as 655 ± 8.

$$\begin{aligned} A &= 430 \pm 5 \ (= \pm 1.163\%) \\ B &= 225 \pm 3 \ (= \pm 1.333\%) \\ \hline C &= 655 \pm 8 \ (= \pm 1.221\%) \end{aligned}$$

Here one important point is to be noted that in case of addition of two quantities, the percentage of doubt in the original figures A and B does not differ from the percentage of doubt in the final result C.

If the same two numbers taken above are subtracted, then we find a very interesting comparison between addition and subtraction with respect to range of doubt

$$\begin{aligned} A &= 430 \pm 5 \ (= \pm 1.163\%) \\ B &= 225 \pm 3 \ (= \pm 1.333\%) \\ \hline \text{Difference } D &= 205 \pm 8 \ (= \pm 3.902\%) \end{aligned}$$

In subtraction also, the doubtful parts are added for the same reason as in addition. Now comparing the results of addition and subtraction of the same numbers A and B, note that the precision of the result, when expressed in percentage, differs greatly. The final result after subtraction shows a large increase in percentage doubt compared to the percentage doubt after addition. The percentage doubt increases even more when the difference between the numbers is relatively small.

So one should avoid measurement technique depending on subtraction of experimental results because the range of doubt in the final result may be greatly increased.

2.6 TYPES OF ERRORS

No measurement can be carried out with complete accuracy, so a study of errors is necessary in the study of measuring process. A study of errors is a first step in finding ways to reduce them and such a study also allows to determine accuracy of the final test results.

To study the matter of accuracy, it is very important to know the different types of errors that may enter during measurement of any quantity. Errors may originate in a variety of ways, but it can be grouped in three main categories.

1. *Gross Errors.* In this category, errors occur because of mistakes in reading or using instruments and in recording and calculating measurement results. These errors are usually because of human mistakes and these may be of any magnitude and cannot be subjected to mathematical treatment.

The complete elimination of gross errors is not possible, but one can minimize them. Some errors are easily detected while others may be elusive.

One common gross error frequently committed during measurement is improper use of measuring instrument. Any indicating instrument changes conditions to some extent when connected into a complete circuit so that the reading of measured quantity is altered by the method used. For example in Figs. 2.1 (*a*) and 2.1 (*b*) two possible connections of pressure coil and current coil of wattmeter are shown.

The connection shown in Fig. 2.1 (*a*) is used when applied voltage is high and current flowing in the circuit is low while the connection shown in Fig. 2.1 (*b*) is used when applied voltage is low and current flowing in the circuit is high. If these connections of wattmeter are used in opposite order then error is liable to enter in wattmeter readings. Another example of this type of error is in the use of a well calibrated voltmeter for the measurement of voltage across a resistance of very high value. The same voltmeter, when connected in a low resistance circuit may give a more dependable reading. This shows that the voltmeter has a loading effect on the circuit, which alters the original situation during the measurement.

Errors caused by loading effect of the meters can be avoided by using them intelligently. For example, when measuring a low resistance by voltmeter-ammeter method a high resistance voltmeter, such as a VTVM or TVM, should be used.

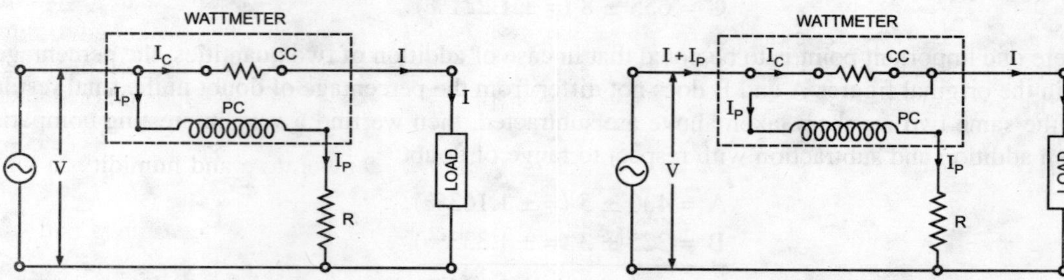

(*a*) *Wattmeter Connections* (*b*) *Wattmeter Connections*

Fig. 2.1

Gross errors are also contributed by the other factors such as improper reading of an instrument, failure to eliminate parallax or recording the result different from the actual reading taken or adjusting the instrument incorrectly.

For example a multirange instrument has different scale for each range. During measurements operator may use a scale which does not correspond to the setting of the range selector of the instrument. Gross error may also be there because of improper setting of zero before the measurement and this will affect all the readings taken during measurement.

As mentioned above that these errors cannot be treated mathematically so a great care should be taken during reading and recording the data to avoid these errors. Apart from this, at least two, three or more readings should be taken of the quantity under measurement, preferably by a different observer at different reading points. Then if the readings differ by an unreasonably large amount, the situation can be investigated and the bad reading eliminated.

2. *Systematic Errors.* These are errors that remain constant or change according to a definite law on repeated measurement of the given quantity. These errors can be evaluated and their influence on the results of measurement can be eliminated by the introduction of proper corrections.

These errors are sometimes referred to as *bias*, and they influence all measurements of a quantity alike. A constant uniform deviation of the operation of an instrument is known as a systematic error. There are basically three types of systematic errors: (*a*) instrumental, (*b*) environmental and (*c*) observational.

(*a*) *Instrumental Errors*. Such errors are inherent in the measuring instruments because of their mechanical structure and calibration or operation of the apparatus used. For example in the D'Arsonval movement, friction in bearings of various components may cause incorrect readings. Improper zero adjustment has a similar effect. Poor construction, irregular spring tensions, variations in the air gap may also cause instrumental errors. Calibration errors may also result in the instrument reading either being too low or too high.

These errors may be detected by checking for erratic behaviour, stability and reproducibility of results. A quick and easy way to check an instrument is to compare it with another instrument of the same characteristics or compare it with a comparatively more accurate instrument.

Such errors may be avoided by: (*a*) selecting a proper measuring device for the particular application, (*b*) applying correction factors after determining the magnitude of instrumental errors, and (*c*) calibrating the measuring device or instrument against a standard.

(*b*) *Environmental Errors*. Such errors are much more troublesome as these change with time in an unpredictable manner. These errors are introduced due to use of an instrument in different conditions than in which it was assembled and calibrated.

Change in temperature is the major cause of such errors as temperature affects the properties of materials in many ways, including dimensions, resistivity, spring effect and others. Other environmental changes also affect the results given by the instruments such as humidity, altitude, earth's magnetic field, gravity, stray electrostatic and magnetic fields etc. These errors can be eliminated or reduced by taking the following precautions:

(*i*) Using instrument in controlled conditions of pressure, temperature and humidity in which it was originally assembled and calibrated.

(*ii*) If above is not possible then deviations in local conditions must be determined and suitable corrections to instrument readings applied.

(*iii*) Using equipment which is immune to these effects. For example, variation in resistance with temperature can be minimized by using resistance materials of very low resistance temperature coefficient.

(*iv*) Using techniques that eliminate the effects of these disturbances. For example the effect of humidity, dust etc., can be entirely eliminated by hermetically sealing the equipment.

(*v*) The effects of external electrostatic or magnetic fields can be avoided by providing electrostatic or magnetic shields.

(*vi*) Altogether new calibrations may be made in the new conditions.

(*c*) *Observational Errors*. Such errors are introduced by the observer. The most common error is the *parallax error* introduced in reading a meter scale, and the error of estimation when obtaining a reading from a meter scale.

Parallax error is caused by the observer by not having his line of sight on the pointer exactly at right angles to the plane of the scale. Such an error can be eliminated by providing a mirror beneath the scale and a knife-edged pointer.

Tests have been conducted in which a number of persons read fractional divisions on an ammeter scale under carefully controlled conditions and consistent individual differences were found. In measurements involving the timing of an event, one observer may tend to anticipate the signal and read too soon. Very considerable differences are likely to appear in determination of light intensities or sound levels. Important readings which may be subject to this type of error should be shared by two or more observers to minimize the possibility of a constant bias.

Modern electrical instruments have digital display of output which completely eliminates the errors due to human observational or sensing powers as the output is in the form of digits.

Systematic errors can also be subdivided into *static* or *dynamic* errors. Static errors are caused due to limitations of the measuring device or the physical laws governing its behaviour. A static error is introduced in a micrometer when excessive pressure is applied in torquing the shaft.

Static error is defined as the difference between the measured value and true value of a quantity *i.e.*

$$\delta A = A_m - A$$

where δA is error, A_m is measured value of quantity and A is true value of quantity.

Dynamic errors are caused by the instruments not responding fast enough to follow the variations in a measured variable.

In this case, the instrument readings lag behind the measured value in time.

3. *Random (or Accidental) Errors.* These are errors that remain even after systematic errors have been substantially reduced or at least accounted for. Random errors are generally an accumulation of a large number of small effects and may be of real concern only in measurements requiring a high degree of accuracy.

These errors are of variable magnitude and sign and do not obey any known law. The presence of random errors becomes evident when different results are obtained on repeated measurements of one and the same quantity. The effect of random errors is minimized by measuring the given quantity many times under the same conditions and calculating the arithmetical mean of the values obtained. The mean value can just be considered as the most probable value of the measured quantity since random errors of equal magnitude but opposite sign are of approximately equal occurrence when making a great number of measurements.

The problem of random errors is treated mathematically as one of the probability and statistics, and is beyond the scope of this book. The following observations may, however, be useful.

If the errors are truely random, a plot of the number of readings of each value as ordinates, against the reading values as abscisae gives a curve of a particular shape known as the *error curve*. If the plot does not give such a curve then the errors are not truely random. This test cannot, of course be applied to a set of as few, half a dozen.

2.7 MEASUREMENT ERROR COMBINATION

When a quantity is computed from measurement made on two or more instruments, it is to be assumed that the errors due to instrument inaccuracy combine in the worst possible way. The resulting error is then larger than the error in any one instrument.

Provided that the errors are small, their effect on the final result is readily obtained from the simple rules outlined below:

(i) Sum of Two or More Quantities. Let the final result *y* be the sum of measured quantities, *u*, *v* and *z*, each of which is subjected to possible systematic error ± δ*u*, ± δ*v*, ± δ*z*.

Then the nominal result is $y = u + v + z$

The relative increment of the function is given by

$$\frac{dy}{y} = \frac{d(u+v+z)}{y} = \frac{du}{y} + \frac{dv}{y} + \frac{dz}{y}$$

Expressing the result in terms of relative increment of the component quantities

$$\frac{dy}{y} = \frac{u}{y}\cdot\frac{du}{u} + \frac{v}{y}\cdot\frac{dv}{v} + \frac{z}{y}\cdot\frac{dz}{z}$$

Since the errors in the component quantities are represented by ± δ *u*, ± δ *v* and ± δ*z* then corresponding limiting error δ*y* in *y* is given by

$$\frac{\delta y}{y} = \pm\left(\frac{u}{y}\cdot\frac{\delta u}{u} + \frac{v}{y}\cdot\frac{\delta v}{v} + \frac{z}{y}\cdot\frac{\delta z}{z}\right) \qquad \text{...(2.6)}$$

The above expression shows that the resultant systematic error is equal to the sum of the products formed by multiplying the individual systematic errors by the ratio of each term to the function.

Since no approximation has been made in working out this particular result, it is true for all values of the errors and is not restricted to the case of small errors.

(ii) Difference of Two Quantities.

Let $y = u - v$

$$\frac{dy}{y} = \frac{du}{y} - \frac{dv}{y}$$

Expressing the result in terms of relative increments of component quantities

$$\frac{dy}{y} = \frac{u}{y}\cdot\frac{du}{u} - \frac{v}{y}\cdot\frac{dv}{v}$$

If the errors in u and v are $\pm\,\delta u$ and $\pm\,\delta v$ respectively, the signs may be interpreted to give the worst possible discrepancy *i.e.*, when the error in u is $+\,\delta u$ and the error in v is $-\,\delta v$ and vice-versa, then the corresponding relative limiting error δy in y is given by

$$\frac{\delta y}{y} = \pm\left(\frac{u}{y}\cdot\frac{\delta u}{u} + \frac{v}{y}\cdot\frac{\delta v}{v}\right) \qquad \text{...(2.7)}$$

This expression is the same as obtained in first case. It may, however, be mentioned that in this case when u and v are almost equal in magnitude then the relative error in y would be very large.

(iii) Product of Two or More Quantities.

Let $y = u\,v\,z$

$$\log_e y = \log_e u + \log_e v + \log_e z$$

Differentiating with respect to y, we get

$$\frac{1}{y} = \frac{1}{u}\cdot\frac{du}{dy} + \frac{1}{v}\cdot\frac{dv}{dy} + \frac{1}{z}\cdot\frac{dz}{dy}$$

or
$$\frac{dy}{y} = \frac{du}{u} + \frac{dv}{v} + \frac{dz}{z}$$

Representing the errors in u, v, and z as $\pm\,\delta u$, $\pm\,\delta v$ and $\pm\,\delta z$ respectively, the error δy in y is given by

$$\frac{\delta y}{y} = \pm\left(\frac{\delta u}{u} + \frac{\delta v}{v} + \frac{\delta z}{z}\right) \qquad \text{...(2.8)}$$

From the above equation we conclude that the relative limiting error of the product of the terms is equal to the sum of the relative errors of the terms.

(iv) Quotient of Two Quantities.

Let $y = \dfrac{u}{v}$

$$\log_e y = \log_e u - \log_e v$$

Differentiating with respect to y, we have

$$\frac{1}{y} = \frac{1}{u}\cdot\frac{du}{dy} - \frac{1}{v}\cdot\frac{dv}{dy}$$

$$\text{or } \frac{dy}{y} = \frac{du}{u} - \frac{dv}{v}$$

Representing the errors in u and v as $\pm\,\delta u$ and $\pm\delta v$ respectively, the relative error in y is given by

$$\frac{\delta y}{y} = \pm\frac{\delta u}{u} \mp \frac{\delta v}{v}$$

The maximum possible error occurs when $\frac{\delta u}{u}$ is + ve and $\frac{\delta v}{v}$ is – ve or vice-versa.

∴ Relative limiting error in y is given by the equation

$$\frac{\delta y}{y} = \pm\left(\frac{\delta u}{u} + \frac{\delta v}{v}\right) \quad ...(2.9)$$

(v) *Power of a Factor*

Let $y = u^n$ where n may be + ve or – ve, integral or fractional

$$\log_e y = n \log_e u$$

Differentiating with respect to y, we get

$$\frac{1}{y} = n\cdot\frac{1}{u}\cdot\frac{du}{dy}$$

$$\text{or } \frac{dy}{y} = n\frac{du}{u}$$

Hence the relative limiting error of y is

$$\frac{\delta y}{y} = \pm\, n\frac{\delta u}{u} \quad ...(2.10)$$

(vi) *Composite Factors*

Let $y = u^n v^m$

$$\log_e y = n \log_e u + m \log_e v$$

Differentiating with respect to y we get

$$\frac{1}{y} = \frac{n}{u}\frac{du}{dy} + \frac{m}{v}\frac{du}{dy}$$

$$\text{or } \frac{dy}{y} = n\frac{du}{u} + m\frac{dv}{v}$$

∴ Relative limiting error of y is

$$\frac{\delta y}{y} = \pm\left(n\frac{\delta u}{u} + m\frac{\delta v}{v}\right) \quad ...(2.11)$$

Example 2.15. Two capacitors 150 ± 2.4 μF and 120 ± 1.5 μF are connected in parallel. Determine the limiting error of the resultant capacitance in μF and in percentage.

Solution: We have $u = 150 \pm 2.4$ μF

$v = 120 \pm 1.5$ μF

When the two capacitors are connected in parallel, the resultant capacitance is

$$y = u + v = (150 \pm 2.4) + (120 \pm 1.5) = (270 \pm 3.9)\ \mu F$$

Therefore, the limiting error is ± 3.9 μF **Ans.**

$$\text{Relative limiting error is } \frac{\delta y}{y} = \pm \frac{3.9}{270} = \pm 0.0144 \text{ or } \pm 1.44\% \textbf{ Ans.}$$

Example 2.16. Two resistors R_1 and R_2 are connected in parallel. The values of resistances are:

$$R_1 = 1\ k\Omega \pm 1\% \qquad R_2 = 500\ \Omega \pm 1\%$$

Assuming given errors as uncertainty, calculate the uncertainty in the combined resistance.

[*U.P. Technical Univ. Elec. Measurements and Measuring Instruments 2005-06*]

Solution: The equivalent resistance is given by the equation

$$\frac{1}{R} = \frac{1}{R_1} + \frac{1}{R_2} = \frac{R_1 + R_2}{R_1 R_2}$$

$$\text{or } R = \frac{R_1 R_2}{R_1 + R_2} = \frac{1{,}000 \times 500}{1{,}000 + 500} = 333.33\ \Omega$$

$$\text{Let } R = \frac{X}{Y}$$

$$\text{The } X = R_1 \times R_2 = 1{,}000 \times 500 = 500{,}000$$

$$\text{and } Y = R_1 + R_2 = 100 + 500 = 1{,}500$$

$$\text{Error in } X = \frac{\delta R_1}{R_1} + \frac{\delta R_2}{R_2} = 1 + 1 = \pm 2\%$$

$$\text{Error in } Y = \frac{\delta R_1}{Y} + \frac{\delta R_2}{Y} = \frac{R_1}{Y} \times \frac{\delta R_1}{R_1} + \frac{R_2}{Y} \times \frac{\delta R_2}{R_2} = \frac{1{,}000}{1{,}500} \times 1 + \frac{500}{1{,}500} \times 1 = \pm 1\%$$

So percentage error (maximum possible) in equivalent parallel resistance

$$= \text{Error in X + error in Y} = \pm 2\% \pm 1\% = \pm 3\% \textbf{ Ans.}$$

Error (maximum possible) in equivalent parallel resistance

$$= \pm 333.33\ \Omega \times \frac{3}{100} = 10\ \Omega$$

Example 2.17. Three resistors have the following ratings:

$$R_1 = 200\ \Omega \pm 5\%,\ R_2 = 100\ \Omega \pm 5\%,\ R_3 = 50\ \Omega \pm 5\%$$

Determine the magnitude of resultant resistance and limiting errors in percentage and ohms, if the above resistances are connected in (*a*) series and (*b*) parallel. [*Pb Univ. Elec. Measurements-I June 1993*]

Solution: (*i*) When the resistances are connected in series

$$\text{Equivalent resistance, } R_{se} = R_1 + R_2 + R_3 = 200 + 100 + 50 = 350\ \Omega \textbf{ Ans.}$$

Relative limiting error of series resistances in percentage

$$= \frac{R_1}{R_{se}} \cdot \frac{\delta R_1}{R_1} + \frac{R_2}{R_{se}} \cdot \frac{\delta R_2}{R_2} + \frac{R_3}{R_{se}} \cdot \frac{\delta R_3}{R_3}$$

$$= \frac{200}{350} \times 5 + \frac{100}{350} \times 5 + \frac{50}{350} \times 5 = \pm 5\% \quad \textbf{Ans.}$$

Relative limiting error of series equivalent resistance in ohms

$$= 350 \times \frac{5}{100} = \pm 17.5\ \Omega \quad \textbf{Ans.}$$

(*ii*) When the resistances are connected in parallel

The equivalent resistance is given by the equation

$$\frac{1}{R_p} = \frac{1}{R_1}+\frac{1}{R_2}+\frac{1}{R_3} = \frac{R_2R_3+R_1R_3+R_1R_2}{R_1R_2R_3}$$

$$R_p = \frac{R_1R_2R_3}{R_2R_3+R_1R_3+R_1R_2} = \frac{200\times100\times50}{100\times50+200\times50+200\times100} = 28.57\ \Omega$$

$$\text{Let } R_p = \frac{X}{Y}$$

then $X = R_1\,R_2\,R_3 = 200 \times 100 \times 50 = 1{,}000{,}000$

and $Y = R_1\,R_2 + R_2\,R_3 + R_3\,R_1 = y_1 + y_2 + y_3 = 200 \times 100 + 100 \times 50 + 50 \times 200 = 35{,}000$

$$\text{Error in X} = \frac{\delta R_1}{R_1}+\frac{\delta R_2}{R_2}+\frac{\delta R_3}{R_3} = 5+5+5 = \pm 15\%$$

$$\text{Error in } y_1 = \frac{\delta R_1}{R_1}+\frac{\delta R_2}{R_2} = 5+5 = 10\%$$

$$\text{Error in } y_2 = \frac{\delta R_2}{R_2}+\frac{\delta R_3}{R_3} = 5+5 = 10\%$$

$$\text{Error in } y_3 = \frac{\delta R_3}{R_3}+\frac{\delta R_1}{R_1} = 5+5 = 10\%$$

$$\text{Percentage error in Y} = \left(\frac{y_1}{Y}\frac{\delta y_1}{y_1}+\frac{y_2}{Y}\frac{\delta y_2}{y_2}+\frac{y_3}{Y}\frac{\delta y_3}{y_3}\right)\times 100$$

$$= \frac{20{,}000}{35{,}000}\times10+\frac{5{,}000}{35{,}000}\times10+\frac{10{,}000}{35{,}000}\times10 = \pm 10\%$$

∴ Percentage error (maximum possible) in equivalent parallel resistance = 15 + 10 = ± 25% **Ans.**

Error (maximum possible in equivalent parallel resistance in ohms)

$$= 28.57 \times \frac{25}{100} = 7.1425\ \Omega \textbf{ Ans.}$$

Example 2.18. A voltmeter reading of 70 V on its 100 V range and an ammeter reading of 80 mA on its 150 mA range are used to determine the power dissipation in a resistor. Both these instruments are guaranteed to be accurate within ± 1.5% at full-scale deflection. Determine power and limiting error of the power. *[U.P. Technical Univ. Measurements and Instrumentation 2002-03]*

Solution: The magnitude of limiting error of the voltmeter, $\delta V = \varepsilon_r V = 0.015 \times 100 = 1.5$ V

The magnitude of voltage being measured = 70 V

$$\text{The percentage limiting error at this voltage} = \frac{1.5}{70}\times 100 = 2.143\%$$

The magnitude of limiting error of the ammeter, $\delta I = \varepsilon_r I = 0.015 \times 150$ mA = 2.25 mA

The magnitude of current being measured = 80 mA

$$\text{The percentage limiting error at this current} = \frac{2.25\text{ mA}}{80\text{ mA}}\times 100 = 2.813\%$$

$$\text{Power, } P = V \times I = 70 \times \frac{80}{1{,}000} = 5.6\text{ W}$$

$$\text{Relative limiting error in power measurement, } \frac{\delta P_x}{P_x}\times 100 = \left[\frac{\delta V}{V}+\frac{\delta I}{I}\right]\times 100$$

$$= 2.143 + 2.813 = 4.956\% \quad \textbf{Ans.}$$

Example 2.19. The wattmeter is used to measure power in the circuit with the help of the following equation $P^2 = \frac{E^2}{R}$, where limiting values of voltage and resistance are, E = 200 V + 1% and R = 1,000 Ω ± 5%.

Calculate (*i*) the nominal power consumed (*ii*) the limiting error of power in watts and percent.

[U.P. Technical Univ. Elec. Measurements and Measuring Instruments 2004-05]

Solution: Normal power consumed, $P = \frac{E^2}{R} = \frac{(200)^2}{1{,}000} = 40$ W **Ans.**

Relative limiting error in measurement of power, $\frac{\delta P}{P} = \pm\left[2\frac{\delta E}{E} + \frac{\delta R}{R}\right] \times 100 = \pm (2 \times 1 + 5) = \pm 7\%$ **Ans.**

Limiting error of power $= P \times \frac{\delta P}{P} = 40 \times \frac{7}{100} = 2.8$ watts **Ans.**

Example 2.20. The resistance of an unknown resistor is determined by a Wheatstone bridge. The solution for the unknown resistance is stated as $R_4 = \frac{R_1 R_2}{R_3}$. The limiting values of various resistances are:

$$R_1 = 500 \pm 1\%, R_2 = 615 \pm 1\%, R_3 = 100 \pm 0.5\%$$

Calculate (*i*) the nominal value of unknown resistance, (*ii*) its limiting error in per cent, and (*iii*) its limiting error in ohm. *[U.P.S.C. I.E.S. Elec. Engineering-I, 1993]*

Solution : (*i*) Unknown resistance, $R_4 = \frac{R_1 R_2}{R_3} = \frac{500 \times 615}{100} = 3{,}075\ \Omega$ **Ans.**

(*ii*) Relative limiting error of unknown resistance,

$$\frac{\delta R_4}{R_4} = \pm\left[\frac{\delta R_1}{R_1} + \frac{\delta R_2}{R_2} + \frac{\delta R_3}{R_3}\right] \times 100 = \pm (1 + 1 + 0.5) = \pm 2.5\% \quad \textbf{Ans.}$$

(*iii*) Limiting error in ohms $= \pm R_4 \times \frac{\delta R_4}{R_4} = \pm 3{,}075 \times \frac{2.5}{100} = 76.875\ \Omega$ **Ans.**

Example 2.21. Calculate the relative error in power factor if relative error in power, current, and voltage are respectively 0.5%, 10% and 1%. Quantities have SI units.

[Pb. Technical Univ. Electrical Measurements and Measuring Instruments May 2008]

Solution: Power factor is given by

$$PF = \frac{P}{VI}$$

Relative limiting error, $\frac{\delta PF}{PF} = \mp \frac{\delta P}{P} \mp \frac{\delta V}{V} \mp \frac{\delta I}{I} = \mp 0.5 \mp 1 \mp 1 = \mp 2.5\%$ **Ans.**

Example 2.22. The unknown inductance is determined by Anderson bridge and is given by the equation

$$L_x = \frac{CP[r(Q+S) + Q \cdot S]}{S}$$

where C = 1 μF ± 1.0% ; P = 1,000 Ω ± 0.4%; Q = 2,000 Ω ± 1.0%; *r* = 200 Ω ± 0.5% and S = 2,000 Ω ± 0.5%.

Determine the magnitude of unknown inductance in Henry and limiting error in percent.

[U.P. Technical Univ. Elec. Measurements & Measuring Instruments 2001-02]

Solution: Unknown inductance, $L_x = \frac{CP}{S}[r\,(Q + S) + QS]$

$$= \frac{1 \times 10^{-6} \times 1{,}000}{2{,}000} \times [200\ (2{,}000 + 2{,}000) + 2{,}000 \times 2{,}000]$$

$$= 0.5 \times 10^{-6} \times [0.8 \times 10^6 + 4 \times 10^6] = 2.4 \text{ Henry} \quad \textbf{Ans.}$$

Let $u = Q + S = 2{,}000 + 2{,}000 = 4{,}000\ \Omega$

Percentage error in $u = \left[\frac{Q}{u}\cdot\frac{\delta Q}{Q} + \frac{S}{u}\cdot\frac{\delta S}{S}\right] = \left[\frac{2{,}000}{4{,}000}\times 1.0 + \frac{2{,}000}{4{,}000}\times 0.5\right] = 0.75\%$

Let $v = r(Q + S) = r\,u = 200 \times (2{,}000 + 2{,}000) = 0.8 \times 10^6$

Percentage error in $v = \frac{\delta r}{r} + \frac{\delta u}{u} = 0.5 + 0.75 = 1.25\%$

Let $x = QS = 2{,}000 \times 2{,}000 = 4 \times 10^6$

Percentage error in $x = \frac{\delta Q}{Q} + \frac{\delta S}{S} = 1.0 + 0.5 = 1.5\%$

Let $y = r(Q + S) + QS = v + x = 0.8 \times 10^6 + 4 \times 10^6 = 4.8 \times 10^6$

Percentage error in $y = \left[\frac{v}{y}\cdot\frac{\delta v}{v} + \frac{x}{y}\cdot\frac{\delta x}{x}\right] = \left[\frac{0.8\times 10^6}{4.8\times 10^6}\times 1.25 + \frac{4\times 10^6}{4.8\times 10^6}\times 1.5\right] = 1.458\%$

Percentage error in inductance $L_x = \frac{\delta C}{C} + \frac{\delta P}{P} + \frac{\delta S}{S} + \frac{\delta y}{y} = 1.0 + 0.4 + 0.5 + 1.458 = 3.358\%$ **Ans.**

Example 2.23. The impedance of R-L circuit operating on ac is given by

$$Z = \sqrt{R^2 + \omega^2 L^2}$$

The resistance R is known to be 100 Ω with an uncertainty of 5%, L is known to be 2H with an uncertainty of 10% and ω is known exactly 2π × 50. Determine the uncertainty in the measurement of Z.

[*U.P. Technical Univ. Elec. Measurements & Measuring Instruments 2005-06*]

Solution: Let $u = R^2 = 100^2 = 10{,}000$

Percentage error in $u = \frac{2\,\delta R}{R} = 2 \times 5 = 10\%$

Let $v = \omega^2 L^2 = (2\pi \times 50)^2 \times 2^2 = 394{,}784$

Percentage error in $v = \frac{2\,\delta L}{L} = 2 \times 10 = 20\%$

Let $x = u + v = R^2 + \omega^2 L^2 = (100)^2 + (2\pi \times 50)^2 \times 2^2 = 404{,}784$

Percentage error in $x = \frac{u}{x}\cdot\frac{\delta u}{u} + \frac{v}{x}\cdot\frac{\delta v}{v} = \left[\frac{10{,}000}{404{,}784}\times 10 + \frac{394{,}784}{404{,}784}\times 20\right] = 0.24 + 19.5 = 19.74\%$

Now $Z = x^{1/2}$

So $\frac{\delta Z}{Z} = \pm\frac{1}{2}\frac{\delta x}{x} = \pm\frac{1}{2} \times 19.74 = 9.873\%$ **Ans.**

2.8 STATISTICAL ANALYSIS

No measurement is made with 100 percent accuracy and, therefore, there is always some error, which varies from one determination to another, and gets introduced in the value of the quantity under measurement. It is a function of statistics to separate, as far as possible, the truth from error by narrowing and defining the region of doubt. But statistical study is mainly concerned with precision of measurement and so it cannot remove systematic errors from set of data. So systematic errors should be small as compared with residual or random errors.

To make statistical methods and interpretations meaningful, a large number of measurements is usually required.

Sometimes simple approach is required for describing and summarizing the results of the measurements. Some of these methods are described below.

(i) Arithmetic Mean. The most probable value of a measured variable is the arithmetic mean of the number of readings taken. Theoretically the best approximate value will be obtained when number of observations of the quantity under measurement is infinite but in practice, only a finite number of observations can be made. The arithmetic mean is given by the following equation

$$\bar{x} = \frac{x_1 + x_2 + x_3 + ... x_n}{n} = \frac{\Sigma x}{n} \qquad ...(2.12)$$

where $\bar{x}$ is arithmetic mean and $x_1, x_2, x_3, ... x_n$ are the readings taken and n is the number of readings taken.

(ii) Deviation from the Mean. The deviation of a reading is the amount by which it differs from the mean. If we have a set of readings x_1, x_2, x_3 ... with mean $\bar{x}$, the deviations of the individual readings are

$$\begin{aligned} \text{Deviation of } x_1 &= d_1 = x_1 - \bar{x} \\ \text{Deviation of } x_2 &= d_2 = x_2 - \bar{x} \end{aligned} \qquad ...(2.13)$$

Deviation from the mean may have a + ve or – ve value but the algebraic sum of all the deviations is always zero.

(iii) Average Deviation. Average deviation is the sum of the scaler values (without sign) of the deviations divided by the number of readings. Average deviation may be expressed as

$$D = \frac{|d_1| + |d_2| + |d_3| + ... |d_n|}{n} = \frac{\Sigma |d|}{n} \qquad ...(2.14)$$

Average deviation gives an indication of the precision of the instruments used in carrying out measurements. Low average deviation between readings shows that instruments used for measurements are highly precise.

(iv) Standard Deviation. The standard deviation of an infinite number of data is the square root of the sum of all the individual deviations squared, divided by the number of readings.

$$\text{Standard deviation, } \sigma = \sqrt{\frac{d_1^2 + d_2^2 + d_3^2 + ... d_n^2}{n}} = \sqrt{\frac{\Sigma d^2}{n}} \qquad ...(2.15)$$

The standard deviation is also known as *root mean square deviation*, and is the most important factor in the statistical analysis of measurement data. Reduction in this quantity effectively means improvement in measurement.

In practice, the possible number of observations is finite. When the number of readings exceeds 20, the standard deviation is denoted by σ but if it is less than 20 the symbol '*s*' is used to denote the same. The standard deviation of a finite number of observations is given as

$$s = \sqrt{\frac{d_1^2 + d_2^2 + d_3^2 + ... d_n^2}{n-1}} = \sqrt{\frac{\Sigma d^2}{n-1}} \qquad ...(2.16)$$

Another expression for essentially the same quantity is the *variance* or *mean square deviation*, which is same as the standard deviation except that the square root is not extracted.

$$\text{So, variance, V} = \text{Mean square deviation} = \sigma^2 \qquad ...(2.17)$$

Variance is a convenient quantity for use in many computations because variances are additive. The standard deviation, however, has the advantage of being of the same units as the variable, making easy to compare magnitudes. Nowadays most scientific results are expressed in terms of standard deviation.

Example 2.24. Calculate the standard deviation for the following data:
$x_1 = 49.7, x_2 = 50.1, x_3 = 50.2, x_4 = 49.6$ and $x_5 = 49.7$

Solution: Arithmetic mean,
$$\bar{x} = \frac{x_1 + x_2 + x_3 + x_4 + x_5}{5}$$
$$= \frac{49.7 + 50.1 + 50.2 + 49.6 + 49.7}{5} = \frac{249.3}{5} = 49.86$$

Deviation from the mean,
$$d_1 = x_1 - \bar{x} = 49.7 - 49.86 = -0.16$$
$$d_2 = x_2 - \bar{x} = 50.1 - 49.86 = +0.24$$
$$d_3 = x_3 - \bar{x} = 50.2 - 49.86 = +0.34$$
$$d_4 = x_4 - \bar{x} = 49.6 - 49.86 = -0.26$$
$$d_5 = x_5 - \bar{x} = 49.7 - 49.86 = -0.16$$

Since the observations are lesser in number,

Standard deviation,
$$s = \sqrt{\frac{d_1^2 + d_2^2 + d_3^2 + d_4^2 + d_5^2}{n-1}}$$
$$= \sqrt{\frac{0.16^2 + 0.24^2 + 0.34^2 + 0.26^2 + 0.16^2}{5-1}} = 0.27 \quad \textbf{Ans.}$$

2.9 CHARACTERISTICS OF EXPERIMENTAL DATA

During measurement of any quantity, scattered data is obtained and this variation can be controlled by taking all care in all manipulations and by holding conditions as steady as possible during the period of measurement. But even with maximum care an unavoidable uncertainty remains. So no measurement can be carried out with absolute definiteness and as the measurements are made closer to limits, presence of smaller disturbances becomes more evident.

All known errors from data such as known systematic effects, calibration etc., should be removed first, before applying statistical methods as they are based on laws of chance, and not on consistent factors. Statistical analysis allows us to determine the best value possible from the given data and set the limits of uncertainty inherent in the scattering of the data.

The distribution of data in a set of readings may be presented in several ways, one of which is a block diagram or *histogram*. Table 2.1 shows a set of 60 current readings, that were taken at small intervals and recorded to the nearest of hundredth of an ampere. The nominal value of measured current is 10.00 A.

Table 2.1 *Tabulation of Current Readings*

Current Reading in Amperes	*Number of Readings*
9.97	1
9.98	3
9.99	13
10.00	23
10.01	15
10.02	4
10.03	1
Total	60

In Fig. 2.2, a histogram for Table 2.1 is shown, in which the number of readings are plotted against each observed current reading.

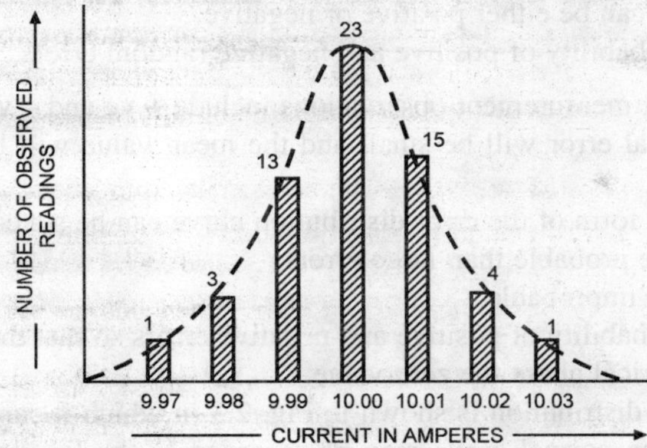

Fig. 2.2

From the figure it is obvious that the largest number of readings (23) occurs at the central value of 10.00 A while the other readings are placed more or less symmetrically on either side of central value. If more readings are taken at smaller increments of 0.005 A intervals (200 readings), then the distribution of observations will remain approximately symmetrical about the central value and the shape of the histogram will be approximately the same as before. With more and more data taken at smaller and smaller increments, the contour of the histogram will finally become a smooth curve, as shown in Fig. 2.2 by the dashed line. The bell shaped curve is known as a *Gaussian curve*. The sharper and narrower the curve, the more definitely an observer may state that the most probable value of the true reading is the central value or mean value.

The normal, or Gaussian, law of errors is the basis for the major part of study of random effects.

2.10 NORMAL LAW OF ERROR

When measurement of a quantity is carried out, the determinations are always finite and limited in number. Random effects cancel each other completely in an infinite set of measurements but it is not true for a small set of measurements. Hence in a limited set of measurements, mean of sample is not necessarily the mean of the larger set, and the standard deviation as obtained from the limited set of measurements may not be the standard deviation of a larger sample or of the universe. Generally precision increases with the size of the sample but the means and the standard deviation of various samples have the property of scatter around the universe values.

The distribution between limited set and universe is a basic concept in statistical study. It was considered formerly that reliable results could be obtained only from a large sample but advances have been made in the field of small sample theory, so that more precise conclusions can be drawn from a small sample.

A probability distribution expresses the likelihood of a particular event or of a deviation of a particular amount. A type of variation may be characterised by a curve or equation which gives the relative probability of a particular result, or of a deviation of a given amount. There are several types of variations that occur in statistical work for different purposes and 'normal law of error' is one of them and is the basis for the major part of the study of random effects.

The assumptions made for the normal law of error are as follows:

(*i*) All observations include a large number of small random disturbing effects or random errors.
(*ii*) These random errors can be either positive or negative.
(*iii*) There is an equal probability of positive and negative random errors.

So it can be expected that measurement observations include + ve and – ve errors in more or less equal amounts so that the total error will be small and the mean value will be the true value of the measured variable.

The possibilities as to the form of the error distribution curve can be stated as follows:

(*i*) Small errors are more probable than large errors.
(*ii*) Large errors are very improbable.
(*iii*) There is an equal probability of positive and negative errors so that the probability of a given error will be symmetrical about the zero value.

A normal curve for error distribution is shown in Fig. 2.3 in which it can be seen that errors are distributed symmetrically. The normal curve may be regarded as the limiting form of a histogram shown in Fig. 2.2, when the data become increasingly large, gathered in smaller and smaller bands.

One form of the equation for the normal law of error may be written in terms of a simple equation. This equation is

$$y = \frac{h}{\pi} e^{-h^2 x^2} \quad \text{...(2.18)}$$

where x is the magnitude of deviation from mean, y is the probability of occurrence of deviation x (number of readings at any deviation x) and h is a constant often referred as *precision index.*

The above Eq. (2.18) leads to a curve of the form shown in Fig. 2.3, which may be seen in a general way to match the requirements of the probability behaviour as outlined in the assumptions given above. The abscissa represents the magnitude of deviation from the central value, and the ordinate is the probability of the occurrence in an infinite set of an observation with a deviation of this magnitude. The presence of the square, x^2 in the exponent provides symmetry for plus and minus deviations, and provides flatness of slope at the origin.

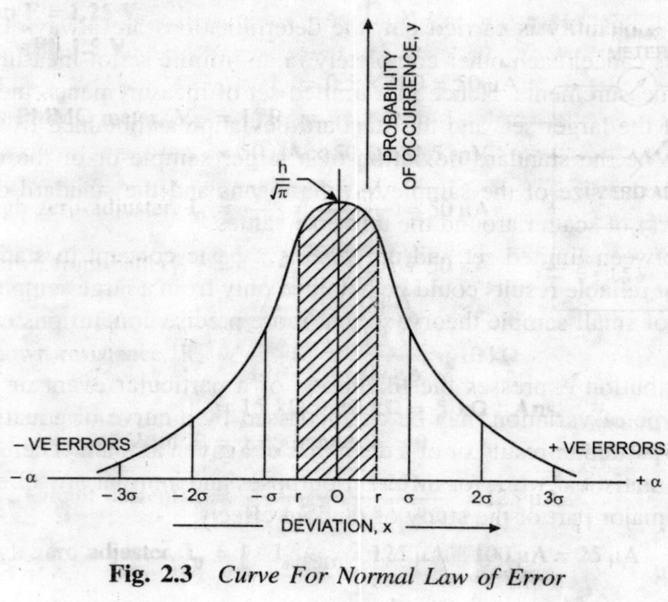

Fig. 2.3 *Curve For Normal Law of Error*

The parameter h reveals the spread or scatter of the measured values about O. The larger the value of h, the greater peak the distribution curve has. A larger h implies higher precision and consistency in measurement and hence it is treated as *precision index*. For $x = 0$, the ordinate $y = h\sqrt{\pi}$. If the two curves shown in Fig. 2.4 are compared, it is noticed that the curve with the larger value of h has a greater central probability, and drops off more rapidly with increase of x. Curve II has a lower h, a lower central value, and a slower drop-off at the sides. (Both curves I and II have the same unit area under them). In summary, a larger value of h means a more closely grouped set of observations, with smaller dispersion, and should, therefore, be considered as a better set, with respect to the control of random effects.

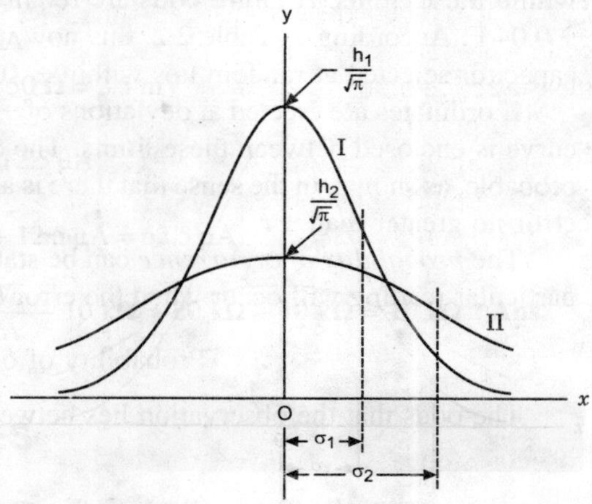

Fig. 2.4 *Comparison of Two Frequency Distribution Curves With Different Degrees of Dispersion*

The value of h is given as

$$h = \frac{1}{\sigma\sqrt{2}} \qquad \text{...(2.19)}$$

If we substitute $h = \frac{1}{\sigma\sqrt{2}}$ in Eq. (2.18), we have

$$y = \frac{1}{\sqrt{2}\,\pi\sigma} e^{-x^2/2\sigma^2} \qquad \text{...(2.20)}$$

This form of equation is particularly useful, as σ is the quantity we ordinarily know and are interested in. σ is a quantity of the same units as the observed quantity and its mean. This makes it easy to visualize the dispersion in comparison with the mean.

2.11 PROBABLE ERROR

The area under the Gaussian probability curve shown in Fig. 3.3 within certain limits represents the number of cases among the observations within those deviation limits, expressed as a fraction of the total cases. The area between the limits $-\infty$ to $+\infty$ represents the entire number of observations and taken as unity.

The area under the curve between $-\sigma$ and $+\sigma$ limits represents the number of observations that differ from the mean by no more than the standard deviation. This area can be determined by integration in series, and for normally dispersed data, following the Gaussian distribution, the value is found to be 0.68, so 68 percent of the cases for the normally dispersed data lie between the limits of $\pm\sigma$. Corresponding values of other deviations, expressed in terms of σ, are given in Table 2.2.

Table 2.2

Deviation (±)	*Fraction of Total Area Included*
0.6745 σ	0.5000
1.0000 σ	0.6828
2.0000 σ	0.9546
3.000 σ	0.9972
1.96 σ	0.9500

If, for example, a large number of capacitors having nominal value of 10 μF are measured and the mean value is found to be 10.00 μF with a standard deviation of 0.02 μF, then we know that on an average 68% of all the capacitors have values lying between limits of ± 0.02 μF of the mean. There is then approximately a two to one chance that any capacitor, selected from the lot at random, will lie within these limits. If larger odds are required, then deviation may be extended to a limit of $\pm 2\sigma =$ ± 0.04 F. According to Table 2.2, this now includes 95% of all the cases, giving 21 to 1 odds and any capacitor selected at random lies within ± 0.04 μF of the mean value of 10.00 μF.

If ordinates are erected at deviations of $\pm r = 0.6745\,\sigma$, as shown in Fig. 2.3, half the area under the curve is enclosed between these limits. The quantity r is called the *probable error (PE)*. This value is probable, as shown, in the sense that there is an even chance that any one observation will have a random error no greater than $\pm r$.

The *probability of occurrence* can be stated in terms of *odds* which is the number of chances that a particular reading will occur when the error limit is specified. The odds can be determined as follows:

$$\text{Probability of occurrence} = \frac{\text{Odds}}{\text{Odds} + 1} \quad \text{...(2.21)}$$

The odds that the observation lies between $\pm\sigma$ limits are

$$\frac{\text{Odds}}{\text{Odds} + 1} = 0.6828$$

$$\text{or Odds} = 2.15 : 1$$

Probable error has been used in experimental work to some extent but standard deviation is more convenient in statistical work and is preferred.

2.12 MEASUREMENT DATA SPECIFICATION

After making statistical analysis of multisample data, the results of the measurements are to be specified. The results are expressed as deviations about a mean value. The deviations may be expressed as follows:

1. *Standard Deviation.* The result is expressed as $\overline{x} \pm \sigma$. The error limit in this case is the standard deviation which means that 68.28% (or about two-thirds) of all the readings have values which lie between limits of $\pm\sigma$ and the odds are 2.15 to 1. Thus there is approximately a two to one chance (or possibility) that a new observation will be within these limits.

2. *Probable Error.* The result is expressed as $\overline{x} \pm r$ or $\overline{x} \pm 0.6745\,\sigma$. It means that 50% or half of all the readings lie within these limits and odds are 1 to 1. There is an even chance that any one observation will be within these limits.

3. ± 2σ *Limits.* The result is expressed as $\overline{x} \pm 2\sigma$. In this case, probability range is increased, approximately 95 per cent of all the readings fall within these limits and odds are 21 to 1.

4. ± 3σ *Limits.* The result is expressed as $\overline{x} \pm 3\sigma$. The probability in this case is 0.9972 which means that 99.72% of all the readings fall within these limits *i.e.* practically all the readings are included in these limits. The odds of any observation falling within these limits are 356 to 1.

Example 2.25. The following 10 observations were recorded when measuring a voltage:

1	**2**	**3**	**4**	**5**	**6**	**7**	**8**	**9**	**10**
41.7	**42**	**41.8**	**42**	**42.1**	**41.9**	**42.5**	**42**	**41.9**	**41.8**

Find (*i*) mean (*ii*) standard deviation (*iii*) probable error of one reading.

[UP.S.C. I.E.S. Elec. Engineering-I, 2003]

Solution: (*i*) Arithmetic mean, $\bar{x} = \dfrac{x_1 + x_2 + x_3 + x_4 + x_5 + x_6 + x_7 + x_8 + x_9 + x_{10}}{10}$

$$= \frac{41.7 + 42 + 41.8 + 42 + 42.1 + 41.9 + 42.5 + 42 + 41.9 + 41.8}{10}$$

$$= \frac{419.7}{10} = 41.97 \quad \textbf{Ans.}$$

Deviation from the mean

$$d_1 = 41.7 - 41.97 = -0.27$$
$$d_2 = 42 \quad - 41.97 = +0.03$$
$$d_3 = 41.8 - 41.97 = -0.17$$
$$d_4 = 42 \quad - 41.97 = +0.03$$
$$d_5 = 42.1 - 41.97 = +0.13$$
$$d_6 = 41.9 - 41.97 = -0.07$$
$$d_7 = 42.5 - 41.97 = 0.53$$
$$d_8 = 42 \quad - 41.97 = +0.03$$
$$d_9 = 41.9 - 41.97 = -0.07$$
$$d_{10} = 41.8 - 41.97 = -0.17$$

(*ii*) Standard deviation,

$$\sigma = \sqrt{\frac{d_1^2 + d_2^2 + d_3^2 + d_4^2 + d_5^2 + d_6^2 + d_7^2 + d_8^2 + d_9^2 + d_{10}^2}{10 - 1}}$$

$$= \sqrt{\frac{(0.27)^2 + (0.03)^2 + (0.17)^2 + (0.03)^2 + (0.13)^2 + (0.07)^2 + (0.53)^2 + (0.03)^2 + (0.07)^2 + (0.17)^2}{9}}$$

$$= \sqrt{\frac{0.4410}{9}} = 0.221 \quad \textbf{Ans.}$$

(*iii*) Probable error of one reading,

$$r = 0.6745 \times \sigma = 0.6745 \times 0.221 = 0.149 \quad \textbf{Ans.}$$

Example 2.26. A set of independent 10 measurements were made to determine the weight of a lead shot. The weights in gramme were:

1.570, 1.597, 1.591, 1.562, 1.577, 1.580, 1.564, 1.586, 1.550, 1.575.

Determine the (*i*) arithmetic mean, (*ii*) average deviation, (*iii*) standard deviation, (*iv*) variance, (*v*) probable error of one reading, (*vi*) probable error of the mean.

[*Pb. Univ. Elec. Measurements-I, July 1993*]

Solution: (*i*) Arithmetic mean, $\bar{x} = \dfrac{1.570 + 1.597 + 1.591 + 1.562 + 1.577 + 1.580 + 1.564 + 1.586 + 1.550 + 1.575}{10}$

$= 1.5752$ gramme **Ans.**

Deviation from the mean

$$d_1 = 1.570 - 1.5752 = -0.0052$$
$$d_2 = 1.597 - 1.5752 = +0.0218$$
$$d_3 = 1.591 - 1.5752 = +0.0158$$
$$d_4 = 1.562 - 1.5752 = -0.0132$$
$$d_5 = 1.577 - 1.5752 = +0.0018$$

$d_6 = 1.580 - 1.5752 = +0.0048$

$d_7 = 1.564 - 1.5752 = -0.0112$

$d_8 = 1.586 - 1.5752 = +0.0108$

$d_9 = 1.550 - 1.5752 = -0.0252$

$d_{10} = 1.575 - 1.5752 = -0.0002$

(*ii*) Average deviation,

$$D = \frac{0.0052 + 0.0218 + 0.0158 + 0.0132 + 0.0018 + 0.0048 + 0.0112 + 0.0108 + 0.0252 + 0.0002}{10}$$

$= 0.011$ gramme **Ans.**

(*iii*) Standard deviation, $\sigma = \sqrt{\dfrac{(0.0052)^2 + (0.0218)^2 + (0.0158)^2 + (0.0132)^2 + (0.0018)^2 + (0.0048)^2 + (0.0112)^2 + (0.0108)^2 + (0.0252)^2 + (0.0002)^2}{10-1}}$

$= 0.01426$ gramme **Ans.**

(*iv*) Variance, $= (\sigma)^2 = (0.01426)^2 = 2.033 \times 10^{-4}$ gramme^2 **Ans.**

(*v*) Probable error, $r = 0.6745\ \sigma = 0.6745 \times 0.01426 = 0.0096$ gramme **Ans.**

(*vi*) Probable error of mean, $r_m = \dfrac{r}{\sqrt{n-1}} = \dfrac{0.0096}{\sqrt{10-1}} = 0.0032$ gramme **Ans.**

EXERCISES

1. Explain the terms "accuracy", "sensitivity" and "resolution" as used for indicating instruments. **[Pb. Univ. Elec. Measurement-I, June 1992; July 1991]**
2. Define random errors and explain how are they analysed statistically. **[Pb. Univ. Elec. Measurement-I, December 1991]**
3. Explain the different types of errors that may occur in measurements. **[U.P.S.C. I.E.S. Elec. Engineering-I, 1993]**
4. What are the instrumental, environmental and observational errors ? Briefly explain about each one of them.
5. Give an account of the various errors involved in measurement. Classify them into broad categories.
6. What are various types of instrumental errors? How these errors can be minimized ?
7. Differentiate clearly between systematic and random errors in measurements. **[U.P. Technical Univ. Elec. Measurements and Measuring Instruments 2002-03; 2005-06]**
8. Discuss random errors and explain how are they analyzed analytically. **[Pb.Univ. Elec. Measurements-I, June 1993]**
9. Explain different types of errors that occur in measuring instruments. How do you compute error limits?
10. Discuss how limiting errors, in y can be computed from measurement of two quantities u, v each having limiting errors, when (*i*) $y = u + v$, (*ii*) $y = \dfrac{u}{v}$.
11. Define probable and random errors. What do you understand by resolution of an instrument? **[Pb. Univ. Elec. Measurements-I, June 1993]**
12. Differentiate giving suitable examples:
    (*i*) Probable errors (*ii*) random errors, (*iii*) systematic errors. **[Pb. Univ. Elec. Measurements-I, December 1992]**

13. Define probable errors and explain how they are statistically dealt with.

[Pb. Univ. Elec. Measurements-I, January 1991]

14. Give the meanings of the following terms:

(*i*) Precision (*ii*) Accuracy (*iii*) Standard deviation, and (*iv*) Probable error.

[U.P.S.C. I.E.S. Elec. Engineering-I, 1994]

SHORT ANSWER TYPE QUESTIONS WITH ANSWERS

Q. 1. What is meant by an absolute error of measurement ?

Ans. The difference between the measured value A_m and the true value A of the unknown quantity or measurand is known as the absolute error of measurement δA *i.e.*

$$\delta A = A_m - A$$

Q. 2. What is meant by reading correction and how is it related to absolute error ?

Ans. The difference of true value A and measured value A_m of the measurand is known as the reading correction. Absolute error and reading correction are of the same magnitude but are of opposite sign.

Q. 3. What is difference between absolute error and relative error?

Ans. Absolute error δA is equal to the difference of the measured value A_m and the true value A of the measurand.

$$i.e.\ \delta A = A_m - A.$$

But relative error is equal to the ratio of absolute error to the true value of the quantity under measurement.

$$i.e.\ \text{Relative error, } \varepsilon_r = \frac{\delta A}{A} = \frac{\varepsilon_o}{A} = \frac{\text{Absolute error}}{\text{True value}}.$$

Q. 4. How is relative error expressed ?

Ans. Relative error is defined as the ratio of absolute error to the true value of the measurand and is expressed in fraction or percentage.

Q. 5. What is meant by limiting error ?

Ans. The limits of the deviations from the specified value, as mentioned by the manufacturer of the equipment/apparatus is known as limiting error.

Q. 6. What is meant by resolution of an instrument ?

Ans. The resolution of any instrument is the smallest change in the input signal (quantity under measurement) which can be detected by the instrument. It may be expressed as an actual value or as a fraction or percentage of the full-scale value.

Q. 7. How do accuracy and precision differ from each other ?

Ans. Accuracy is defined as the degree of exactness (closeness) of a measurement compared to the expected (desired) value, whereas precision is a measure of the consistency or repeatability of measurements, *i.e.* successive readings do not differ. Precision is the consistency of the instrument output for a given value of input.

In brief, accuracy can be defined as *conforming to truth* and precision can be defined as *sharply* or *closely defined*.

Q. 8. Name the categories of static errors.

Ans. Static errors can be grouped in three categories viz. gross errors, systematic errors and random or accidental errors. Systematic errors may further be categorised as instrumental errors, environmental errors and observational errors.

Q. 9. What are systematic errors ?

Ans. Systematic errors remain constant or change according to a definite law on repeated measurement of the given quantity. These errors can be evaluated and their influence on the results of measurement can be eliminated by the introduction of proper corrections.

Q. 10. How do random errors differ from systematic errors ?

Ans. Systematic errors remain constant or change according to a definite law on repeated measurement of the given quantity whereas random errors, also called the accidental errors, are of variable magnitude and sign and do not obey any known law.

Q. 11. What is meant by "arithmetic mean", "average deviation" and "standard deviation" ?

Ans. *Arithmetic mean* is the sum of the number of observations divided by the number of observations *i.e.*

$$\overline{x} = \frac{x_1 + x_2 + ... + x_n}{n} = \frac{\sum x}{n}$$

where $\overline{x}$ is the arithmetic mean and $x_1, x_2, ..., x_n$ are the readings taken and n is the number of readings taken.

Average deviation is the sum of the scaler values (without sign) of the deviations divided by the number of readings *i.e.*

$$\text{Average deviation, D} = \frac{|d_1|+|d_2|+...+|d_n|}{n} = \frac{\sum d}{n}$$

where $d_1, d_2 ... d_n$ are the deviations of the readings from the mean.

Standard deviation, also known as *root mean square deviation*, of an infinite number of data is the square root of the sum of all the individual deviations squared, divided by the number of readings

$$\textit{i.e. } \text{Standard deviation, } \sigma = \sqrt{\frac{d_1^2 + d_2^2 + d_3^2 + ... + d_n^2}{n}}.$$

Q. 12. What is difference between "standard deviation" and "variance" ?

Ans. Standard deviation is the root mean square deviation whereas variance is the mean square deviation. So variance, V = Mean square deviation = σ^2.

Q. 13. What is probable error ?

Ans. The probable error is given as $r = 0.6745\ \sigma$, where σ is the standard deviation. The quantity r is called the probable error in the sense that there is an even chance that any one observation will have a random error not greater than $\pm r$.

Q. 14. What is the expression for probable error if the readings of measurement follow Gaussian distribution?

Ans. The probable error is given as $r = 0.6745\ \sigma$, where σ is the standard deviation.

PROBLEMS

1. The measured value of a resistance is 111 Ω whereas its true value is 110 Ω. Determine absolute error of measurement and relative error. **[Ans.** 1 Ω; 1%]
2. The capacitance of a capacitor is specified as 200 μF ± 5 percent by manufacturer. Determine the limits of capacitance between which it is guaranteed. **[Ans.** 200 ± 10 μF]
3. A moving coil voltmeter has a uniform scale with 100 divisions and gives full-scale reading of 200 V. The instrument can read upto 1/5th of a scale division with fair degree of certainty. Determine the resolution of the instrument in volt. **[Ans.** 0.4 V]
4. The value of resistance r was determined by measuring current I flowing through the resistance with an error $\varepsilon_1 = \pm 1.5$ percent and power loss P in it with an error $\varepsilon_2 = \pm 1$ percent.

Determine the maximum possible relative error to be expected on measuring resistance r calculated from formula

$$r = \frac{P}{I^2} \qquad \text{[Ans. } \pm 4\%]$$

5. Three resistors have the following ratings

$$R_1 = 47\ \Omega \pm 4\%,\ R_2 = 65\ \Omega \pm 4\%;\ R_3 = 55\ \Omega \pm 4\%$$

Determine the magnitude and limiting errors in ohms and in percentage of the resistance of these resistors connected in series. **[Ans.** 167 Ω, ± 6.68 Ω; ± 4%]

6. A voltmeter and an ammeter are to be used to determine the power dissipated in a resistor. Both the instruments are guaranteed to be accurate within ± 1% at full-scale deflection. If the voltmeter reads 80 V on its 150 V range and the ammeter reads 70 mA on its 100 mA range, determine the limiting error for the power calculation. **[Ans.** ± 3.3%; ± 0.185 watts]

7. The unknown resistance is determined by Wheatstone bridge and is given by the equation

$$R_x = \frac{R_2 R_3}{R_1}$$

where $R_1 = 100 \pm 0.4\%\ \Omega$; $R_2 = 800 \pm 0.7\%\ \Omega$ and $R_3 = 825 \pm 0.5\%\ \Omega$

Determine the magnitude of the unknown resistance and the limiting error in percent and in ohms for the unknown resistance. **[Ans.** 6,600 Ω; ± 1.6%; 105.6 Ω]

8. Find the value of $(R_1 + R_2)$, considering the errors in their values as (*i*) limiting error (*ii*) standard deviation when

$$R_1 = (100 \pm 2\%)\text{ ohms},\ R_2 = (200 \pm 2.5\%)\text{ ohms}$$

[U.P.S.C. I.E.S. Electrical Engineering-I, 1997]

[Ans. Limiting error = ± 7 Ω; ± 2.333%; 9.8995 Ω]

9. The measurement of resistance of a resistor gives the following results:

101.2, 101.7, 101.3, 101.0, 101.5, 101.3, 101.2, 101.4, 101.3, 101.1 ohms.

Assuming that the random errors are present calculate (*i*) arithmetic mean (*ii*) the standard deviation of the readings, (*iii*) the probable error of average of 10 readings. **[Pb. Univ. Elec. Measurements-I, December 1999]**

[Ans. 101.3, 0.2 Ω, 0.04497 Ω]

10. Following readings were obtained in respect of measurement of a resistor:

0.903, 0.895, 0.892, 0.899, 0.901, 0.890, 0.906, 0.897, 0.912, 0.892, 0.898 and 0.902 Ω.

Determine (*a*) the arithmetic mean (*b*) the standard deviation of the readings (*c*) the probable error of mean value. **[Ans.** (*a*) 0.8989 Ω (*b*) 0.00637 (*c*) 0.138%]

11. Given the following set of voltage measurements taken from a voltmeter, find their (*i*) average value (*ii*) average deviation (*iii*) standard deviation and (*iv*) probable error.

Data: 153 V, 162 V, 157 V, 161 V, 155 V. **[Ans.** (*i*) 157.6 V (*ii*) 3.12 V (*iii*) 3.847 Ω (*iv*) 2.595 Ω]

12. Two resistors R_1 and R_2 are connected in series and then in parallel. The value of resistances are:

$$R_1 = 100.0 \pm 0.1\ \Omega;\ R_2 = 50 \pm 0.05\ \Omega$$

Calculate the uncertainty in the combined resistance for both series and parallel arrangements.

[U.P.S.C. I.E.S. Elec. Engineering-I, 1994]

[Ans. (*i*) (150 ± 0.1118) Ω; (*ii*) (33.333 ± 0.02485) Ω]

❑❑❑

3

PMMC Instruments

3.1 INTRODUCTION

The permanent magnet moving coil (PMMC) instrument is a direct reading instrument. It consists basically of a circular or rectangular coil of very fine insulated wire with many turns usually wound on an aluminium frame or bobbin and suspended between the poles of a permanent magnet. When current flows through the coil, a uniform magnetic field is produced. This field interacts with the field produced due to the permanent magnet, resulting in partial rotation of the coil. A pointer connected to the coil deflects over a calibrated scale, indicating the magnitude of current flowing through the coil.

The PMMC instrument is essentially a milliammeter or microammeter. However, with the use of parallel-connected resistors, it can be employed for measuring a wide range of direct current magnitude. This instrument can be converted into a voltmeter by connecting appropriate value resistor in series with the coil. Ohmmeters can be made from precision resistors, PMMC instruments, and batteries.

3.2 PERMANENT MAGNET MOVING COIL (PMMC) INSTRUMENT

3.2.1. Principle of Operation. A deflection instrument makes use of a pointer that moves over a calibrated scale to indicate the magnitude of *measurand* (quantity under measurement). For this to occur, three torques are operating on the electromechanical mechanism (or movement) inside the instrument: a *deflecting torque*, a *controlling torque* and a *damping torque*. The *deflecting torque* causes the moving system of the instrument to move from its zero position when the instrument is connected in an electrical circuit to measure the electrical quantity.

The magnitude of the movement of the moving system would be somewhat indefinite under the influence of deflecting torque unless some controlling torque existed. This torque opposes the deflecting torque and increases with the increase in deflection of the moving system, thus limits the movement and ensures that the magnitude of the deflection is always the same for a given value of quantity to be measured. Under the influence of the controlling torque the pointer (or moving system) will return to its zero position on removing the source producing the deflecting torque. Without controlling system, the pointer would swing over its maximum deflected position irrespective of magnitude of current and moreover, once deflected it would not return to its zero position on removing the current.

Damping force or torque is also necessary to avoid oscillations of the moving system about its final deflected position owing to the inertia of the moving parts and to bring the moving system to rest in its final deflected position quickly.

In the absence of damping, the moving system of an instrument would oscillate about the position at which the deflecting and restoring torques are equal. The function of damping is to absorb energy from the oscillating system and to bring it to rest promptly in its equilibrium position so that its indication may be observed.

3.2.2. Construction. A permanent magnet moving coil instrument is shown in Fig. 3.1. Main components, the instrument consists of, are described below:

1. *Moving System.* The moving element essentially consists of a light rectangular former, on which a multi-turn rectangular coil of 0.02 to 0.2 mm diameter insulated copper or aluminium wire is wound. The ratio of length to breadth of the former is kept between 1.3 and 1.5 in order to have best compromise between high weight-efficiency and high torque-efficiency. The width of the winding across the former is usually kept as narrow as possible, 6 mm to 10 mm, as the broader coil will cause trouble to occur from fouling the poles and difficulty of getting it into place over the iron core without serious distortion. Formers may be of copper or of aluminium but aluminium is usually preferred on account of weight and inertia. In some cases former of phosphor bronze or German silver is also used in order to avoid overdamping. The former is lightly insulated by lacquering or serving it with a thin layer of insulation before winding, and the conductors are laid over it as evenly as possible. Copper wire, insulated with enamel, silk, nylon etc., is widely used but in order to have reduced weight and inertia aluminium wire is preferred nowadays. The ampere-turns in various instruments vary over a wide range; the lowest value seems to be approximately 0.2 and the highest 1.5. Most of the instruments have the ampere-turns between 0.4 and 0.7.

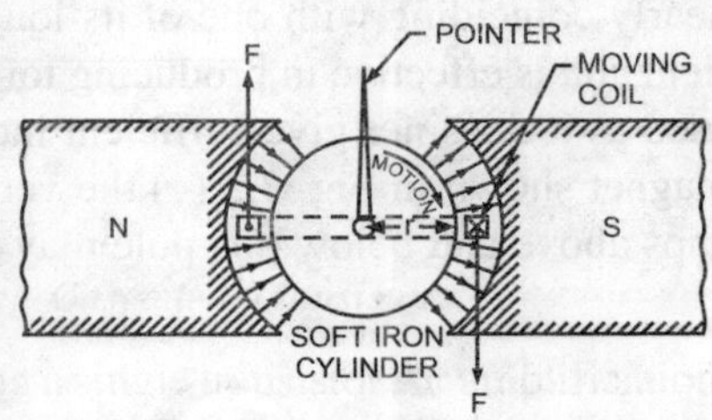

(a) Permanent Magnet Moving Coil Instrument

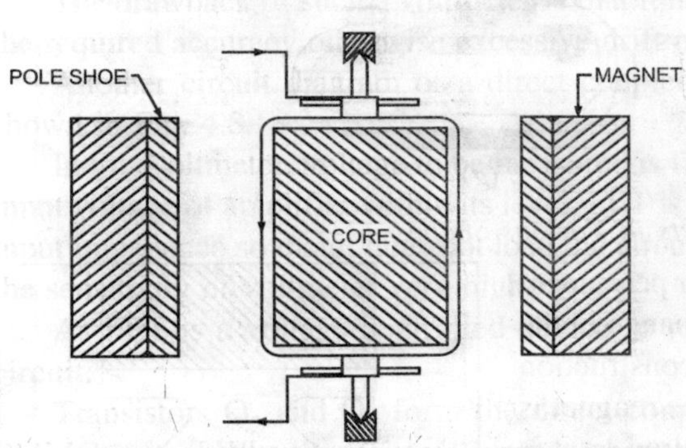

(b) Permanent Magnet Moving Coil Instrument

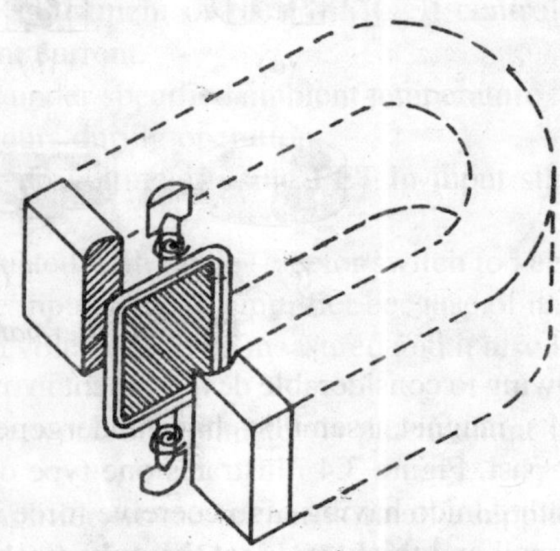

(c) Permanent Magnet Moving Coil Instrument

Fig. 3.1

The coil is so pivoted on jewel bearings that its sides lie in the air-gaps between the poles of a permanent magnet and a soft-iron cylinder and the coil can rotate freely. The V-jewel, shown in Fig. 3.2 (*a*), is widely used in instrument bearings. The pivot, bearing in the pit in the jewel, may have a radius at its tip from 0.01 mm to 0.02 mm, depending on the weight of the moving element and the vibration the instrument will encounter. The radius of the pit in the jewel is kept slightly larger than the pivot radius in order to have contact area circular, only a few microns across. Such a design has the least friction of any practical type of instrument bearing. Specially protected instruments employ the spring-back (in cabloc) jewel bearing, shown in Fig. 3.2 (*b*). It is located in normal position by the spring and is free to move axially when the mechanism is subjected to severe shock.

(a)

(b)

Fig. 3.2 *Instrument Bearings*

2. *Magnet System.* The permanent magnets, used by different manufacturers for construction of permanent magnet moving coil instruments, in various shapes are shown in Fig. 3.3. Simple U-shaped permanent magnet is most widely used magnet and is shown in Fig. 3.3 (*a*). The expanded horse shoe type magnet shown in Fig. 3.3 (*b*), circular one shown in Fig. 3.3 (*c*) and Siemens and Halske type shown in Fig. 3.3 (*d*) have definite advantages over U-shaped magnet but are more costly to produce. In a single gap magnet shown in Fig. 3.3 (*e*) the polar extensions curve over one another, as shown, and the coil is pivoted on an axis parallel to and nearly coinciding with one of its long sides, so that only the other long side is in field and is effective in producing torque. The magnet shown in Fig. 3.3 (*f*) is not used as it does not give sufficient induction in the gap and is not well designed in permanence. In the magnet shown in Fig. 3.3 (*g*) the vertical gap flux links with the two horizontal sides, so that the two gaps above and below this pole may be considered as in parallel.

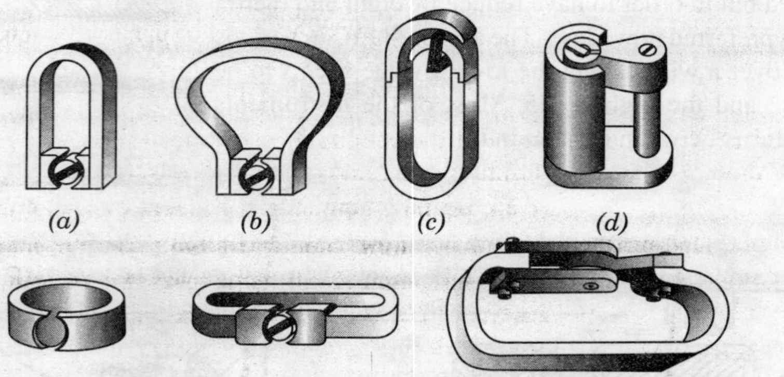

(*a*) (*b*) (*c*) (*d*) (*e*) (*f*) (*g*)

Fig. 3.3 *Permanent Magnets in Various Shapes*

Owing to considerable development in materials for permanent magnets, magnet assemblies have undergone a lot of change in the recent past. Figure 3.4 illustrates one type of magnet construction in which Alnico having high coercive force is used. The magnet is very small and thick, and that the only machining required on it is to grind the ends flat and parallel to receive the soft iron poles.

Thus in small instruments it is possible to employ a small coil having small number of turns and so reduction in volume. In this design the movement of the coil is restricted and so angular span of the scale is restricted to approximately 90º.

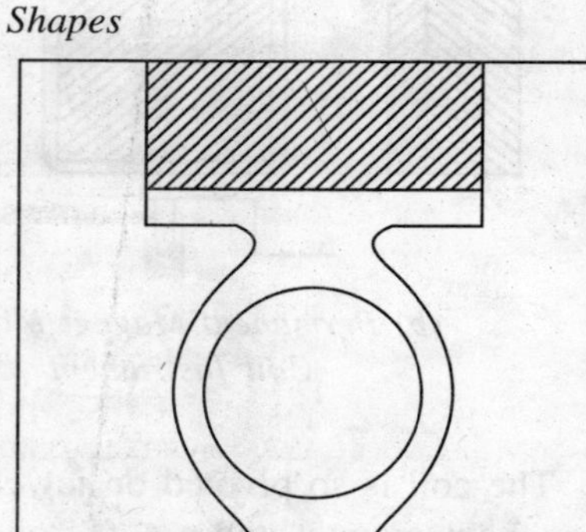

Fig. 3.4

Another design utilizing the special properties of Alnico is the concentric structure shown in Fig. 3.5. Two sector-shaped Alnico magnets, having ground cylindrical faces, are placed inside a soft-steel ring and the assembly is die-cast together. The moving element and soft-iron core are mounted in a separate die-cast frame which fits between the inner Alnico pole faces. The soft-steel ring, surrounding the Alnico magnets provides magnet shielding to the moving coil against stray magnetic fields. In this construction the Alnico magnet is magnetized radially, i.e. the entire inner surface of the ring shaped magnet is one pole, and the soft-steel core is the other pole. Only the outer side of the moving coil is active in developing torque, as the inner side is inside the middle core and has no field. This type of construction gives coil travel of about 240º, a long scale in an instrument having small overall dimensions. Such a construction is used in many panel type instruments and some portable instruments.

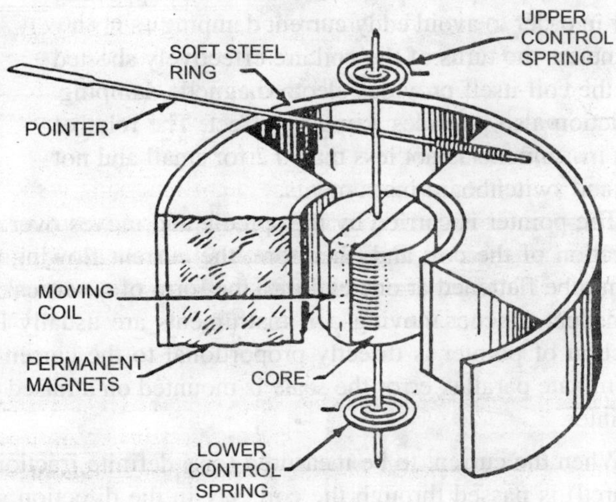

Fig. 3.5 *Concentric Magnet Instrument*

In Fig. 3.6 a design is shown in which the magnet is inside the moving coil, and the outside member is a soft steel ring to complete the magnetic circuit. This construction is very compact and light and have the obvious advantage of being relatively unaffected by external magnetic fields, eliminating the magnetic shunting effects in steel panel construction, where many meters operating side by side may affect each other's readings. The requirement for magnetic shielding, in the form of iron cases, is also eliminated by this construction.

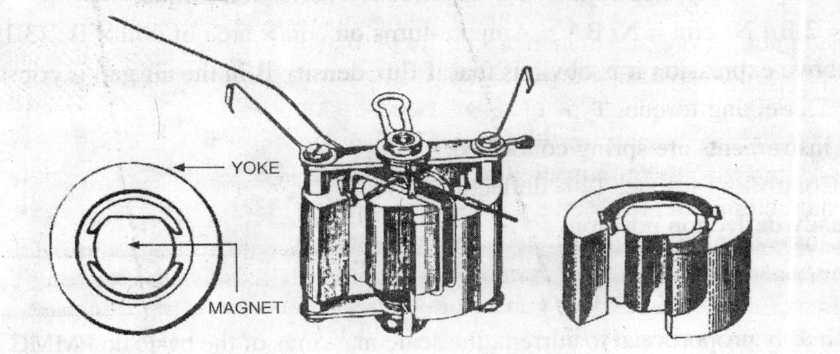

Fig. 3.6 *Core Magnet Construction*

3. *Control System.* The controlling torque is provided by two phosphor bronze hair springs, either helical or spiral, coiled in opposite directions, normally equal in strengths. Constancy of spring performance is essential in order to maintain accuracy. The spring thickness is accurately controlled in manufacture to avoid permanent set of the spring.

The control springs also conduct the operating current into and out of the moving coil. Figure 3.7 shows how the current under measurement flows from one of the instrument terminals to the second terminal through the spiral spring, coil winding and the second spring.

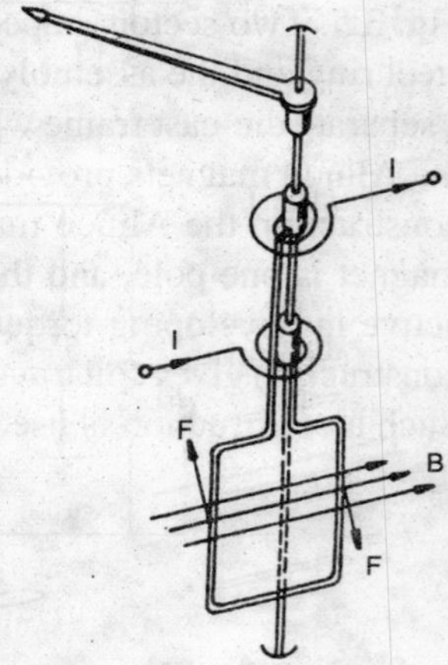

Fig. 3.7

4. *Damping System.* Usually eddy current damping is employed in this type of instrument. In microammeters the moving coil is made self supporting *i.e.*, without a metal former in order to avoid eddy current damping as it may be too strong. In microammeters the turns of the coil are effectively shorted by the ammeter shunt and the coil itself provides electromagnetic damping. In addition, the bearing friction also provides damping effect. The relative damping employed in such instruments is not less than 0.2 for small and not less than 0.45 for portable and switchboard instruments.

5. *Pointer and Scale.* The pointer is carried by the spindle and moves over a graduated scale and indicates the angular deflection of the coil and, therefore, the current flowing through the coil. The pointer is a light aluminium tube flattened at one end into the form of a vertical knife edge. The scale markings of basic dc permanent magnet moving coil instruments are usually linearly spaced as the deflecting torque or deflection of pointer is directly proportional to the current flowing through the moving coil. In order to eliminate parallax error the scale is mounted on a raised platform and a mirror is provided beneath the pointer.

3.2.3. Torque Equation. When the current to be measured (or a definite fraction of it or proportional to the voltage to be measured) is passed through the coil, say in the direction shown in Fig. 3.2 (*a*), deflecting torque is produced on account of reaction of the permanent magnetic field with the coil magnetic field. The direction of deflecting torque can be determined by applying Fleming's left hand rule.

If i is the current in amperes flowing through the coil of turns N and length l metres and B is the flux density in teslas in the air gap,

Then deflecting force, $F = BilN$ Newtons

If r is the distance in metres between the centre of the coil and the force F

Since force acts on each side of the coil, the effective deflection torque,

$T_d = 2\,Fr = 2\,BilNr$ Nm $= NiBA$ = Ampere-turns on coil × area of coil × B ...(3.1) $\quad \because A = 2\,lr$

From the above expression it is obvious that if flux density B in the air gap is constant, then

Deflecting torque, $T_d \propto i$

Since such instruments are spring controlled,

Controlling torque, $T_c \propto$ deflection θ

Since in steady deflection position

$$T_c = T_d,$$

$$\theta \propto i \qquad ...(ii)$$

Since θ is directly proportional to current, the scale markings of the basic dc PMMC instrument are usually linearly spaced. If low frequency ac is applied to the moving coil, the pointer will deflect up-scale for one half-cycle of the input waveform and down-scale for the next half cycle. At power frequencies of 50 Hz and above, the pointer will not be able to follow the rapid variations in direction and will quiver slightly around the zero mark, seeking the average value of ac *i.e.*, zero. That is why, PMMC instrument is unsuitable for ac measurements, unless the current is rectified before applying it to the moving coil.

The deflecting torque equation indicates several things of interest from the design stand point. In case of a large instrument with a long and heavy pointer it is necessary to provide larger deflecting torque to overcome increased friction and also to maintain a satisfactory speed of response. The larger deflecting torque is usually obtained by increasing the dimensions of a moving coil. The larger torque can be obtained from an increase in the flux density. The flux densities used in PMMC instruments vary from 0.1 T to 1.0 T. Thus for smaller instruments it is possible to use a small coil with small number of turns.

The power requirement for PMMC movement to give full-scale deflection is surprisingly small ranging from 25 μW to 200 μW depending upon the size of the movement. These values may be further reduced by using improved magnetic materials.

Accuracy of PMMC instrument is usually of the order 2–5% of full-scale reading.

3.2.4. Errors in PMMC Instruments. The errors that usually occur in PMMC instruments are as follows:

(*i*) Frictional error, (*ii*) Temperature error, (*iii*) Error owing to weakening of permanent magnet, (*iv*) Stray magnetic field error, (*v*) Thermoelectric error.

(*i*) *Frictional Error*. The frictional error in PMMC instruments is usually very small with proper pivoting and balancing because of high torque-weight ratio. The mechanical error, which is more troublesome, may occur between core and coil owing to slight warping of the former on which the coil is wound. Such a trouble gradually increases with the time. It may be mainly due to too small a gap, too light a former, too great a winding tension, or an unsuitable and insufficiently hard setting cement. This trouble can be avoided by winding of coil carefully.

(*ii*) ***Temperature Error.*** This type of error, apart from the change in room temperature, is mainly due to heating of the working coils and other resistance coils connected internally in the instrument casing by the operating current. As a result there is a change in resistance of the working coils and other resistance coils. Such an error is not serious in case of an ammeter but it is very important in case of a voltmeter which is of high resistance and where constancy of resistance is very important. For reducing such an error the power lost in a voltmeter should be kept very small and arrangement for ventilation and cooling should be provided. Series resistance and shunts housed internally in instrument cases should be well separated from working elements and well ventilated. Other errors resulting from heating may be caused by expansion of the control spring or of other parts of the instruments but such errors are usually negligibly small.

The PMMC instruments are not inherently insensitive to temperature, but can be *temperature compensated* by the appropriate use of series and shunt resistors of copper and manganin.

The change in temperature affects the coil resistance, tension of spring and field strength of the magnets. With the increase in temperature the spring becomes more flexible by about 0.04 percent per °C rise and the pointer indicates high. The magnets become weaker at high temperature, which nullifies the effect of springs. The net effect of spring and magnets at high temperature is that the movement gives reading high by approximately 0.013 to 0.02 percent per °C rise. If the voltmeter circuit is made entirely of copper, the instrument will give indication low at high temperature by approximately 0.4 percent per °C rise. The *uncompensated instrument* tends to read low by approximately 0.2 percent per °C rise in temperature. Temperature compensation may be affected by connecting a series resistor, called the *swamping resistor*, in series with the moving coil. The swamp resistor is made of manganin, which is hardly affected by temperature variations, combined with copper in the ratio of 20:1 to 30:1. The total resistance of the moving coil and swamping resistor increases slightly with a rise in temperature, but only just enough to overcome the effect of springs and magnets, so that the overall temperature effect is zero. Manufacturers design these instruments so that the temperature error is not more than half the guaranteed tolerance for a temperature change of 10ºC.

Instruments in the millivolt range may be compensated, as shown in Fig. 3.8. Here the total resistance of the circuit increases slightly with a rise in temperature, due to copper moving coil and the copper shunt resistor. By correct proportioning of the copper and manganin components in the circuit, the temperature effect may be completely nullified.

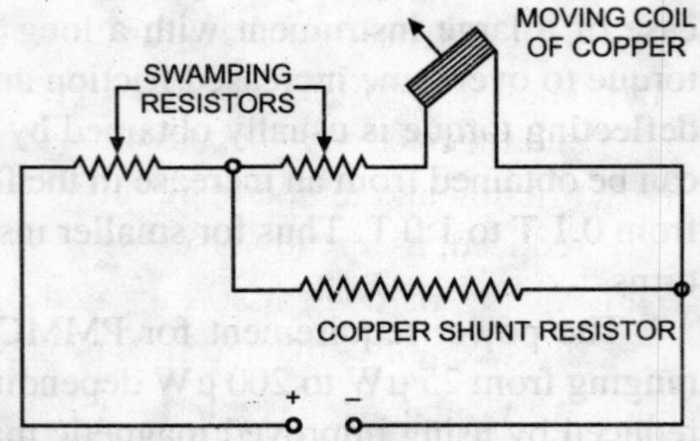

Fig. 3.8

(*iii*) *Error Owing To Weakening of Permanent Magnet.* This error is because of weakening of permanent magnet with the time of usage. In good magnet, this occurs very slowly, but in instruments liable to vibration, change of position, and stray magnetic fields *i.e.*, in portable instruments demagnetization may be considerably accelerated. In some portable instruments the error of this type may be as high as 5–10%. However, such an error can be reduced by proper attention to ageing and reasonable care in use. The effect, however, can be eliminated completely by weakening of mechanical control and attaching a symmetrical iron or steel needle to the coil. The magnetic needle is so arranged that magnetic control provided by it diminishes as the magnet becomes weaker.

(*iv*) *Stray Magnetic Field Error*. It is said that the PMMC instruments are not affected by the external fields owing to its strong field. But this is not strictly true as due to presence of iron in the working parts the effect of external magnetic fields is increased, specially when the disturbing field coincides with, or opposes the field in the gap of the instrument. This type of error can be avoided by using an iron case for ordinary type instruments. Unshielded instruments of ordinary type require careful consideration and the close proximity of two instruments may cause an appreciable error.

(*v*) *Thermoelectric Error*. This error, of course, pertains chiefly to ammeters using shunts, and is due to uneven temperature distribution in the shunt, arising from unequal cooling, position of mounting, bad contacts at the current terminals, and the Peltier effect. The error of this type may be detected by switching off the current flowing through the shunt after it has attained its maximum temperature and leaving the instrument connected. In case of presence of such an error the pointer will not return immediately to zero, but will gradually approach that position as the temperature of the shunt decreases.

The best method of eliminating such an error is by using an alloy of small thermoelectric power to copper such as manganin, tarnac, therlo, but such materials require careful treatment as regards soldering and protection. So materials of constantan class are used owing to cheapness, despite their higher thermoelectric properties. One method of compensating for this error is by connecting the instrument to the shunt through strips of the same material as that used for the shunt plates, making these tags sufficiently long to ensure that the outer ends, to which the copper leads are soldered, are at the same temperature.

3.2.5. Merits and Demerits of PMMC Instruments. *Merits.* (*i*) Uniform scale.

(*ii*) Low power consumption (say from 25 microwatt to 200 microwatt) because of small driving power.

(*iii*) No hysteresis loss as the former is of copper or aluminium.

(*iv*) Very effective and reliable eddy current damping.

(*v*) High torque-weight ratio resulting in high accuracy (2–5% of full-scale reading).

(*vi*) Use of single instrument for measurement of currents and voltages by employing shunts and multipliers of different resistances.

(*vii*) No effect of stray magnetic fields because of use of intensive polarised or unidirectional field.

(*viii*) The instrument using core magnet is very suitable in aircraft and aerospace applications, where a multiplicity of instruments are mounted in close proximity to each other. This is because of self-shielding property of core magnets.

Demerits. (*i*) These instruments cannot be used for ac measurements.

(*ii*) These are costlier in comparison to moving iron instruments because of delicate construction and the necessary accurate machining and assembly of various parts.

(*iii*) Friction and temperature might introduce errors, as in case of other instruments, which can be reduced, as already discussed.

(*iv*) Ageing of control springs and of the permanent magnets might cause errors which can be considerably obviated by careful choice of material and preageing during manufacture.

Example 3.1. The moving coil of a PMMC instrument has 100 turns and dimensions of 4 cm × 3 cm. The flux density in the air gap is 0.05 T. Determine the torque developed by each coil when it carries a current of 20 mA.

Solution: Number of turns on the coil, $N = 100$

Area of coil, $A = 4 \times 3 = 12 \text{ cm}^2 = 12 \times 10^{-4} \text{ m}^2$

Current carried by the coil, $i = 20 \text{ mA} = 0.02 \text{ A}$

Flux density in the air gap, $B = 0.05 \text{ T}$

Torque developed by the coil, $T = NiBA = 100 \times 0.02 \times 0.05 \times 12 \times 10^{-4} = 12 \times 10^{-5} \text{ Nm}$ **Ans.**

Example 3.2. The moving coil of a PMMC instrument has 125 turns and is of dimensions 4 cm × 2.5 cm. The flux density in the air-gap is 0.06 T. Determine the deflection when the coil carries a current of 25 mA. The spring control provides a deflection of one degree for a torque of 25 × 10⁻⁷ Nm.

Solution: Number of turns on the coil, $N = 125$

Area of coil, $A = 4 \times 2.5 = 10 \text{ cm}^2 = 1 \times 10^{-3} \text{ m}^2$

Current carried by the coil, $i = 25 \text{ mA} = 0.025 \text{ A}$

Flux density in the air-gap, $B = 0.06 \text{ T}$

Torque developed by the coil, $T_d = NiBA = 125 \times 0.025 \times 0.06 \times 1 \times 10^{-3} \text{ Nm} = 18.75 \times 10^{-5} \text{ Nm}$

Controlling torque, $T_c = 25 \times 10^{-7} \, \theta \text{ Nm}$

where θ is the deflection in degrees produced by 25 mA current

Since for steady deflection state, $T_c = T_d$,

$$25 \times 10^{-7} \theta = 18.75 \times 10^{-5}$$

$$\text{or } \theta = \frac{18.75 \times 10^{-5}}{25 \times 10^{-7}} = 75° \quad \textbf{Ans.}$$

Example 3.3. The resistance of a moving coil voltmeter is 12,000 Ω. The moving coil has 100 turns and is 4 cm long and 3 cm wide. The flux density in the air gap is 6 × 10⁻² Wb/m². Find the deflection produced by 300 V if the spring control gives a deflection of one degree for a torque of 25 × 10⁻⁷ Nm.

[Allahabad Univ. Electrical Engineering 1974]

Solution: Voltage applied, $V = 300 \text{ V}$

Coil resistance, $R = 12{,}000 \ \Omega$

Current flowing through the coil, $i = \frac{V}{R} = \frac{300}{12,000} = 0.025$ A

Deflecting torque, $T_d = NiBA = 100 \times 0.025 \times 6 \times 10^{-2} \times 0.04 \times 0.03 = 18 \times 10^{-5}$ Nm

Controlling torque, $T_c = 25 \times 10^{-7}\,\theta$ Nm

where θ is the deflection in degrees produced by 300 V

Since for steady deflection state $T_c = T_d$,

$$\therefore\ 25 \times 10^{-7}\,\theta = 18 \times 10^{-5}$$

$$\text{or}\quad \theta = \frac{18 \times 10^{-5}}{25 \times 10^{-7}} = 72° \ \textbf{Ans.}$$

3.3 GALVANOMETER

The term 'galvanometer' is commonly applied to any current-indicating instrument which is not calibrated, but in the majority of the cases an instrument operating with a small current is implied. Galvanometers are often used in deflection methods, but their principal application is as indicators of zero current in bridge and potentiometer measurements. In many of the applications the circuit is adjusted to give a zero deflection or "null" on the galvanometer, so the function of the galvanometer is to show the presence or absence of current with no requirement to measure the actual magnitude. It is merely necessary for the instrument to show a readable deflection for the smallest current that is significant in the particular application. Thus a galvanometer besides being sufficiently sensitive, should have a stable zero, a short periodic time, and nearly critical damping, so that the movement comes to rest quickly at a definite position.

There are a few places in measurement work in which deflections of a galvanometer are read, but such use is avoided as much as possible, for the accuracy in reading a deflection is seriously limited.

Robustness and portability are obviously desirable but may be unobtainable in the more sensitive instruments.

A *galvanometer* is essentially a PMMC instrument designed to be sensitive to extremely small currents. The torque equation for a galvanometer is exactly as discussed in Art. 3.2.3.

The main advantage of permanent magnet moving coil galvanometers, frequently known as D'Arsonval type instrument, lies in its freedom from the effects of stray fields and the ease with which a long and uniform scale may be attained. These are the features that have caused it to be adopted, in a modified form, for direct current ammeters and voltmeters.

The essential features of such a galvanometer are shown in Fig. 3.9.

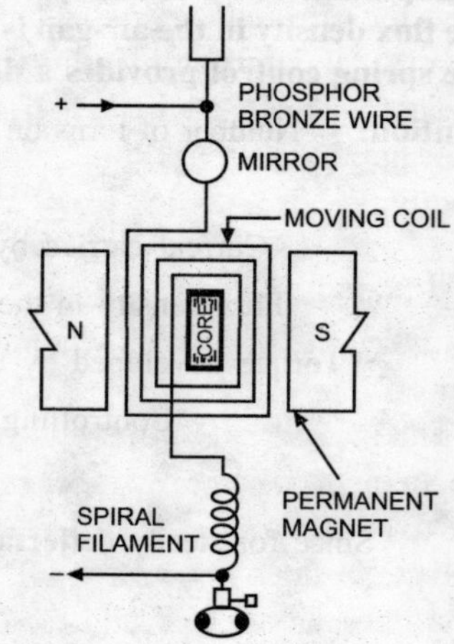

Fig. 3.9 *D'Arsonval Galvanometer*

The instrument consists of a circular or rectangular coil of very fine insulated wire with many turns usually wound on an aluminium frame or bobbin and suspended between the poles of a permanent magnet. The coil is suspended by a phosphor bronze filament which acts as one lead-in for the coil. The other lead-in wire is a very flexible spiral wire at the bottom of the coil. The elasticity of the filament sets up a moderate torque in opposition to the rotation of the coil. The poles of the magnet are usually cylindrical in shape and a cylindrical soft iron core is placed between the poles and coaxial with them. The cylindrical form of the poles faces and core has got two advantages. The length of the air gap is reduced so that the amount of flux linking the coil is increased, thus increasing the sensitivity of the instrument; and the flux between the pole faces and core is practically radial. In radial field the deflection of the coil is directly proportional to the current in the coil and, therefore, a uniform scale is obtained.

When current flows through the coil, a deflecting force proportional to the flux density, the coil current and the dimensions of the coil makes the coil to rotate on its vertical axis. The deflecting force is opposed by the restraining force of the suspension filament so that coil does not continue to rotate as in a motor but turns until the deflecting force is balanced by the restraining force of the suspension filament. Since the deflecting torque is directly proportional to the coil current, the amount of deflection of the coil indicates the magnitude of current flowing in the coil.

The amount of deflection and hence the amount of current flowing through the coil may be indicated by a pointer attached to the moving element and moving over a calibrated scale.

A torsion head is provided for adjustment of position of coil and also for zero adjustment.

On the sensitive galvanometer a small mirror is attached to the moving coil to indicate the position of the coil. It may be used in two ways. The first way is to mount a telescope and scale in front of the galvanometer, so that the observer reads the scale reflected by the mirror. The second arrangement which is more common and less tiring to the user, consists of a lamp and scale device in which a spot of light is projected on the mirror and reflected to the scale. The scale is usually mounted 1 metre away from the mirror, although 0.5 m is used sometimes for compactness.

The aluminium frame, upon which the coil is wound, provides the *damping force* in the instrument. Whenever the frame moves in the magnetic field, currents are induced flowing round the closed circuit formed by the frame. The induced currents oppose the motion that produces them, thereby bringing the moving element immediately to a standstill. Damping may also be obtained by connecting a low resistance across the terminals of the galvanometer. Damping torque depends upon the resistance and so we can have critical damping by adjustment of resistance.

This type of galvanometer requires careful levelling so that the coil hangs straight and centrally in the air gap without rubbing on poles or cylinder. Some galvanometers, intended to be portable and not to require exact levelling have *"taut suspensions"* consisting of straight flat strips, top and bottom, kept under tension.

With modern refinements, the suspension galvanometer is still used in certain high sensitivity laboratory measurements when the delicacy of the instrument is not objectionable and portability is not required.

The instrument of this type, and other instruments based on this construction, operate only on dc. If such an instrument is operated on ac supply, the magnet poles of its coil reverses rapidly. The coil is too large to swing back and forth 50 times per second (that's is a normal supply voltage frequency in India); therefore, it does not turn at all. An instrument intended for one ampere dc would not be damaged by one ampere ac but it would read zero on ac.

An ordinary simple galvanometer, by itself, is of very limited use. As a current indicator, only a very small current can be allowed to flow in the fine wire of the moving coil. And, since the coil is of very low resistance, only a very small voltage can be applied to the moving coil.

The most useful galvanometers are scaled as milliammeters or microammeters indicating how many thousandths or millionths of an ampere flow through the instrument.

The relation between the angular deflection θ of the coil and the linear displacement of the light beam on the scale can be derived as follows:

As the mirror deflects from its initial position by an angle θ, the reflected beam is displaced by an angle 2θ, as depicted in Fig. 3.10, and the displacement can approximately be given as

$$d = r \times 2\theta$$

where r is the distance of scale from mirror. If the scale is 1 m from the mirror *i.e.* $r = 1$ m, then

$$d = 2{,}000\,\theta \text{ mm} \quad ...(3.2)$$

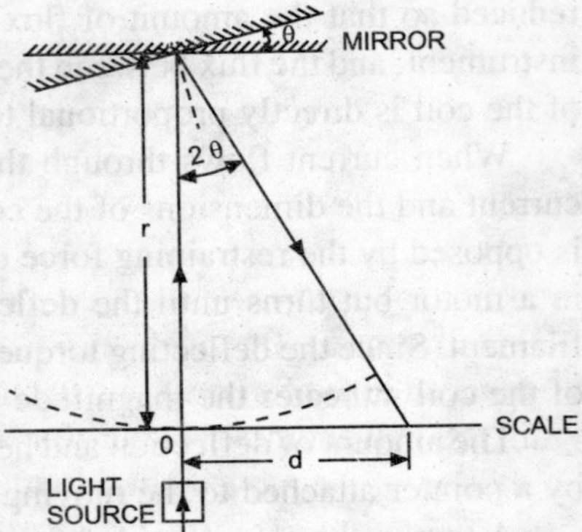

Fig. 3.10

Although the suspension type galvanometer is neither a practical nor portable instrument, the principles governing its operation apply equally to its more modern version, the PMMC movement, depicted in Fig. 3.11. This mechanism also consists of a coil, suspended in a magnetic field of a permanent magnet in the form of a horse-shoe. The coil is so suspended that it can rotate freely in the magnetic field. When current flows through the coil, an electromagnetic torque is developed causing the coil to rotate. The torque is proportional to the magnitude of a current being measured and is counter balanced by the mechanical torque of control springs attached to moving coil. The balance of torques, and therefore, the angular position of the moving-coil, is indicated by a pointer against a fixed reference, called a scale. The PMMC movement, also called the *D'Arsonval movement*, provides an instrument with very low power consumption and low current required for full-scale deflection.

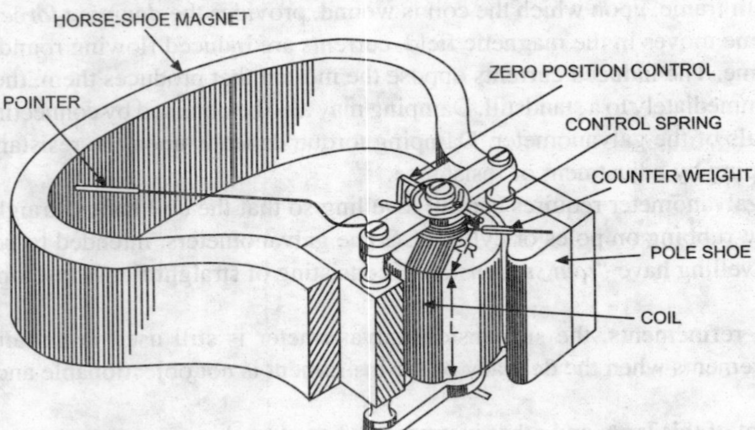

Fig. 3.11 *Constructional Details of External Magnet PMMC Movement*

If i is the current flowing through the coil of turns N and length l metres and B is the flux density is teslas (Wb/m^2) in the air gap, then deflecting force,

$$F = Bil\,N \text{ newtons} \quad ...(3.3)$$

If r is the distance between the centre of coil and the force F, then deflecting torque,

$$T_d = F \times 2r = 2Bil\,N\,r = Gi \text{ newton-metres} \quad ...(3.4)$$

where G is called the *displacement constant* of the galvanometer and

$$= BN \times (2\,lr) = BNA \text{ Nm/A} \quad ...(3.5)$$

If C is controlling or restoration constant of spring in Nm/radian, then controlling torque exerted by suspension,

$$T_c = C\theta \text{ where } \theta \text{ is the final deflection in radians.} \quad ...(3.6)$$

Since in steady-state deflected position

$$T_d = T_c$$

$$2\,BNilr = C\theta$$

$$\text{or } Gi = C\theta$$

$$\text{or } i = \frac{C\theta}{G} \qquad ...(3.7)$$

Damping Due to Coil Circuit Resistance. When the coil rotates in a magnetic field, a voltage is induced in the coil resulting in circulation of a current through the coil and the external resistor connected in the galvanometer circuit for this purpose. It causes an opposing or retarding torque, that damps the motion of the movement. Frequently, a *critical damping resistance* value is stated, which provides just sufficient damping to allow the pointer to settle down quickly with only a very small short lived oscillation. The external resistance required for critical damping is called CDRX or ECDR (critical damping external resistance).

Galvanometer Sensitivity. The sensitivity of a galvanometer may be defined in several different ways such as on the basis of current, voltage and mega ohms.

(a) Current Sensitivity. The deflection caused by unit current is known as *current sensitivity*. In case of a galvanometer attached with a mirror it may be defined as deflection in mm produced on a scale at a distance of one metre from the galvanometer mirror by unit current.

That is current sensitivity, $S_i = \dfrac{d}{i \times 10^6}$ mm per microampere where d is the deflection on scale in mm.

Sometimes the *current sensitivity* is defined as the current required to cause a deflection of one scale division.

The scale has usually 1 mm divisions and is placed at a distance of one metre from the galvanometer. If the deflection of the coil is θ, beam falling on scale will be deflected by an angle 2θ, as it is the reflected beam from the mirror.

$$\therefore\ d = 1{,}000 \times 2\theta \text{ mm}$$

and current sensitivity, $S_i = \dfrac{2{,}000\,\theta}{i \times 10^6} = \dfrac{\theta}{500\,i}$ mm/μA $= \dfrac{G}{500\,C}$ mm/μA...(3.8) $[\because Gi = C\theta]$

A term 'figure of merit', expressed in μA/mm, is also sometimes used in connection with a galvanometer. It has meaning opposite to that of current sensitivity.

(b) Voltage Sensitivity. The voltage sensitivity is defined in terms of deflection in scale divisions caused by unit voltage applied across galvanometer.

In case of galvanometer attached with a mirror the voltage sensitivity is defined as the deflection in mm produced on a scale at a distance of 1 m from the mirror by a unit voltage impressed on the galvanometer.

i.e. Voltage sensitivity, $S_V = \dfrac{d}{V \times 10^6}$ mm/μV $= \dfrac{d}{iR \times 10^6}$ mm/μV

For a galvanometer with scale 1 m away and one scale division = 1 mm,

$$S_V = \frac{G}{500\,CR} \text{ mm/}\mu\text{V} = \frac{S_i}{R} \text{ mm/}\mu\text{V} \qquad ...(3.9)$$

where R is the total galvanometer resistance in ohms.

(c) Megohm Sensitivity. The resistance required in MΩ in the circuit to cause a deflection of one scale division on impressing 1 V is called the *megohm sensitivity*.

In case of a galvanometer attached with a mirror the megohm sensitivity is defined as the resistance in MΩ placed in series with the galvanometer through which one volt of impressed voltage will cause a deflection of 1 mm on a scale at a distance of 1 m from the mirror. In fact megohm sensitivity is the current sensitivity expressed in μA per scale division.

$$S_o = \text{Resistance in M}\Omega \text{ to give a deflection of one scale division with 1 V}$$

$$= \frac{d}{1 \times 10^6} \text{ M}\Omega \text{ per scale division.}$$

For a galvanometer with scale 1 m away and one scale division = 1 mm

$$S_o = \frac{G}{500\ C} \text{ M}\Omega \text{ per mm} \qquad ...(3.10)$$

Galvanometer Shunts. A protective device of some kind must be employed with a sensitive galvanometer in the process of balancing a bridge, potentiometer or other network where the potential difference in the galvanometer branch may be relatively high. If the galvanometer has sufficient sensitivity for the final stages of balancing, the large unbalance that probably exist at the start is likely to damage the coil or suspension, or, if not this serious, it sends the light spot far off scale.

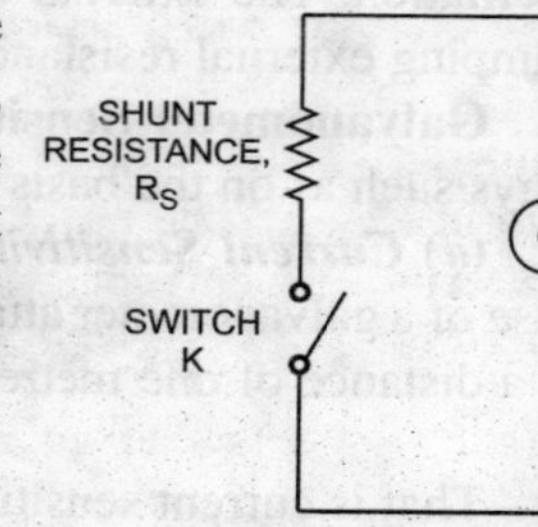

Fig. 3.12

The simplest procedure would be to use either a high resistance in series with the galvanometer to limit the current caused by available emf or a low-resistance shunt across the terminals of the galvanometer to bypass most of the current. Protection provided by connecting an adjustable resistance across the instrument is shown in Fig. 3.12. When the shunt resistance R_S is zero, all of the circuit current flows through the shunt. With the increase in shunt resistance above zero, an increasing amount of current flows through the coil of the galvanometer.

Example 3.4. Determine the angle through which coil turns when a deflection of 42 mm is observed on the scale of a galvanometer placed at a distance of 0.6 m from the mirror.

Solution: From Eq. (3.2) $\quad d = 2\,\theta\, r$

$$\text{or} \quad \theta = \frac{d}{2r} = \frac{0.042}{2 \times 0.6} = 0.035 \text{ radians or } 2^\circ \quad \textbf{Ans.}$$

Example 3.5. A moving coil instrument has a coil 1.5 cm wide and 1.2 cm long. The flux density in the air gap is 1.8 × 10⁻³ Wb/m². The spring constant is 1.4 × 10⁻⁷ Nm/radian. Determine the number of turns required to produce an angular deflection of 90° when current of 5 m A is flowing through the coil.

Solution: Controlling torque for deflection of 90° (or $\frac{\pi}{2}$ radians)

$$= K\,\theta = 1.4 \times 10^{-7} \times \frac{\pi}{2} = 2.2 \times 10^{-7} \text{ N-m}$$

If N is the number of turns for 90° deflection, then deflecting torque

$$T_d = B\,i\,NA = 1.8 \times 10^{-3} \times 5 \times 10^{-3} \times N \times 1.5 \times 10^{-2} \times 1.2 \times 10^{-2}$$

$$= 16.2 \times 10^{-10} \text{ N-m}$$

Since for steady deflection state, $T_d = T_c$,

$$16.2 \times 10^{-10}\ N = 2.2 \times 10^{-7}$$

$$\text{or Number of turns, } N = \frac{2.2 \times 10^{-7}}{16.2 \times 10^{-10}} = 136 \textbf{ Ans.}$$

Example 3.6. The coil of a moving coil meter has 200 turns, wound on a non-inductive former, its width being 2.0 cm and height 2.5 cm. It works in a constant field of 0.1 T. The moment of inertia of the moving parts is 3×10^{-7} kg-m^2 and the control spring produces a torque of 100×10^{-7} Nm per radian. Calculate the current in the coil to produce a deflection of 120°.

Solution: Controlling torque for 120° $\left(\frac{2\pi}{3} \text{ radians}\right)$ deflection

$$= C\theta = 100 \times 10^{-7} \times \frac{2\pi}{3} = 209.33 \times 10^{-7} \text{ N-m}$$

If the current required to produce a deflection of 120° is i amperes then deflecting torque

$$T_d = Bi\,NA = 0.1 \times i \times 200 \times 2 \times 10^{-2} \times 2.5 \times 10^{-2} = 0.01\, i \text{ Nm}$$

Since for steady deflection state, $T_d = T_c$,

$$0.01\, i = 209.33 \times 10^{-7}$$

$$\text{or } i = \frac{209.33 \times 10^{-7}}{0.01} = 2.0933 \text{ mA } \textbf{Ans.}$$

Example 3.7. A moving coil galvanometer gives a deflection of 150 mm on a scale distant 2.5 m for a current of 2.5 µA. The resistance of the moving coil of galvanometer is 250 Ω.

Determine (*i*) current sensitivity (*ii*) voltage sensitivity (*iii*) megohm sensitivity and (*iv*) the deflection produced in radians by a current of 5 µA.

Solution: (*i*) Current sensitivity, $S_i = \frac{d \text{ in mm}}{i \text{ in } \mu A} = \frac{150}{2.5} = 60 \text{ mm}/\mu A$ **Ans.**

(*ii*) Voltage sensitivity, $S_V = \frac{d \text{ in mm}}{V \text{ in } \mu V} = \frac{d \text{ in mm}}{R \times i \text{ in } \mu A} = \frac{150}{200 \times 2.5} = 0.3 \text{ mm}/\mu V$ **Ans.**

(*iii*) Megohm sensitivity, $S_0 = \frac{1}{\frac{1}{60} \times 10^{-6}} = 60 \text{ M}\Omega/\text{mm}$ **Ans.**

(*iv*) Deflection produced by 5 µA = 60 × 5 = 300 mm

$$\text{Deflection in radians, } \theta = \frac{d}{2r} = \frac{300}{2 \times 2.5 \times 1{,}000} = 0.06 \text{ radian } \textbf{Ans.}$$

3.4 DC AMMETER

The *ammeter* carries the current to be measured or a definite fraction of it, and this current or its definite fraction produces the deflecting torque. Ammeter, which is connected in series with the circuit carrying the current under measurement, must be of *very low resistance* so that the voltage drop across the ammeter and power absorbed from the circuit are as low as possible.

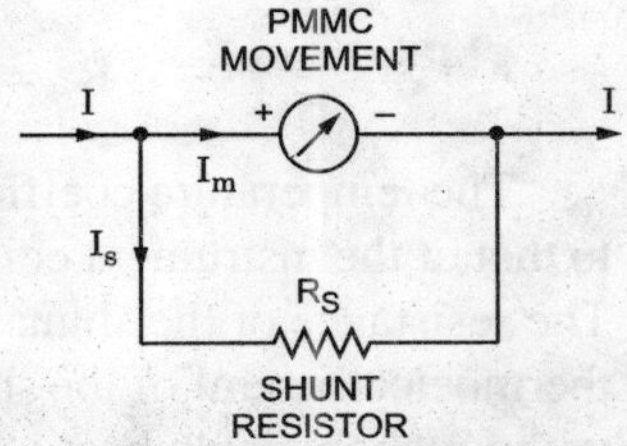

Fig. 3.13

The basic PMMC instrument discussed so far is in reality a milliammeter or microammeter, as their operation depends on the current flowing through the moving coil and current carrying capacity of springs used as current leads is limited. The maximum current, that can be safely carried is about 20 mA. The PMMC instruments are designed for full-scale deflection current ranging from 5 µA to 20 µA and voltage drop of approximately 50 mV to 100 mV.

However, the PMMC instrument can be used to measure higher currents by connecting a small resistor, called the *shunt*, across the meter coil in order to bypass the excessive current (Fig. 3.13). In low range ammeters, especially in portable instruments, the shunt may be mounted inside the case whereas in high range instruments the shunt is mounted separately and connected to the instrument by calibrated leads. The range for dc ammeters is 0–5 μA and upto 0–20 mA without shunts, 0–200 A with internal shunts and 0–5,000 A with external shunts.

3.4.1. Ammeter Shunts. An ammeter shunt is merely a low resistance that is placed in parallel with the coil circuit of the instrument in order to measure fairly large currents. The greater part of the current in main circuit is then diverted around the coil through the shunt.

The connection diagram for a shunt and milliammeter for measuring large currents is shown in Fig. 3.14. The shunt has four terminals, two of large current carrying capacity known as *current terminals* for inserting it in series with the main circuit and the other two of smaller size, known as *potential terminals,* for connecting the ammeter across it.

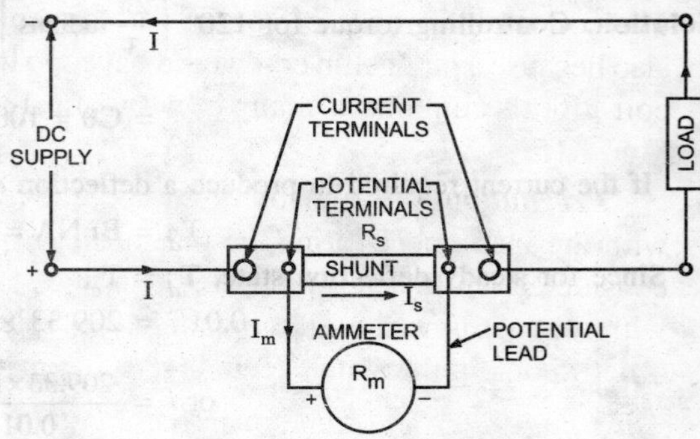

Fig. 3.14 *Diagram Showing The Method of Connecting a Shunt and Milliammeter For Measuring High Currents*

Theory. Let I be the current of the circuit to be measured, R_s be the resistance of the shunt and R_m be the resistance of the ammeter plus its leads to the potential terminals.

The main circuit current will be split into two parts, I_m, the current flowing through the ammeter and I_s, the current through the shunt.

$$\text{Thus } I = I_m + I_s \quad \text{...(3.11)}$$

Since the voltage drop across the shunt and the instrument is same,

$$I_m R_m = I_s R_s$$

$$\text{or } R_s = \frac{I_m R_m}{I_s} = \frac{I_m R_m}{I - I_m} = \frac{R_m}{I/I_m - 1} \quad \text{...(3.12)}$$

Since from Eq. (3.11) $I_s = I - I_m$

The ratio of the total current to the instrument current (*i.e.* I/I_m) is known as the 'multiplying power' of the shunt. Denoting the ratio by N, we have shunt resistance.

$$R_s = \frac{R_m}{N - 1} \quad \text{...(3.13)}$$

The temperature coefficient of resistance of the shunt must be very low and as far as possible equal to that of the instrument coil so that multiplying power of the shunt shall be independent of temperature. The resistance of the shunt should also not vary with the time of usage. Other requirements are that the thermoelectric emf of the shunt should be very small and it must be capable of carrying required current without appreciable heating.

The shunt material is usually manganin, in order that the resistance change due to change in temperature may be as small as possible. Another advantage of manganin is its low thermoelectric emf with copper. The ends of manganin strips are soldered to two large copper blocks. Each copper block carries two terminals, one current terminal and other potential terminal. The potential terminals are separated from current terminals by a sufficient distance to ensure that they are positioned at points where the current flow has become uniform over the cross-section of the shunt. The strips which form the shunt are spaced from each other to promote a good circulation of air and thus efficient cooling.

If the shunt is of manganin, or other low temperature-coefficient material, it is essential that the moving-coil circuit should also be of low temperature coefficient; ideally its temperature coefficient should be the same as that of the shunt, in order that the current may divide correctly. The coil itself is made of copper for convenience of manufacture and also because a manganin coil would have too high total resistance, so the coil circuit contains a manganin swamp resistance R, as shown in Fig. 3.15.

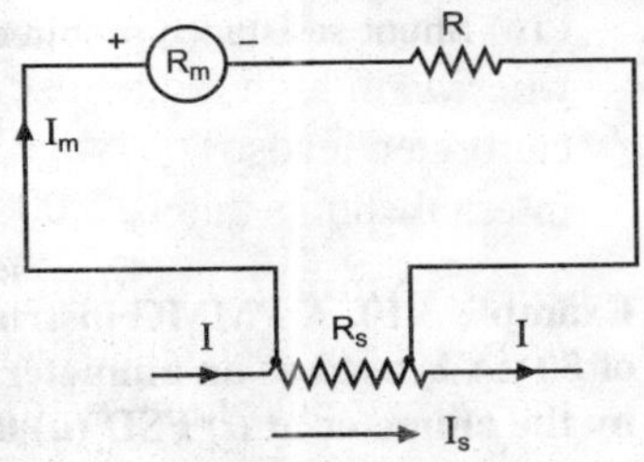

Fig. 3.15

Precautions. As already discussed, the resistance of the leads used with the instrument forms the part of the resistance of the instrument circuit and, therefore, must be considered while determining the required shunt resistance. A special pair of leads is usually supplied with the instrument by the manufacturer intended to be used. These leads should always be used in order to obtain the stated accuracy, as the instrument is calibrated with these leads in use.

Advantages. (*i*) The same instrument can be used for measurement of currents of different ranges by employing different shunts.

(*ii*) The shunt enables the ammeter to be mounted remote from the main circuit, which results in reduction of the possible effect of stray field and have the advantage of remote indication.

Example 3.8. A moving coil ammeter has a full-scale deflection of 50 μA and a coil resistance of 1,000 Ω. What will be the value of the shunt resistance required for the instrument to be converted to read a full-scale reading of 1 A ?

Solution: Full-scale deflection current, $I_m = 50 \times 10^{-6}$ A

Instrument resistance, $R_m = 1{,}000\ \Omega$

Total current to be measured, I = 1 A

$$\text{Resistance of ammeter shunt required, } R_s = \frac{R_m}{\frac{I}{I_m}-1} = \frac{1{,}000}{\frac{1}{50\times10^{-6}}-1} = 0.0500025\ \Omega \textbf{ Ans.}$$

Example 3.9. The coil of a measuring instrument has a resistance of 1.0 Ω and the instrument has a full-scale deflection of 250 V when a resistance of 4,999 Ω is connected in series with it, find:

(*a*) The current range of the instrument when used as an ammeter with the coil connected across a shunt of $\frac{1}{4{,}999}$ Ω, and (*b*) the value of the shunt resistance for the instrument to give a full-scale deflection of 50 A.

Solution: Instrument resistance, $R_m = 1.0\ \Omega$

Series resistance, $R_{se} = 4{,}999\ \Omega$

If I_m is the full-scale deflection current of the instrument

then full-scale deflection voltage $= I_m(R_m + R_{se})$

$$\text{or Full-scale deflection current, } I_m = \frac{\text{Fullscale deflection voltage}}{R_m + R_{se}} = \frac{250}{1+4{,}999} = 0.05 \text{ A}$$

(*a*) Current range of the instrument when used as an ammeter with the coil connected across a shunt of $\frac{1}{4{,}999}$ Ω (*i.e.*, $R_s = \frac{1}{4{,}999}$ Ω),

$$I = I_m + I_s = I_m + \frac{I_m R_m}{R_s} = I_m\left(1+\frac{R_m}{R_s}\right) = 0.05\left(1+\frac{1}{\frac{1}{4{,}999}}\right) = 250 \text{ A } \textbf{Ans.}$$

(*b*) Shunt resistance required for full-scale deflection current of 50 A (*i.e.*, I = 50 A),

$$R_s = \frac{R_m}{N-1} = \frac{R_m}{\frac{I}{I_m}-1} = \frac{1.0}{\frac{50}{0.05}-1} = \frac{1}{999}\ \Omega\ \textbf{Ans.}$$

Example 3.10. A PMMC instrument with a coil resistance of 49 Ω and full-scale deflection (FSD) current of 50 μA is used as an ammeter. Shunt resistance, R_S is 1 Ω. Determine the main circuit current measured by the ammeter at (*i*) FSD (*ii*) 0.5 FSD and 0.1 FSD.

Solution: The main circuit current, $I = I_m + I_s + I_m + \frac{I_m R_m}{R_s} = I_m\left(1+\frac{R_m}{R_s}\right) = I_m\left(1+\frac{49}{1}\right) = 50\,I_m$

(*i*) Main circuit current at FSD, $I_1 = 50\,I_m = 50 \times 50 \times 10^{-6} = 2.5$ mA **Ans.**

(*ii*) Main circuit current at 0.5 FSD, $I_2 = 50 \times \frac{1}{2} \times 50 \times 10^{-6} = 1.25$ mA **Ans.**

(*iii*) Main circuit current at 0.1 FSD, $I_3 = 50 \times \frac{1}{10} \times 50 \times 10^{-6} = 0.25$ mA **Ans.**

3.4.2. Multirange Ammeters. The combination of a millivoltmeter and shunt employed as an ammeter is readily adaptable to multirange construction either by separate, interchangeable, single-range shunts or multirange shunts. A method of operation with multirange shunts is illustrated in Fig. 3.16. The shunt is selected (*i.e.* the current passes through one or more shunt sections, depending on the position of selector switch) to provide the necessary voltage drop for the millivoltmeter. The swamping resistance in series with the moving coil may be nearly constant, assuming the shunt resistance to be low in comparison to swamping resistance. Of course, the switch contacts must be constructed to carry the entire current under measurement.

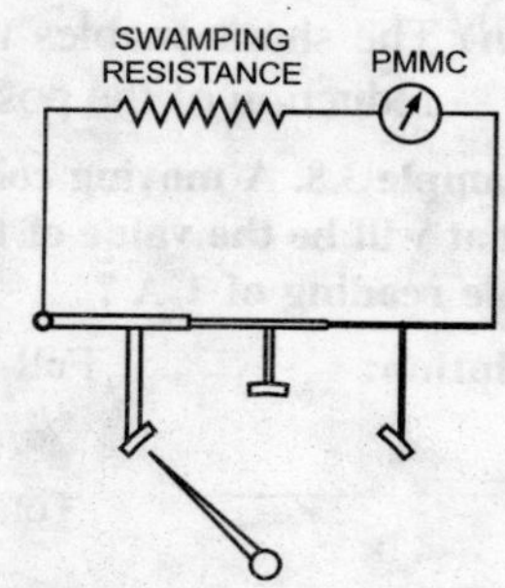

Fig. 3.16 *Multirange Ammeter*

A schematic diagram of a multirange ammeter is shown in Fig. 3.17 (*a*). The circuit has five shunts R_{S_1}, R_{S_2}, R_{S_3}, R_{S_4} and R_{S_5} which can be placed in parallel with the meter movement to give five different current ranges I_1, I_2, I_3, I_4 and I_5. Let N_1, N_2, N_3, N_4 and N_5 be the multiplying power of the shunts for currents I_1, I_2, I_3, I_4 and I_5 respectively.

$$\text{Now } R_{S_1} = \frac{R_m}{N_1-1};\ R_{S_2} = \frac{R_m}{N_2-1};\ R_{S_3} = \frac{R_m}{N_3-1};\ R_{S_4} = \frac{R_m}{N_4-1}$$

$$\text{and } R_{S_5} = \frac{R_m}{N_5-1}$$

Low range ammeters use a multiposition make-before break switch provided on the instrument case. Such switch is essential so that the meter movement is not damaged when changing from the current range one to another. If an ordinary switch is provided, the meter will remain without a shunt and as such it may get damaged when the range is changed.

Multirange ammeters are employed for ranges from 1 to 50 A. When using a multirange ammeter first use the highest current range, then decrease the current range until good up-scale reading is obtained.

The *universal shunt* or Ayrton shunt, shown in Fig. 3.17 (*b*) is also used for multirange ammeters. The advantage of an Ayrton shunt is that it eliminates the possibility of the meter being in the circuit without a shunt. But this advantage is gained at the cost of a higher meter resistance.

For the arrangement (4-range) shown in Fig. 3.17 (*b*), we have

$$\text{For switch at position 1, } R_1 = \frac{R_m}{N-1}$$

$$\text{For switch at position 2, } I_m (R_1 - R_2 + R_m) = (I_2 - I_m) R_2 \text{ or } R_2 = \frac{R_1 + R_m}{N_2}$$

$$\text{For switch at position 3, } I_m (R_1 - R_3 + R_m) = (I_3 - I_m) R_3 \text{ or } R_3 = \frac{R_1 + R_m}{N_3}$$

$$\text{For switch at position 4, } I_m (R_1 - R_4 + R_m) = (I_4 - I_m) R_4 \text{ or } R_4 = \frac{R_1 + R_m}{N_4}$$

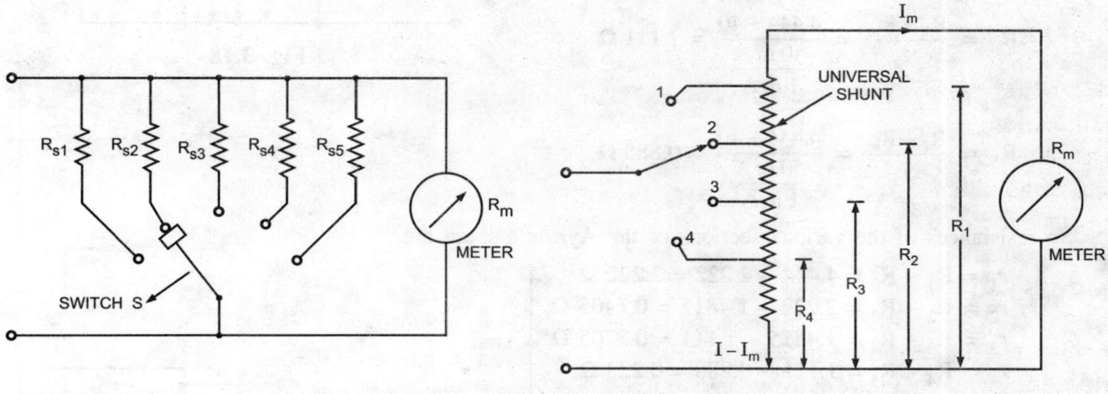

(*a*) *Multirange Ammeter* (*b*) *Multirange Ammeter With Universal Shunt*

Fig. 3.17

Thus the values of different sections of resistances *i.e.*, $(R_1 - R_2)$, $(R_2 - R_3)$, $(R_3 - R_4)$ and R_4 may be determined.

The precautions, while using ammeters, to be observed are:

1. Ammeters are always connected in series with the load capable of limiting the current. It should never be connected across the source; if connected it would draw, because of its low resistance, a high current and the movement may get damaged.
2. In dc ammeters, the correct polarity must be observed while making connections. Reverse polarity causes the meter to deflect against the mechanical stopper, which may damage the pointer.
3. In case of multirange ammeters, highest current range must be used first and then it is decreased to the range until substantial deflection is obtained.

Example 3.11. Calculate the necessary values of resistors to be provided using Ayrton's multiranging method so that a dc milliammeter having a basic movement with R_m = 40 Ω and full-scale current of 1 m A can be used for measuring currents in the range of 10 m A and 50 m A.

[*U.P. Technical Univ. Elec. Measurements and Measuring Instruments 2002-03*]

Solution: Let the current ranges be 10, 20, 30, 40 and 50 m A

For switch at position 1 (lowest range of 10 m A),

$$R_1 = \frac{R_m}{N_1 - 1} = \frac{R_m}{\frac{I}{I_m} - 1} = \frac{40}{\frac{10 \text{ m A}}{1 \text{ m A}} - 1} = 4.444\ \Omega$$

For switch at position 2 (range of 20 m A)

$$R_2 = \frac{R_1 + R_m}{N_2} = \frac{4.444 + 40}{\frac{20 \text{ m A}}{1 \text{ m A}}} = 2.222\ \Omega$$

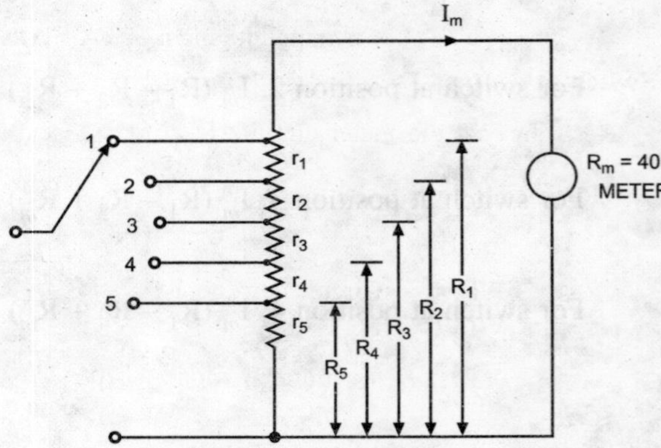

Fig. 3.18

For switch at position 3 (range of 30 mA)

$$R_3 = \frac{R_1 + R_m}{N_3} = \frac{4.444 + 40}{\frac{30 \text{ m A}}{1 \text{ m A}}}$$

$$= 1.4815\ \Omega$$

$$R_4 = \frac{R_1 + R_m}{N_4} = \frac{4.444 + 40}{\frac{40 \text{ m A}}{1 \text{ m A}}} = 1.111\ \Omega$$

$$R_5 = \frac{R_1 + R_m}{N_5} = \frac{4.444 + 40}{\frac{50 \text{ m A}}{1 \text{ m A}}} = 0.888\ \Omega$$

So resistances of the various sections of the Ayrton's shunt are:

$$\left.\begin{aligned} r_1 &= R_1 - R_2 = 4.444 - 2.222 = 2.222\ \Omega \\ r_2 &= R_2 - R_3 = 2.222 - 1.4815 = 0.7405\ \Omega \\ r_3 &= R_3 - R_4 = 1.4815 - 1.111 = 0.3705\ \Omega \\ r_4 &= R_4 - R_5 = 1.111 - 0.888 = 0.223\ \Omega \\ r_5 &= R_5 = 0.888\ \Omega \end{aligned}\right\} \textbf{Ans.}$$

Example 3.12. A PMMC instrument has a three resistor Ayrton shunt connected across it to make an ammeter, as in Fig. 3.19. The full-scale deflection (FSD) current is 1 mA. Calculate the three ranges of the ammeter.

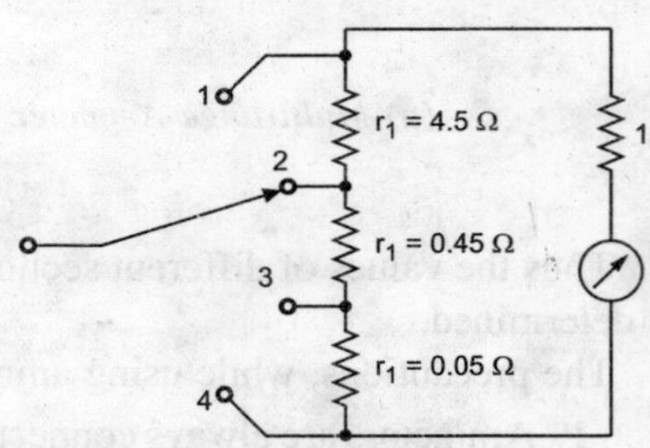

Fig. 3.19

Solution: Full-scale deflection current of PMMC instrument

$$I_m = 1 \text{ mA}$$

For switch at contact 1

Resistance in meter circuit, $R_{m_1} = R_m = 1\ \text{k}\Omega = 1{,}000\ \Omega$

Resistance in shunt circuit, $R_{s_1} = r_1 + r_2 + r_3 = 0.05 + 0.45 + 4.5 = 5\ \Omega$

$$\text{Circuit current, } I_1 = I_m \left(\frac{R_{m_1}}{R_{s_1}} + 1\right) = 1 \text{ mA} \times \left(\frac{1000}{5} + 1\right) = 201 \text{ mA}$$

Ammeter range ~ 200 mA **Ans.**

For switch at contact 2

Resistance in the meter circuit, $R_{m_2} = R_m + r_1 = 1{,}000 + 4.5 = 1{,}004.5\ \Omega$

Resistance in shunt circuit, $R_{s_2} = r_2 + r_3 = 0.45 + 0.05 = 0.5\ \Omega$

$$\text{Circuit current, } I_2 = I_m\left(\frac{R_{m_1}}{R_{s_2}}+1\right) = 1 \text{ mA}\left(\frac{1{,}004.5}{0.5}+1\right) = 2.01 \text{ A}$$

Ammeter range ~ 2 A **Ans.**

For Switch at contact 3

Resistance in the meter circuit, $R_{m_3} = R_m + r_1 + r_2 = 1{,}000 + 4.5 + 0.45 = 1{,}004.95\ \Omega$

Resistance in the shunt circuit $R_{s_3} = r_3 = 0.05\ \Omega$

$$\text{Circuit current, } I_3 = I_m\left(\frac{R_{m_3}}{R_{s_3}}+1\right) = 1 \text{ mA} \times \left(\frac{1{,}004.95}{0.05}+1\right) = 20.1 \text{ A}$$

Ammeter range ~ 20 A **Ans.**

3.5 DC VOLTMETER

Voltmeter carries the current proportional to the voltage to be measured and this current produces the deflecting torque. The voltmeter, which is connected in parallel with the circuit across which the voltage is to be measured, must be of *very high resistance* so that the current flowing through the instrument and the power absorbed from the circuit are the minimum possible.

As already mentioned in Art. 3.4, the PMMC instruments are designed for full-scale deflection current ranging from 5 μA to 20 μA and voltage drop of approximately 50 mV to 100 mV. The milliammeter can also be regarded as millivoltmeter that measure current in terms of the potential drop across the shunt.

For using PMMC instrument for measurement of dc voltage exceeding 100 mV a series resistance is connected in the instrument circuit. The series resistance is usually mounted inside the case for voltmeters of moderate range (say upto 500 volts) and mounted separately, in case of high range voltmeters, in order to avoid excessive heating inside the case. The external resistance inserted in the movement circuit for increasing the range is called the *multiplier*. The range for dc voltmeters is 0–100 mV without series resistance and upto 20 kV or 30 kV with external series resistance.

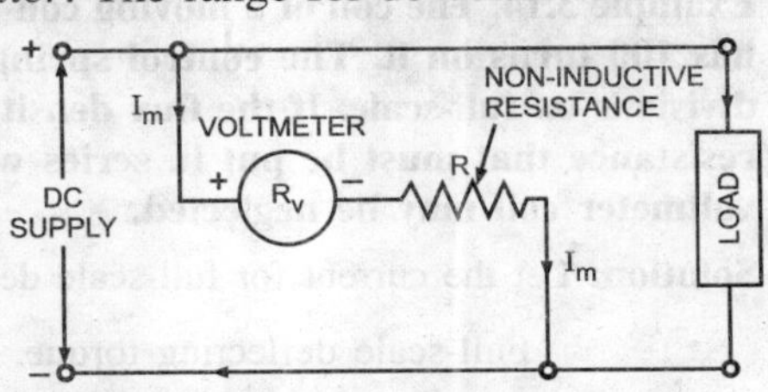

Fig. 3.20

3.5.1. Voltmeter Multipliers. Voltmeter multiplier is a high non-inductive resistance connected in series with the voltmeter coil and is used for increasing the range of the voltmeter. A voltmeter connected in series with the high non-inductive resistance (voltmeter multiplier) for measurement of high voltage is shown in Fig. 3.20.

Theory. Let R_v be the resistance of voltmeter, I_m be the full-scale deflection current, V be the voltage of the circuit to be measured and R be the external series resistance. Now, since voltage across supply leads = voltage drop across the voltmeter + voltage drop across external resistance,

$$\therefore V = I_m R_v + I_m R$$

$$\text{or } R = \frac{V}{I_m} - R_v \qquad \text{...(3.14)}$$

The main requirements of voltage multipliers to be used for dc measurements are that (*i*) resistance should not change with time of usage (*ii*) their temperature coefficient of resistance must be very low so that change in temperature may not affect their resistance values and (*iii*) should have ample provision for cooling to dissipate the heat developed, as an appreciable amount of power is developed in such resistors.

The series resistance is usually of manganin, either in the form of a bobbin or in the form of strip wound on a narrow sheet of bakelite or other insulating material. It may be inside the instrument case or, particularly for higher voltages, in a separate ventilated case.

The question of temperature compensation does not normally arise in the case of a voltmeter, as any slight change in the coil resistance is negligible in comparison with the total resistance, the major portion of which consists of manganin wire. The series resistance, infact acts as a completely effective swamp resistance.

Example 3.13. A moving coil instrument has a resistance of 10 Ω and gives a full-scale deflection when carrying 50 m A. Show how it can be adopted to measure voltage upto 750 V and current 100 A.

[*Pb. Univ. Elec. Measurements-1, Dec. 1992*]

Solution: Instrument resistance, $R_m = 10\ \Omega$

Instrument current, $I_m = 50\text{ m A} = 0.05\text{ A}$

Voltage to be measured, $V = 750\text{ V}$

External resistance required to be connected in series with the instrument to measure voltage up to 750 V,

$$R = \frac{V}{I_m} - R_m = \frac{750}{0.05} - 10 = 14{,}990\ \Omega \textbf{ Ans.}$$

(*ii*) Current to be measured, $I = 100\text{ A}$

$$\text{Multiplying power of shunt, } N = \frac{I}{I_m} = \frac{100}{0.05} = 2{,}000$$

$$\text{Shunt resistance required, } R_s = \frac{R_m}{N-1} = \frac{10}{2{,}000-1} = \frac{10}{1{,}999}\ \Omega \textbf{ Ans.}$$

Example 3.14. The coil of a moving coil permanent magnet voltmeter is 40 mm long and 30 mm wide and has 100 turns on it. The control spring exerts a torque of 120 × 10⁻⁶ N-m when the deflection is 100 divisions on full-scale. If the flux density of the magnetic field in the air gap is 0.5 Wb/m², estimate the resistance that must be put in series with the coil to give one volt per division. The resistance of the voltmeter coil may be neglected.

Solution: Let the current for full-scale deflection be I amperes

Full-scale deflecting torque, $T_d = B\,I\,N\,A\text{ Nm} = 0.5 \times I \times 100 \times 0.04 \times 0.03 = 0.06\,I\text{ N-m}$

Controlling torque, $T_c = 120 \times 10^{-6}\text{ N-m}$

In final deflected position $T_d = T_c$

or $0.06\,I = 120 \times 10^{-6}$

or $I = 0.002\text{ A}$

Full-scale reading of the instrument, $V = 100 \times 1 = 100\text{ V}$

External resistance required to be put in series with the coil,

$$R = \frac{V}{I} - R_m = \frac{100}{0.002} - 0 = 50{,}000\ \Omega \textbf{ Ans.}$$

Example 3.15. A PMMC instrument with FSD of 0.2 mA and a coil resistance of 10 Ω is to be converted into a voltmeter. Determine the required multiplier resistance if the voltmeter is used to measure 100 V at full scale. Also determine the applied voltage when the instrument indicates 0.75, 0.5, 0.25 and 0.1 FSD.

Solution: Full-scale deflection current, $I_m = 0.2 \times 10^{-3}$ A

Instrument resistance, $R_m = 10\ \Omega$

External resistance required for full-scale deflection with 100 V,

$$R = \frac{V}{I_m} - R_m = \frac{100}{0.2 \times 10^{-3}} - 10 = 499.99\ \text{k}\Omega \quad \textbf{Ans.}$$

Instrument current at 0.75 FSD, $I_{m_1} = 0.75 \times 0.2 \times 10^{-3} = 0.15 \times 10^{-3}$ A

Applied voltage, $V_1 = I_{m_1}(R + R_m) = 0.15 \times 10^{-3}\ (499.99 \times 10^3 + 10) = 75$ V **Ans.**

Instrument current at 0.5 FSD, $I_{m_2} = 0.5 \times 0.2 \times 10^{-3} = 0.1 \times 10^{-3}$ A

Applied voltage, $V_2 = I_{m_2}(R + R_m) = 0.1 \times 10^{-3}\ (499.99 \times 10^3 + 10) = 50$ V **Ans.**

Instrument current at 0.25 FSD, $I_{m_3} = 0.25 \times 0.2 \times 10^{-3} = 0.05 \times 10^{-3}$ A

Applied voltage, $V_3 = I_{m_3}(R + R_m) = 0.05 \times 10^{-3}(499.99 \times 10^3 + 10) = 25$ V **Ans.**

Instrument current at 0.1 FSD, $I_{m_4} = 0.1 \times 0.2 \times 10^{-3} = 20 \times 10^{-6}$ A

Applied voltage, $V_4 = I_{m_4}(R + R_m) = 20 \times 10^{-6}(499.99 \times 10^3 + 10) = 10$ V **Ans.**

3.5.2. Multirange Voltmeters. It is a voltmeter which provides a number of voltage ranges. It may be made up with a resistor appropriate to the highest range desired, with tap points appropriate to various lower ranges brought out to binding posts, as illustrated in Fig. 3.21 (*a*) or to a rotary selector switch as illustrated in Fig. 3.21 (*b*).

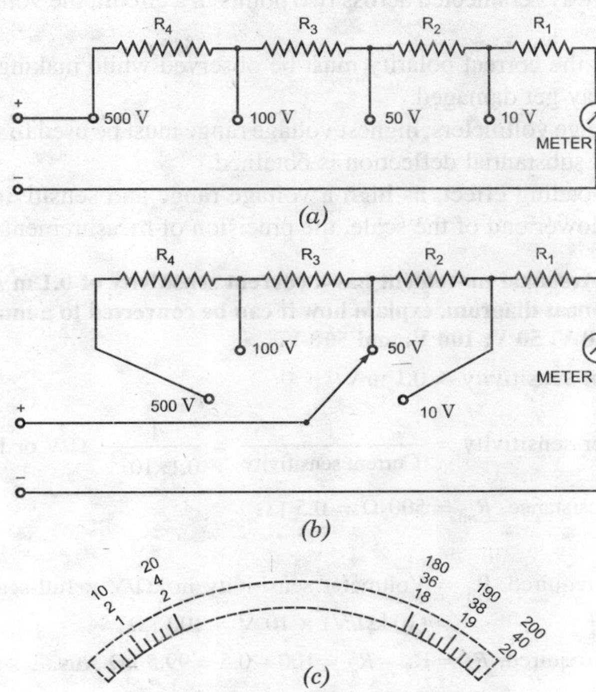

Fig. 3.21

The resistance required for each range can be easily computed by keeping in view the basic fact that the sensitivity of a meter movement is always the same regardless of the range selected. Moreover, the full-scale deflection current is the same for every range. For any range, the total circuit resistance is determined by multiplying the sensitivity by the full-scale voltage for that range. It will be clear from Example 3.16.

If the full-scale values of voltages in the various ranges differ by a factor 2 or 10, the same scale may be used for both ranges with multiple labelling of cardinal points, as illustrated in Fig. 3.21 (*c*). For some other combinations of ranges, double scales, separately labelled, would have to be drawn.

3.5.3. Voltmeter Sensitivity. The sensitivity or ohms per volt rating of a voltmeter is defined as the ratio of the total circuit resistance R_T and the voltage range V. Sensitivity S, is essentially the reciprocal of the full-scale deflection current of the basic movement. Thus

$$\text{Sensitivity, } S = \frac{1}{f_{sd}} \text{ ohms per volt} \quad ...(3.15)$$

The sensitivity 'S' of the voltmeter may be used in calculating the values of multipliers resistors in a dc voltmeter. It will be clear from Example 3.16.

3.5.4. Loading Effect. In selection of a voltmeter for a certain voltage measurement, its sensitivity is an important factor. A low-sensitivity voltmeter may give a correct reading while measuring a voltage in a low-resistance circuit, but it is certain to give unreliable readings in a high-resistance circuit, because, in this case, the meter takes an appreciable current and indicates a lower reading than what existed before the meter was connected. This effect of the voltmeter is called the *loading effect* and is caused by low-sensitivity instruments.

The precautions, while using voltmeters, to be observed are:

1. Voltmeters are always connected across two points in a circuit, the voltage drop across which is to be measured.
2. In dc voltmeters, the correct polarity must be observed while making connections, otherwise the instrument may get damaged.
3. In case of multirange voltmeters, highest voltage range must be used first and then it is decreased to the range until substantial deflection is obtained.
4. For minimising loading effect, as high a voltage range and sensitivity as possible, are used. However, at the lower end of the scale, the precision of measurement decreases.

Example 3.16. A basic D'Arsonval movement has a current sensitivity of 0.1 m A and internal resistance of 500 Ω. With the help of neat diagram, explain how it can be converted to a multirange voltmeter which has the following ranges 10 V, 50 V, 100 V, and 500 V.

Solution : Current sensitivity = 0.1 mA

$$\text{Voltmeter sensitivity} = \frac{1}{\text{Current sensitivity}} = \frac{1}{0.1\times10^{-3}}\ \Omega/\text{V or } 10\ \text{k}\ \Omega/\text{V}$$

$$\text{Meter resistance, } R_m = 500\ \Omega = 0.5\ \text{k}\Omega$$

(1) 0 – 10 V range

$$\text{Total resistance required, } R_T = \text{Voltmeter sensitivity in k}\Omega/\text{V} \times \text{full-scale deflection voltage}$$

$$= (10\ \text{k}\Omega/\text{V}) \times 10\ \text{V} = 100\ \text{k}\Omega$$

$$\text{Additional resistance required, } R_1 = R_T - R_m = 100 - 0.5 = 99.5\ \text{k}\Omega$$ **Ans.**

(2) 0 – 50 V range

Total resistance required = (10 kΩ/V) × 50 V = 500 kΩ

Additional resistance required, R_2 = 500 – 99.5 – 0.5 = 400 k Ω **Ans.**

(3) 0 – 100 V range

Additional resistance required, R_3 = (10 k Ω/V) × 100 – R_2 – R_1 – R_m

= 1,000 – 400 – 99.5 – 0.5 = 500 k Ω **Ans.**

(4) 0 – 500 V range

Additional resistance required, R_4 = (10 k Ω/V) × 500 – 500 – 400 – 99.5 – 0.5 = 4,000 k Ω **Ans.**

The circuit diagram is similar to that shown in Fig. 3.21.

3.6 SERIES OHMMETER

An *ohmmeter* is normally a part of volt-ohm-milliammeter (VOM), or multifunction meter. It does not usually exist as individual instrument.

3.6.1. PMMC as Ohmmeter. The PMMC meter can be used to operate as an ohmmeter by using a small internal battery (dry cell), as illustrated in Fig. 3.22 (*a*).

Meter current will flow through the unknown resistance (R_X). Value of the current, and hence deflection in the meter, will depend on the value of the unknown resistance. The scale for ohmmeter can be calibrated and graduated in ohms. If the resistance is too high, one cell may not be sufficient because current will become too small. Hence provision of a larger cell ($E_2 > E_1$) is made through a switch.

The rheostat R_z is adjusted to ensure that when the unknown resistance is zero (*i.e.* the test probes are shorted with each other), the meter should read zero resistance (full deflection). Resistance scale will show zero at full deflection because zero resistance means maximum current through the meter. Infinite resistance means no current, and so the left most end of the scale (zero current mark) will be marked as ∞ on the resistance scale. Different resistance ranges, say 100 Ω, 10 kΩ and 10 M Ω will be obtained using different resistors with the help of a multi-way switch, as illustrated in Fig. 3.22 (*b*).

For measuring on the lowest resistance range, the shunt resistance will be lowest. For higher ranges, the shunt resistance will increase in value. Hence in Fig. 3.22 (*b*) $R_1 < R_2 < R_3$. R_Z is the zero-adjuster resistance. If the meter is of 1 mA, full-scale deflection (fsd), R_Z will be adjusted so that the circuit draws 1 mA when probes are shorted to each other *i.e.* when $R_X = 0$.

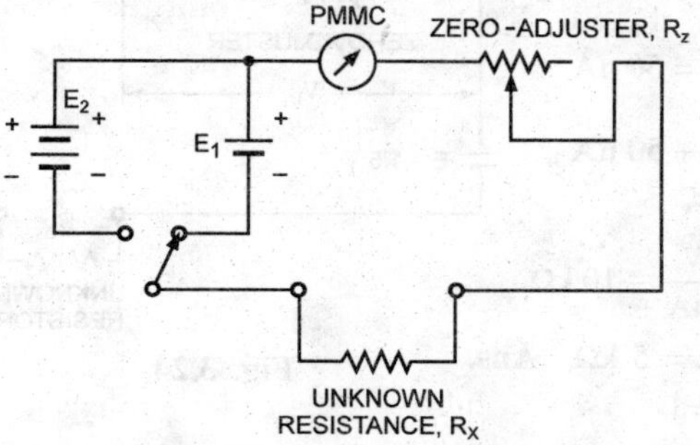

(*a*) *Basic Circuit of an Ohmmeter*

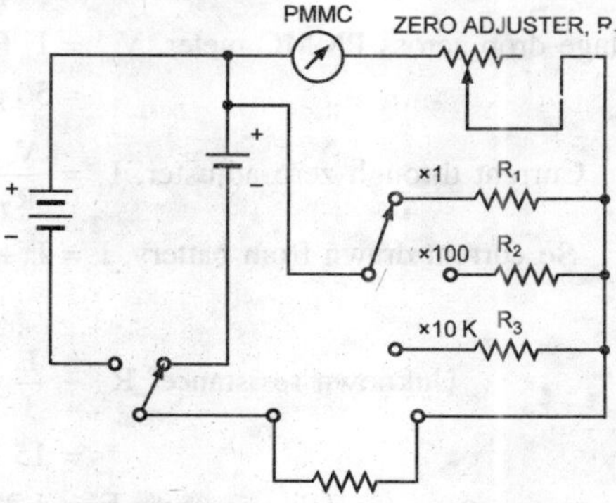

(*b*)

Fig. 3.22

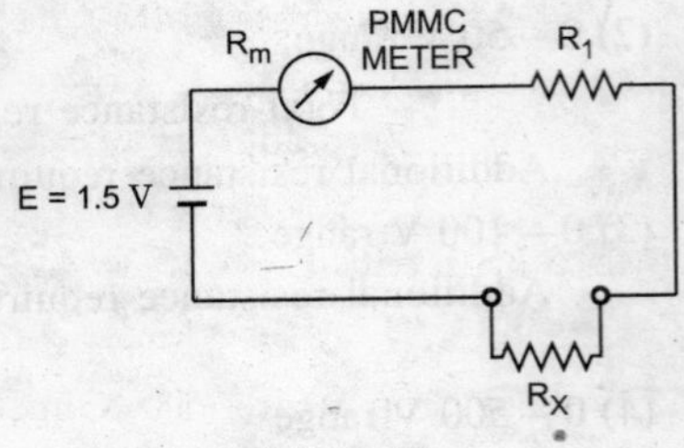

Fig. 3.23

Example 3.17. The series ohmmeter is made up of a 1.5 V battery, a 50 μA meter and a series resistance R_1 which makes $(R_1 + R_m) = 10$ kΩ. (*a*) Determine the instrument indication when resistance under measurement is zero. (*b*) Determine how the scale should be marked at 0.8 FSD, 0.5 FSD, 0.25 FSD and 0.1 FSD.

Solution:

$$\text{Battery emf, } E = 1.5 \text{ V}$$

$$\text{Total resistance in the circuit, } R = R_1 + R_m + R_x = 10 \text{ k}\Omega + 0 = 10 \text{ k}\Omega$$

$$\text{Meter FSD current, } I_m = \frac{E}{R} = \frac{1.5}{10 \text{ k}\Omega} = 150 \ \mu\text{A}$$

At 0.8 FSD

$$I_m = 0.8 \times 150 \ \mu\text{A} = 120 \ \mu\text{A}$$

$$\text{Total resistance in the circuit, } R = \frac{E}{I_m} = \frac{1.5}{120 \times 10^{-6}} = 12{,}500 \ \Omega \text{ or } 12.5 \text{ k}\Omega$$

$$\text{Unknown resistance, } R_x = R - (R_1 + R_m) = 12.5 - 10 = 2.5 \text{ k}\Omega \quad \textbf{Ans.}$$

At 0.5 FSD

$$I_m = 0.5 \times 150 \ \mu\text{A} = 75 \ \mu\text{A}$$

$$\text{Unknown resistance, } R_x = R - (R_1 + R_m) = \frac{E}{I_m} - (R_1 + R_m) = \frac{1.5}{75 \times 10^{-6}} - 10 \text{ k}\Omega = 10 \text{ k}\Omega \quad \textbf{Ans.}$$

At 0.25 FSD

$$I_m = 0.25 \times 150 = 37.5 \ \mu\text{A}$$

$$\text{Unknown resistance, } R_x = \frac{E}{I_m} - (R_1 + R_m) = \frac{1.5}{37.5 \times 10^{-6}} - 10 \text{ k}\Omega = 30 \text{ k}\Omega \quad \textbf{Ans.}$$

At 0.1 FSD

$$I_m = 0.1 \times 150 = 15 \ \mu\text{A}$$

$$\text{Unknown resistance, } R_x = \frac{E}{I_m} - (R_1 + R_m) = \frac{1.5}{15 \times 10^{-6}} - 10 \text{ k}\Omega = 90 \text{ k}\Omega \quad \textbf{Ans.}$$

Example 3.18. In ohmmeter circuit in Fig. 3.24, E = 1.5 V, $R_1 = 10$ kΩ, $R_m = 50$ Ω, $R_2 = 50$ Ω and meter full-scale deflection current is 100 μA. Determine the ohmmeter scale reading at 0.5 FSD, and determine the new resistance value that R_2 must be adjusted to when battery emf falls to 1.25 V. Also, compute the value of R_x at 0.5 FSD when E = 1.25 V.

Solution: At 0.5 FSD, with 1.5 V

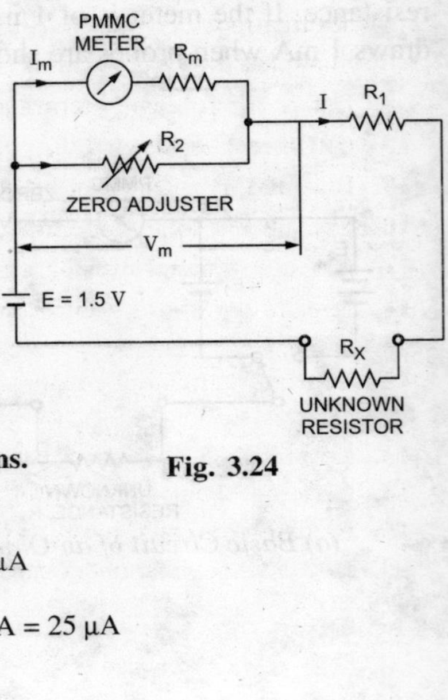

Fig. 3.24

$$I_m = 0.5 \times 100 = 50 \ \mu\text{A}$$

$$\text{Voltage drop across PMMC meter, } V_m = I_m R_m = 50 \ \mu\text{A} \times 50 \ \Omega = 2.5 \text{ mV}$$

$$\text{Current through zero adjuster, } I_0 = \frac{V_m}{R_2} = \frac{2.5 \text{ mV}}{50 \ \Omega} = 50 \ \mu\text{A}$$

$$\text{So current drawn from battery, } I = I_0 + I_m = 50 \ \mu\text{A} + 50 \ \mu\text{A} = 100 \ \mu\text{A}$$

$$\text{Unknown resistance, } R_x \sim \frac{E}{I} - R_1 = \frac{1.5 \text{ V}}{100 \ \mu\text{A}} - 10 \text{ k}\Omega = 15 \text{ k}\Omega - 10 \text{ k}\Omega = 5 \text{ k}\Omega \quad \textbf{Ans.}$$

With E = 1.25 V and $R_x = 0$

$$\text{Circuit current, } I = \frac{E}{R_x + R_1} = \frac{1.25 \text{ V}}{0 + 10 \text{ k}\Omega} = 125 \ \mu\text{A}$$

$$\text{Current through zero adjuster, } I_0 = I - I_{m(FSD)} = 125 \ \mu\text{A} - 100 \ \mu\text{A} = 25 \ \mu\text{A}$$

Voltage drop across PMMC meter, $V_m = I_m R_m = 100\ \mu A \times 50\ \Omega = 5\ mV$

$$\text{Zero adjuster resistance, } R_2 = \frac{V_m}{I_0} = \frac{5\ mV}{25\ \mu A} = 200\ \Omega \quad \textbf{Ans.}$$

At 0.5 FSD, with E = 1.25 V

Meter current, $I_m = 0.5 \times 100 = 50\ \mu A$

Voltage drop across PMMC meter, $V_m = I_m R_m = 50\ \mu A \times 50\ \Omega = 2.5\ mV$

$$\text{Current through zero adjuster, } I_0 = \frac{V_m}{R_2} = \frac{2.5\ mV}{200\ \Omega} = 12.5\ \mu A$$

Circuit current, $I = I_m + I_0 = 50\ \mu A + 12.5\ \mu A = 62.5\ \mu A$

$$\text{Unknown resistance, } R_x \sim \frac{E}{I} - R_1 = \frac{1.25\ V}{62.5\ \mu A} - 10\ k\Omega = 20\ k\Omega - 10\ k\Omega = 10\ k\Omega \quad \textbf{Ans.}$$

EXERCISES

1. List the three torques involved in the moving system of a deflection instrument and explain the function of each torque. How are these torques produced in PMMC instruments ?

2. Describe the principle of operation, construction, main features and applications of a PMMC instrument.

3. Deduce the expression for the deflection of a PMMC type meter in terms of coil parameters, air-gap flux density, spring constant and current flowing through the meter.
State the advantages and disadvantages of a PMMC type meter.

4. Describe with the help of a labelled diagram, the constructional details of a PMMC meter and explain how 'control' and 'damping' forces are obtained ?

5. Sketch the construction of a core-magnet type PMMC instruments and explain.

6. What is galvanometer ? What for is it used ? On what principle does D'Arsonval galvanometer operate ? State the main advantage of PMMC galvanometer.

7. With the help of a neat diagram explain the principle of a moving coil galvanometer. Discuss its applications.
Show how a variable shunt should be used for galvanometer protection. Explain.

8. Define the following terms in reference to galvanometer.
(*i*) Current sensitivity (*ii*) voltage sensitivity and (*iii*) megohm sensitivity and (*iv*) CDRX.

9. How PMMC meter can be converted into an ammeter ? Explain with circuit diagram.

10. Explain the following terms:
(*i*) Four-terminal resistor (*ii*) swamping resistor (*iii*) Ayrton shunt (*iv*) make-before-break switch.

11. Explain the use of shunt as a device for extending the range of ammeters. What are the advantages of shunt ?

12. Explain the multirange ammeter with universal shunt. State the precautions to be observed while using ammeters.

13. How PMMC instrument can be converted into a voltmeter ? Explain with circuit diagram.

14. Give the circuit diagram for a multirange voltmeter and explain.

15. Explain voltmeter sensitivity, voltmeter loading effect and voltmeter swamping resistance.

16. How PMMC instrument can be used as an ohmmeter ? Explain with circuit diagram.

17. Write short notes on
   (*i*) PMMC instrument. (*ii*) Moving coil galvanometer.
   (*iii*) Conversion of PMMC instrument into ammeter. (*iv*) Multirange ammeter.
   (*v*) PMMC instrument as a voltmeter. (*vi*) Multirange voltmeter.
   (*vii*) Series ohmmeter.

SHORT ANSWER TYPE QUESTIONS WITH ANSWERS

Q. 1. How is controlling torque provided in PMMC instruments ?

Ans. In PMMC instruments, controlling torque is provided by two phosphor-bronze hair springs, either helical or spiral, coiled in opposite directions, normally equal in strengths.

Q. 2. What type of damping is provided in PMMC instruments ?

Ans. Usually eddy current damping is used in PMMC instruments.

Q. 3. What is a galvanometer ?

Ans. The galvanometer is a highly sensitive instrument used for measuring current, voltage and quantity of electricity; the scale of the instrument is graduated directly by the observer.

Q. 4. What for galvanometers are used ?

Ans. Galvanometers are used for:

(*i*) measuring currents and voltages of small magnitudes;

(*ii*) determining equality of potentials between two points of an electric circuit (for example, as null detectors in bridge and potentiometer circuits);

(*iii*) determining the equality of two currents (differential galvanometers);

(*iv*) measuring quantity of electricity or current impulse (ballistic galvanometers).

Q. 5. On what principle does D'Arsonval galvanometer operate ?

Ans. When a current carrying conductor is placed in a magnetic field, a mechanical force is experienced on the conductor which tends to move the conductor in a direction given by Fleming's left hand rule. This principle is employed in galvanometers.

Q. 6. What is the main advantage of a PMMC galvanometer ?

Ans. The main advantage of a PMMC galvanometer lies in its freedom from the effects of stray fields and the ease with which a long and uniform scale may be attained.

Q. 7. What is meant by mechanical damping in a D'Arsonval galvanometer ?

Ans. In a D'Arsonval galvanometer, mechanical damping is mainly due to the motion of the coil through the air surrounding it and a small dissipation of energy associated with the flexing of the suspension.

Q. 8. What is meant by electromagnetic damping ?

Ans. Electromagnetic damping is developed because of induced effects in the moving coil when it rotates in the magnetic field and a close path is provided for the flow of current.

Q. 9. What is CDRX ?

Ans. The external resistance required for critical damping is called CDRX or ECDR (critical damping external resistance).

Q. 10. What is the difference between an ammeter and a voltmeter ?

Ans. Ammeter is a low resistance indicating instrument while the voltmeter is high resistance one.

Q. 11. Why should be an ammeter of very low resistance ?

Ans. *Ammeter,* which is connected in series with the circuit carrying the current under measurement, must be of *very low resistance* so that the voltage drop across the ammeter and power absorbed from the circuit are as low as possible.

Q. 12. Why should be a voltmeter of very high resistance ?

Ans. *Voltmeter*, which is connected in parallel with the circuit across which the voltage is to be measured, must be of very *high resistance* so that the current flowing through the voltmeter and the power absorbed from the circuit are minimum possible.

Q. 13. How can a PMMC instrument be converted into an ammeter ?

Ans. PMMC instrument can be converted into an ammeter for measuring higher currents by connecting a small resistor across the meter coil in order to bypass the excessive current.

Q. 14. How can be an ammeter changed into a voltmeter ?

Ans. An ammeter of low range can be converted into a voltmeter by connecting a high resistance in series with it provided the current through the series combination is within the range of the ammeter when connected across the voltage under measurement.

Q. 15. What happens when an ammeter is connected across the circuit ?

Ans. If an ammeter is connected in parallel to the circuit like a voltmeter, a very high current will flow through it which will produce such an excessive heat that the insulation of the wire carrying the current will be destroyed. The wire may itself melt away. Thus the instrument will get damaged.

Q. 16. What happens when a voltmeter is connected in series with the circuit ?

Ans. If a voltmeter is connected in series with the circuit, the circuit resistance will become too large and consequently a very small current will flow through it. The instrument will, however, read almost the same emf acting in the circuit.

Q. 17. What do you understand by ammeter shunt ?

Ans. An ammeter shunt is merely a low resistance that is placed across the coil circuit of the instrument in order to measure fairly large currents (Fig. 3.14).

Q. 18. What do you understand by voltmeter multiplier ?

Ans. Voltmeter multiplier is a high non-inductive resistance connected in series with the voltmeter coil and is used for increasing the range of a voltmeter (Fig. 3.20).

Q. 19. What are the main requirements that must be fulfilled by voltmeter multipliers ?

Ans. The main requirements of voltage multipliers to be used for dc measurements are that (*i*) resistance should not change with time of usage (*ii*) their temperature coefficient of resistance must be very low so that change in temperature may not affect their resistance values and (*iii*) should have ample provision for cooling to dissipate the heat developed, as an appreciable amount of power is developed in such resistors.

Q. 20. What is the advantage of using Ayrton or universal shunt in multirange ammeters ?

Ans. The advantage of using Ayrton or universal shunt is that it eliminates the possibility of the meter being in circuit without a shunt.

PROBLEMS

1. The coil of a moving iron ammeter has a resistance of 6.25 Ω and full-scale reading of 0.012 A. What would be the full-scale reading of the instrument if used as a voltmeter ? **[Ans.** 0.075 V]
2. A permanent magnet moving coil instrument has a coil of dimensions 15 mm × 12 mm. The flux density in the airgap is 1.8×10^{-3} Wb/m^2 and the spring constant is 0.14×10^{-6} Nm/rad. Determine the number of turns required to produce an angular deflection of 90° when a current of 5 mA is flowing through the coil. **[Ans.** 138]
3. A moving coil comprises 100 turns of insulated copper wire wound on a former of length 3 cm and breadth 4 cm. The resistance of the coil is 2,000 Ω. The field strength of the magnet is 0.06 Wb/m^2. The torque exerted by the control spring is 2×10^{-6} Nm per degree. Estimate the deflection of instrument when a voltage of 120 V is applied across it. **[Ans.** 216°]

4. The dimensions of the coil of a permanent magnet moving coil voltmeter are 4 cm × 2.6 cm. The number of turns in the coil are 80 and the flux density in the gap is 0.15 T. The resistance of the instrument is 15,000 Ω. Calculate the deflecting torque produced in the instrument when a voltage of 300 V is applied to its terminals. **[Ans.** 24×10^{-5} Nm]

5. A moving coil meter gives a full-scale deflection with a current of 5 m A. The coil is wound with 100 turns of wire on a rectangular former having a depth of 3 cm and a width of 2.5 cm. The controlling torque supplied by the spring for full-scale deflection is 4.9×10^{-5} N-m. Find the magnitude of the flux density is the air gap in teslas. **[U.P.S.C. I.E.S. Elec. Engg. 1972]**

[Ans. 0.125 T]

6. A moving coil-meter gives a full-scale deflection with a current of 5 mA. If the coil of the instrument has a resistance of 10 Ω, show how it can be adopted to work (*i*) as an ammeter with a range 0–10 A and (*ii*) as a voltmeter with a range of 0–100 V. **[U.P.S.C. I.E.S. Elec. Engg. 1972]**

[Ans. (*i*) $\frac{10}{1,999}$ Ω (*ii*) 19,990 Ω]

7. Pointer of a moving-coil instrument gives full-scale deflection of 20 mA. The potential difference across the meter when carrying 20 mA is 200 mV. This instrument is intended to be used on 200 A for full-scale indication. Find out the shunt resistance required to achieve this. If the instrument is used as a voltmeter for full-scale reading with 1,000 V, find out the series resistance to be connected in it.

[Ans. (*i*) $\frac{10}{9,999}$ Ω; (*ii*) 49,990 Ω]

8. A dc moving coil voltmeter gives full-scale deflection of 100 V when the current flowing through it is 1 mA. It is used for measuring voltage in a circuit containing two equal resistances of 100,000 Ω, each connected in series across a 200 V dc supply. What will be the voltmeter reading when it is used to measure the voltage drop across one of the resistance ? **[Ans.** 66.67]

9. The coil of a moving coil voltmeter is 4 cm × 3 cm wide and has 100 turns wound on it. The control spring exerts a torque of 2.5×10^{-4} N-m when the deflection is 50 divisions on scale. If the flux density on the magnetic field in the air gap is 1 T, estimate the resistance that must be put in series with the coil to give 1 V/division. Resistance of the voltmeter is 10,000 Ω. **[Ans.** 14,000 Ω]

10. A series ohmmeter is made up of the following components: Battery voltage = 3 V, series resistor R_1 = 30 kΩ, meter shunt resistor = 50 Ω, Meter FSD current = 50 μA and instrument resistance, R_m = 50 Ω. Determine the resistance measured at 0, 25, 50 and 75% of full-scale deflection.

[Ans. 0 Ω, 90 kΩ, 30 kΩ, 10 kΩ]

❑❑❑

UNIT-II
Analog Electronic Meters And Digital Meters

CONTENTS IN THIS UNIT

Analog Electronic Meters: Transistor voltmeter circuits, AC electronic voltmeter, Current measurement with electronic instruments, Multimeter probes.

Digital Meters: Digital voltmeter systems, Digital multimeters, Digital frequency meter system.

CHAPTER 4: Analog Electronic Meters.

CHAPTER 5: Digital Meters.

4

Analog Electronic Meters

4.1 INTRODUCTION

In all electronic voltmeter circuits the principle involved is that an indication on a permanent magnet moving coil instrument (normally abbreviated as PMMC or D'Arsonval movement) proportional to the input voltage is obtained by means of amplification in one or more stages with a high input impedance.

Although electronic instruments are usually more costlier than electrical instruments but are becoming more and more popular because of their numerous advantages over conventional ones, as discussed below :

1. A moving coil voltmeter draws a large current, hence a large power from the test circuit during measurement. Generally current drawn by the voltmeter does not make much difference in electrical circuits but when measurement is involved in electronic circuits, current drawn by the voltmeter load the test circuit, and therefore, voltmeter shows an error in the reading.
   In electronic voltmeters power required for deflection of a permanent magnet moving coil (PMMC) movement is not drawn from the test circuit but supplied from the amplifier whose output is proportional to the test circuit voltage *i.e.* input to the amplifier, so power drawn from the test circuit becomes almost negligible. So electronic voltmeters draw very low power from the test circuit and it can be said that it has a very high input impedance. This feature of electronic voltmeters is indispensable for voltage measurements in many high impedance circuits such as encountered in communication of the order of microvolts.
2. Voltages of the order of microvolts are measured in electronic circuits which is not possible with the sensitive PMMC voltmeters but these voltages can be measured with electronic voltmeter by using its amplifying properties.
3. Electronic voltmeters measure voltage both at audio — as well as radio — frequency power level, as the action of thermionic valves or transistors can be made independent of frequency up to 10 Hz–100 MHz or even higher. Hence such an instrument has an extremely wide frequency range (from direct current to frequencies of the order of hundred of MHz), if the circuit is properly designed. The high frequency range may also be attributed to low input capacitance (of the order of a few pF) of most electronic devices. An electronic voltmeter calibrated at power frequency of 50 Hz can be used at radio frequency without any error.

4. Transistorized voltmeters can be designed for measuring very high voltages such as hundreds or thousands of volts.
5. Owing to amplifying properties electronic instruments possess very high sensitivity and because of high sensitivity, their input impedance is increased. It results in reduced loading effect while carrying out measurements.
6. Electronic voltmeters give faster response and flexibility.
7. Such instruments can monitor remote signals. Electronic voltmeters may use either vacuum tube or transistor. Later one is called *transistorized voltmeter* (TVM) and former is called the *vacuum tube voltmeter* (VTVM).

Transistorized voltmeters have numerous advantages over vacuum tube voltmeters and hence replacing them in almost every field of electronics. The advantages of TVMs over VTVMs are given below:

1. A TVM does not require any warm up time because of absence of heating element.
2. Use of transistor makes the instrument portable because of its light weight.
3. A transistorized instrument can be operated on low voltage *i.e.* on battery so TVM is well suited for the field work where power source is not available.
4. VTVM cannot measure current directly because of its high resistance whereas TVM can do.
5. TVM consumes very less power because of absence of heating element.

The only disadvantage with TVM is low input impedance which it offers in comparison to VTVM. But this shortcoming is overcome by using field effect transistor (FET) in input stage of voltmeter because FET offers input impedance almost equal to vacuum tubes.

In electronic voltmeters current proportional to the voltage to be measured is generated by use of amplifier, rectifier etc. and then this current is measured by PMMC.

Electronic voltmeters are of two types namely dc voltmeters and ac voltmeters.

4.2 TRANSISTOR VOLTMETER CIRCUITS

Analog electronic meter basically consists of an amplifier connected to a D'Arsonval meter movement. In meters designed for measuring dc quantities, the amplifier receives the signal under measurement, amplifies it and feeds to the D'Arsonval movement. The D'Arsonval movement converts the amplified signal to an indication by causing a pointer to move along a calibrated scale. If the meters are designed to measure ac quantities, a rectifier is inserted in the circuit along with the amplifier and D'Arsonval movement.

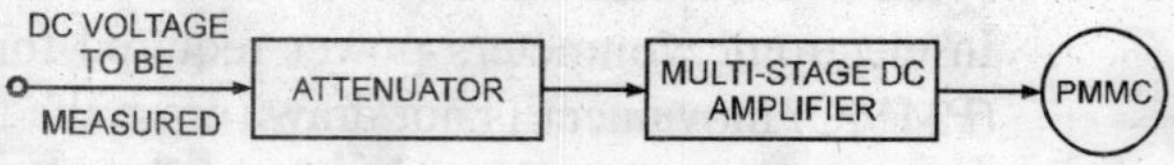

Fig. 4.1 *Block Diagram of DC Voltmeter*

DC Voltmeter mainly consists of dc amplifier apart from the attenuator, as shown in Fig. 4.1.

4.2.1. Direct Coupled Amplifier DC Voltmeter. This type of voltmeter is very common because of its low cost. This instrument can be used only to measure voltages of the order of millivolts owing to limited amplifier gain.

The circuit diagram for a direct coupled amplifier dc voltmeter using cascaded transistors is shown in Fig. 4.2.

An attenuator is used in input stage to select voltage range. A transistor is a current controlled device so resistance is inserted in series with the transistor Q_1 to select the voltage range. It can be seen from Fig. 4.2 that sensitivity of voltmeter is 200 k Ω/volt neglecting small resistance offered by transistor Q_1. Other values of range selecting resistors are also so chosen that sensitivity remains the same for all ranges. So current drawn from the circuit is only 5 μA.

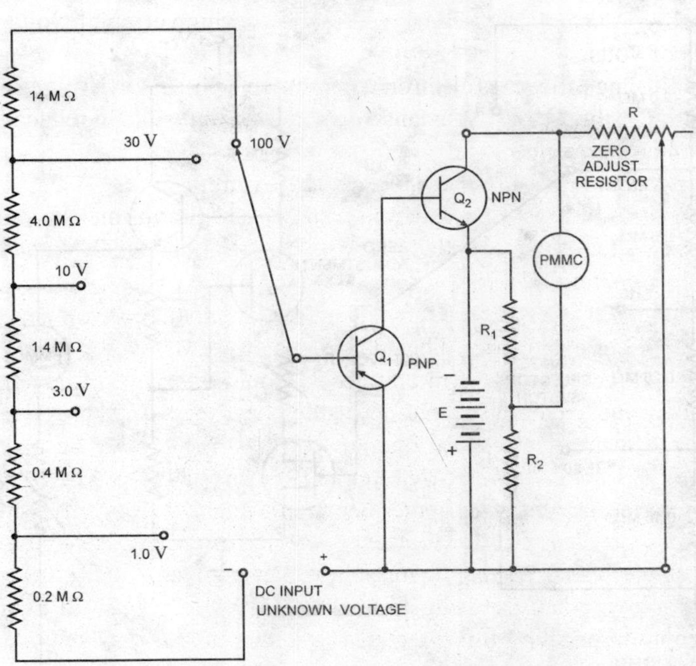

Fig. 4.2 *Direct Coupled Amplifier DC Voltmeter Using Cascaded Transistors*

Two transistors in cascaded connections are used instead of using a single transistor for amplification in order to keep the sensitivity of the circuit high. Transistors Q_1 and Q_2 are taken complement to each other and are directly coupled to minimise the number of components in the circuit. They form a direct coupled amplifier.

A variable resistance R is put in the circuit for zero adjustment of the PMMC. It controls the bucking current from the supply E to buck out the quiescent current.

The drawback of such a voltmeter is that it has to work under specified ambient temperature to get the required accuracy otherwise excessive drift problem occurs during operation.

Another circuit diagram of a direct coupled amplifier dc voltmeter using FET in input stage is shown in Fig. 4.3.

In this voltmeter, voltage to be measured is firstly attenuated with range selector switch to keep the input voltage of amplifier within its level. FET is used in the input stage of amplifier because of its high input impedance so that it does not load the circuit of which voltage is to be measured and it also keeps the sensitivity of voltmeter very high.

As FET is a voltage controlled device so resistance network of attenuator is put in shunt in the circuit.

Transistors Q_2 and Q_3 form the direct coupled dc amplifier whose output is finally supplied to PMMC meter. When transistors work within their dynamic region, the deflection of meter remains proportional to the applied input voltage.

Apart from the high input impedance, this circuit has another advantage that when input voltage exceeds its limit, amplifier gets saturated which limits the current passing through the PMMC meter. So meter does not burn out.

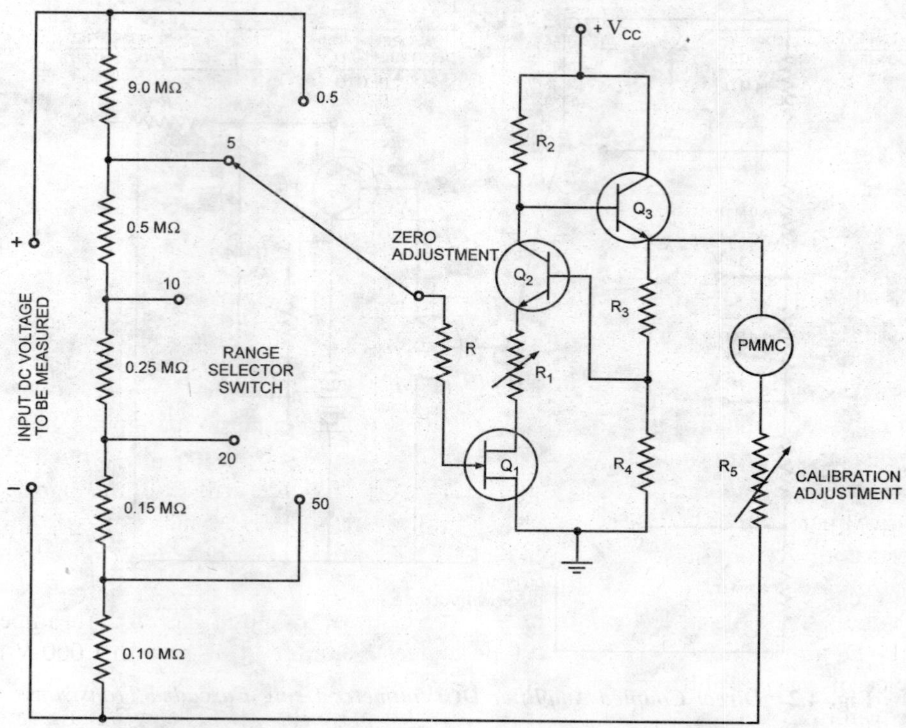

Fig. 4.3 *Direct Coupled Amplifier Transistorized DC Voltmeter Using FET*

The gain of the dc amplifier allows the instrument to be used for measurement of voltages in the milli-volt range. Instruments in the microvolt range of measurement need a high gain dc amplifier to supply sufficient current for driving the meter movement. The drift problem in direct-coupled dc amplifiers restricts the use of a number of stages required for high gains. In order to avoid the drift problems of dc amplifiers, chopper type dc amplifiers are commonly used in high sensitivity voltmeters.

4.2.2. Chopper Type DC Voltmeter. Chopper type dc amplifier is used in highly sensitive dc electronic voltmeters. Its block diagram is shown in Fig. 4.4. Firstly dc input voltage is converted into ac voltage by chopper modulator and then it is supplied to an ac amplifier. Output of amplifier is then demodulated to a dc voltage proportional to the original input voltage. Modulator chopper and demodulator chopper act in antisynchronism.

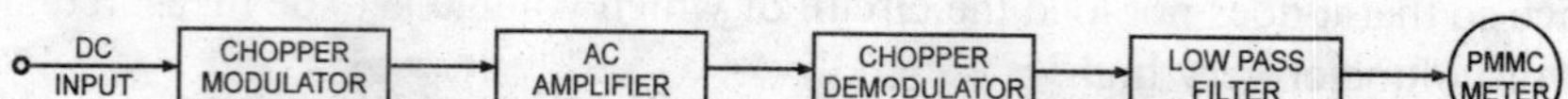

Fig. 4.4 *Block Diagram of Chopper Type DC Voltmeter*

Chopper system may be either mechanical or electronic. Circuit diagram of an electronic chopper employing photodiodes is shown in Fig. 4.5.

Photodiode changes its resistance under different illumination conditions, this property of photodiode is used in chopper amplifier. Its resistance changes from the order of few mega ohms to few hundred ohms when it is illuminated by a light source in the dark place.

Two neon lamps are used in this circuit, these are supplied by an oscillator for alternate half cycles. Two photo-diodes are used in input stage which act as half-wave modulators because of their alternate switching action by the neon lamps at the frequency of oscillator.

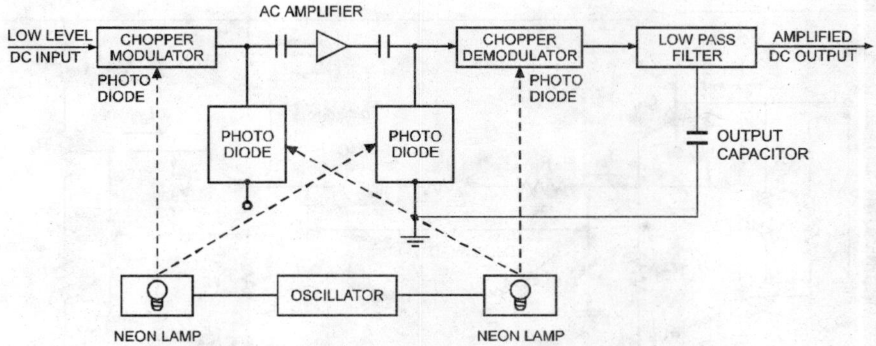

Fig. 4.5 *Circuit Diagram of Chopper Type DC Voltmeter*

Output of chopper modulator is a square-wave voltage (proportional to the input signal) which is supplied to the ac amplifier through a capacitor. Amplified output is again passed through a capacitor and then fed to chopper demodulator. Capacitor is used to remove dc drift from the signal. Chopper demodulator gives a dc output voltage (proportional to the input voltage) which is passed through the low pass filter to remove any residual ac component. Now this dc output voltage is supplied to the PMMC meter for measurement of input voltage.

A commercially available instrument using a photo chopper amplifier has an input impedance of the order or 100 Ω, a resolution of 0.1 μV and input ranges from 3 μV full-scale to 1,000 V full-scale with an accuracy of ± 2% of full-scale deflection.

The drift in an ordinary dc amplifier is of the order of millivolts. The full-scale range of an ordinary dc amplifier is restricted to measuring input signal of 1–100 mV. In a chopper modulating system with the use of ac amplifier, drift can be brought down to one hundredth, thus allowing an input signal range of about 0.01 mV = 10 μV full-scale to be handled.

4.2.3. Emitter Follower Voltmeters. The basic emitter-follower voltmeter circuit is shown in Fig. 4.6. In this circuit, a PMMC meter and a multiplier resistance R_S are connected in series with the transistor emitter. The dc supply V_{CC} is connected between transistor collector and PMMC meter and voltage under measurement is connected between transistor base and the same terminal as the power supply negative. The polarities are shown in the circuit diagram. The transistor base current,

$$I_B \simeq \frac{I_E}{h_{FF}} \simeq \frac{I_m}{h_{FE}} \qquad \text{...(4.1)}$$

where h_{FF} is the transistor current gain

$$\text{Input resistance, } R_{in} = \frac{V}{I_B} = \frac{V_B}{I_m} \times h_{FE} \qquad \text{...(4.2)}$$

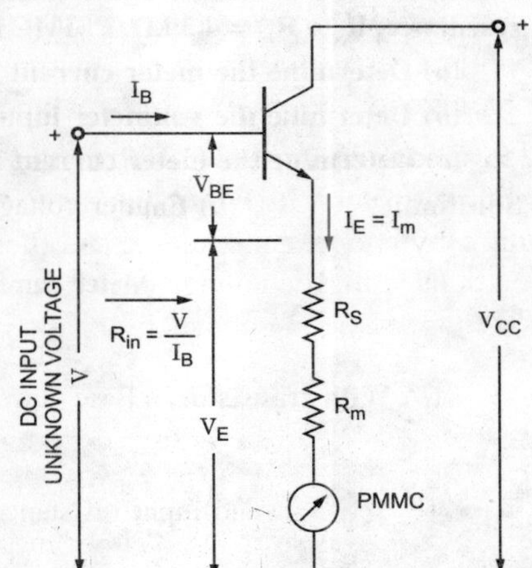

Fig. 4.6 *Basic Emitter-follower Voltmeter Circuit*

Thus, the circuit input resistance is much larger than the meter circuit resistance.

The drawback of the above circuit is that the transistor base-emitter voltage V_{BE} introduces an error. However, this error can be eliminated by using a potential divider and an additional emitter follower, as shown in Fig. 4.7. This circuit uses a dual polarity supply (typically, ± 12 V).

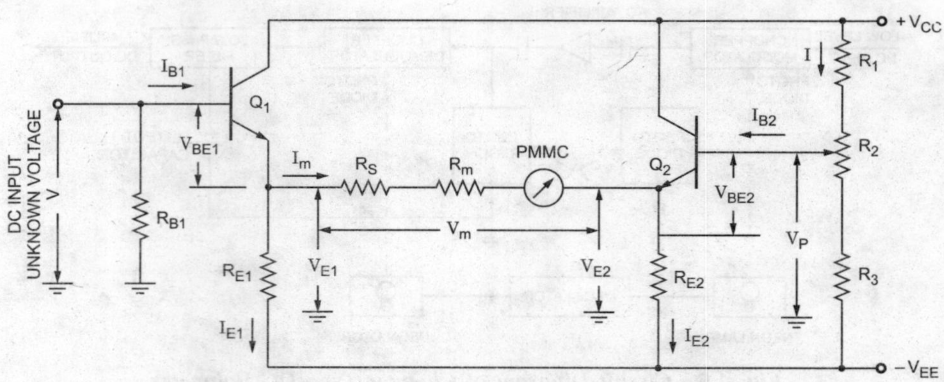

Fig. 4.7 *Practical Emitter Follower Voltmeter Circuit*

The base of transistor Q_1 is biased to ground via resistor R_{B_1} and an adjustable bias voltage V_P is provided to the base of transistor Q_2 by potential divider (R_1, R_2 and R_3). The PMMC meter circuit is connected between the emitters of transistors Q_1 and Q_2. The emitter terminals of transistors Q_1 and Q_2 are connected to the negative supply voltage $-V_{EE}$ through emitter resistors R_{E_1} and R_{E_2} respectively.

The input resistance of the circuit is R_{B_1} in parallel with the input resistance at the transistor base.

When no input voltage is applied *i.e.* when V = 0, the base voltage of transistor Q_2 is adjusted to provide zero current in PMMC meter. This makes bias voltage applied to transistor Q_2, $V_P = 0$, $V_{E_1} = V_{E_2} = V_{BE} = -0.7$ V, and meter circuit voltage $V_m = 0$ V.

When input applied voltage to the base of transistor Q_1 is V volts, the meter voltage, $V_m = V_{E_1} - V_{E_2} = (V - V_{BE_1}) - V_{E_2} = (V - 0.7\text{ V}) - (-0.7\text{ V}) = V$ volts *i.e.* all of the voltage to be measured appears across the meter and no part of its is lost as transistor V_{BE}.

Example 4.1. The simple emitter follower circuit shown in Fig. 4.6 has V_{CC} = 12 V, and meter circuit resistance, $R_S + R_m$ = 4.3 kΩ. PMMC meter current is 1 mA at full scale and transistor current gain is 100.

(*a*) Determine the meter current when dc input voltage is 5 V.

(*b*) Determine the voltmeter input resistance with and without the transistor.

(c) Determine the meter current when the dc input voltage is 2.5 V.

Solution: (*a*) Emitter voltage, $V_E = V - V_{BE} = 5 - 0.7 = 4.3$ V

$$\text{Meter current, } I_m \simeq I_E \simeq \frac{V_E}{R_S + R_m} \simeq \frac{4.3\text{ V}}{4.3\text{ k}\Omega} = 1\text{ mA} \quad \textbf{Ans.}$$

(*b*) **With transistor,** Base current, $I_B \simeq \dfrac{I_m}{h_{FE}} \simeq \dfrac{1\text{ mA}}{100} = 10\ \mu\text{A}$ $\qquad \because h_{FE} = 100$

$$\text{and input resistance, } R_{in} = \frac{V}{I_B} = \frac{5\text{ V}}{10\ \mu\text{A}} = 500\text{ k}\Omega \quad \textbf{Ans.}$$

Without transistor Input resistance, $R_{in} = R_S + R_m = 4.3$ kΩ **Ans.**

(*c*) Meter current when the dc input voltage is 2.5 V,

$$I_m \simeq \frac{V_E}{R_S + R_m} = \frac{V - V_{BE}}{R_S + R_m} \simeq \frac{(2.5 - 0.7)\text{ V}}{4.3\text{ k}\Omega} \simeq \frac{1.8\text{ V}}{4.3\text{ k}\Omega} = 0.42\text{ mA}$$

i.e. meter reads 0.42 mA instead of 0.5 A (half of full-scale reading). This error is introduced due to base-emitter voltage drop, V_{BE}.

Example 4.2. An emitter follower voltmeter circuit shown in Fig. 4.7 has emitter resistances of 3.3 kΩ each and $V_{CC} = \pm 12$ V.

(*a*) Determine emitter currents when input voltage is zero volt.

(*b*) Compute the meter circuit voltage when input dc voltage is (*i*) 2 V and (*ii*) 1 V.

Solution: Emitter voltages, $V_{E_1} = V_{E_2} = V - V_{BE} - V_{EE} = 0 - 0.7 - (-12) = 11.3$ V

Emitter currents, $I_{E_1} = I_{E_2} = \dfrac{V_{E_1}}{R_{E_1}} = \dfrac{11.3 \text{ V}}{3.3 \text{ k}\Omega} \simeq 3.4$ mA **Ans.**

When input dc voltage, V = 2 volts

Emitter voltage, $V_{E_1} = V - V_{BE} = 2 - 0.7 = 1.3$ V

Emitter voltage, $V_{E_2} = V_P - V_{BE} = 0 - 0.7 = -0.7$ V

Meter circuit voltage, $V_m = V_{E_1} - V_{E_2} = 1.3 - (-0.7) = 2$ V **Ans.**

When input dc voltage, V = 1 volt

Emitter voltage, $V_{E_1} = V - V_{BE} = 1 - 0.7 = 0.3$ V

Emitter voltage, $V_{E_2} = V_P - V_{BE} = 0 - 0.7 = -0.7$ V

Meter circuit voltage, $V_m = V_{E_1} - V_{E_2} = 0.3 - (-0.7) = 1$ V **Ans.**

4.2.4. Balanced Bridge Voltmeters. Such voltmeters can be used for measurement of both dc and ac voltages and are very popular. These voltmeters are available in both versions, vacuum tube type and transistorized type and are described below:

1. Balanced Bridge Triode VTVM. The circuit diagram of a balanced bridge triode VTVM is shown in Fig. 4.8.

In this instrument two identical triodes T_1 and T_2 are used. Plate voltage is applied to both of the triodes through two identical resistors R_L and a variable resistance R_0. Resistance R_0 is so adjusted that plate voltages of both of the triodes become equal when there is no voltage applied on the grid of triode T_1. So, in this condition, meter connected between the plates of the two triodes reads zero. The bias battery E_{CC} biases the cathodes of both of the triodes positive and so the bias of the grid of both of the triodes becomes negative with respect to their respective cathodes.

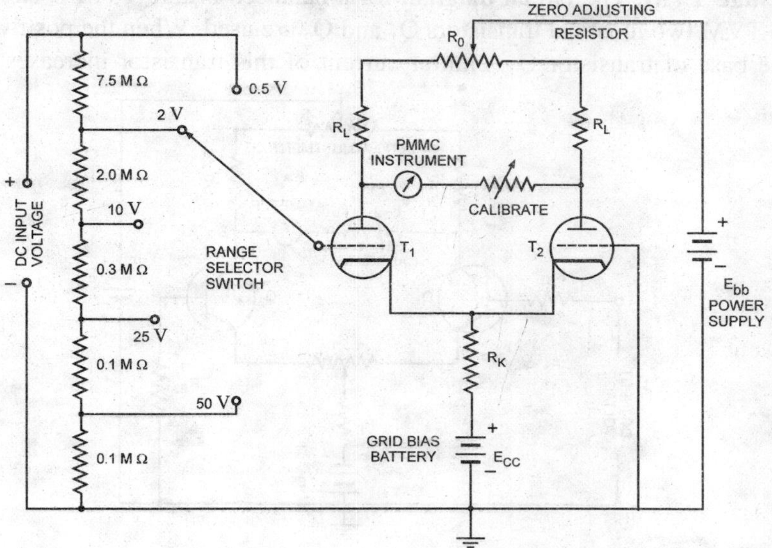

Fig. 4.8 *Circuit Diagram of Balanced Bridge Triode VTVM*

When a positive dc voltage is applied to the grid of triode T_1, potential of grid with respect to cathode increases and so the plate current increases. Due to increase in plate current of triode T_1 the voltage drop across the resistor R_L connected in series with triode T_1 increases and, therefore, plate voltage of triode T_1 falls. The increased current also flows through the resistor R_k which is common to both of the tubes T_1 and T_2. An increase in current through resistor R_k causes increase in voltage drop across it and, therefore, cathode of triode T_2 becomes more positive. It results in increase of – ve grid bias of tube T_2 and, therefore, plate current of triode T_2 is reduced. With this voltage drop across resistor R_L connected with plate of triode T_2 falls and, therefore, plate voltage of triode T_2 increases. Thus with the application of dc voltage on the grid of triode T_1, voltages of the two triode plates do not remain same and a current flows through the PMMC meter, which is proportional to the applied dc voltage. So PMMC meter can be directly calibrated to read the applied input dc voltage.

Range of this instrument can be extended by using a range selector switch and a potential divider network. For measuring ac voltage with this voltmeter, it is first rectified and then applied to this voltmeter. A resistance is connected in series with the meter for the purpose of setting the full-scale reading of the meter accurately.

The balanced bridge VTVM have the following advantages over an ordinary VTVM (*i.e.* a VTVM using a single tube):

1. The effective input impedance of a cathode follower circuit is very high.
2. The bias level applied to the grid of valve T_2 is made variable so that the meter reading can be set to zero when the input to terminal are shorted together. This balances out differences in the tube characteristics.

   Generally 12 A770–dual triode may be used so that the characteristics of the triodes are closely matched.
3. Fluctuations in the power supply have a smaller effect on the measuring circuit.
4. There is a lesser tendency for the meter zero to shift during the meter operation.

The disadvantage of VTVM is that quite a lot of heat is produced and, therefore, the instrument needs to be switched for time to allow the effects of temperature variations inside the case to settle down to a stable level before making any measurement.

2. Balanced Bridge TVM. The circuit diagram for a balanced bridge TVM is shown in Fig. 4.9. In balanced bridge TVM two identical transistors Q_1 and Q_2 are used. When the positive input dc voltage is applied to the base of transistor Q_1, emitter current of this transistor increases which causes the

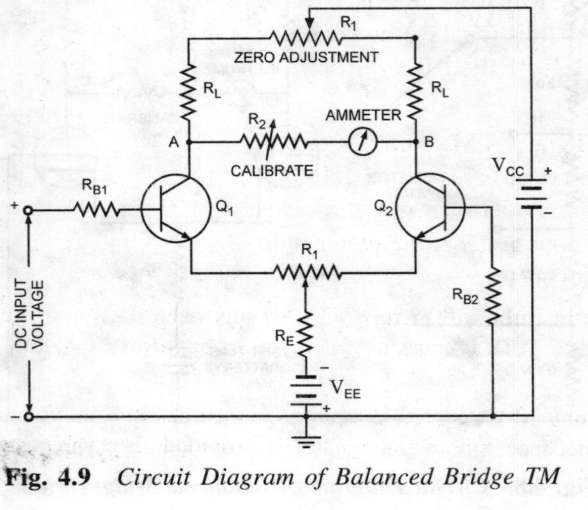

Fig. 4.9 *Circuit Diagram of Balanced Bridge TM*

increase in voltage drop across resistor R_L and, therefore, reduces the potential of collector of transistor Q_1. With the increase in emitter current, voltage drop across resistance R_E increases and so the emitter current of transistor Q_2 is reduced. This increases the potential of the collector of transistor Q_2 and a current proportional to the applied input voltage starts flowing in the ammeter connected between collectors of transistors Q_1 and Q_2.

Resistor R_1 is used for achieving *zero adjustment.* The purpose of this adjustment is to equalise both halves of the difference amplifier so that there is no current through the meter under the absence of dc input voltage.

Resistor R_2 is used for *calibration purpose.* This is to do the accurate setting of the full-scale reading of the ammeter. This adjustment is required because characteristics are not same for all the transistors. For instance, if the meter is made to have 1 V full-scale reading, the calibration is accomplished by measuring an input which is exactly 1 V. The calibration adjustment is made to give reading of exact voltage.

For obtaining several different full-scale ranges, a range switch may be used.

Main advantage of TVM is that if both transistors Q_1 and Q_2 are similar then with the change in ambient temperature, β values of both the transistors will change equally so there will be no effect on collector current in changed ambient temperature condition. Also with the change in ambient temperature, saturation current of both the transistors will change equally and, therefore, the meter reading will remain unaffected.

Resistor R_E also provides negative feedback for transistors Q_1 and Q_2 as any increase in ambient temperature also increases the emitter current which in turn is limited by the increase in the voltage drop across resistor R_E which is normally of high magnitude. Thus accuracy of calibration of TVM is maintained over a wide range of temperature.

Another advantage of TVM is that any fluctuation in power supply V_{CC} and V_{EE} does not affect its performance as it changes the biasing of both transistors equally.

Main disadvantage of balanced bridge TVM is that it offers less input impedance comparatively. But by using FET in the circuit of TVM, a very high input impedance can be achieved.

In the circuit of this TVM, two resistors R_L and the internal resistance of transistors Q_1 and Q_2 form the arms of a balanced bridge and the current flows through the meter when external voltage is applied to the base of transistor Q_1, which disturbs the balance of the bridge. That is why, this is called the *balanced bridge TVM.* This is also called the *difference-amplifier type TVM.*

3. Emitter-Follower Type TVM. Another circuit of a TVM using emitter followers is depicted in Fig. 4.10. If the transistors are identical, the circuit is balanced and for zero input voltage, the meter indication is zero. The adjustments are as described above.

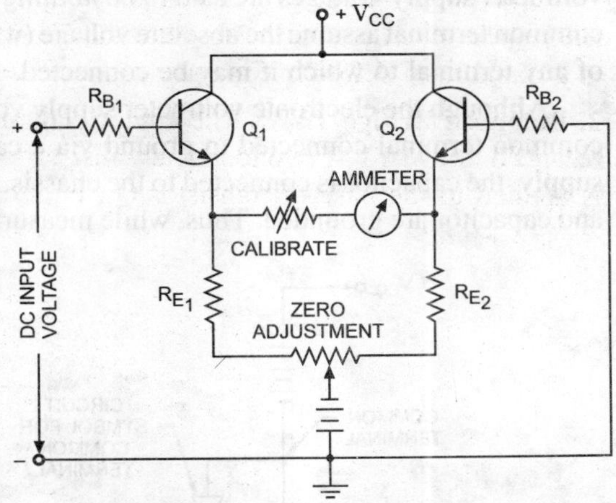

Fig. 4.10 *Emitter Follower Type Voltmeter*

Example 4.3. The vacuum tube voltmeter (VTVM) has been designed for 1.0 V dc output. The input impedance is specified as 1 M Ω. Discuss how can you incorporate 10 V and 100 V multirange facilities in the instrument.

Solution : For incorporating 10 V and 100 V multirange facilities in a VTVM designed for 1.0 V dc input and having 1 M Ω input impedance, suitable attenuator is provided. The values of attenuator resistances should be such that the voltage applied to the instrument is always 1 V and the rest is dropped across the resistances. The detail of attenuator is shown in Fig. 4.11.

Thus when voltage under measurement is 100 V, it will be connected across 100 V point. In this case 99 V will drop across 99 MΩ resistor and 1 V across 1 MΩ resistance connected across which is VTVM.

For measuring 10 V, it will be connected across 10 V point. In this case 9 V will drop across 9 MΩ resistor and 1 V across 1 MΩ input impedance.

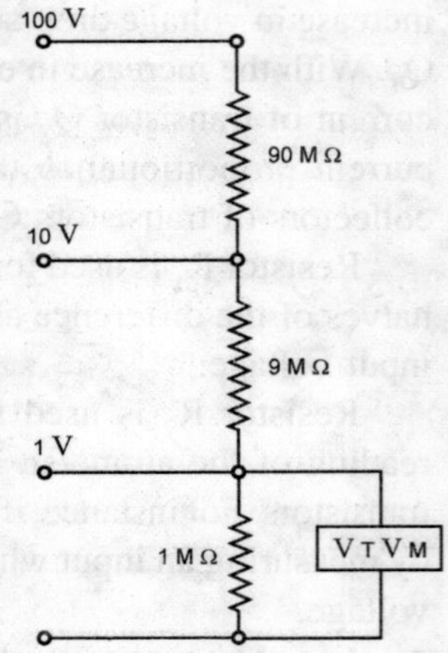

Fig. 4.11

4.2.5. Ground Terminals and Floating Power Supplies. The circuit in Fig. 4.7 shows the input voltage V as being measured with respect to ground. But, this may not always be convenient. For example, if voltage across resistor R_2 were to be measured by a voltmeter with its negative terminals grounded (Fig. 4.12). The voltmeter ground would short-circuit resistor R_3 and seriously effect the voltage and current conditions in the resistor circuit. Obviously, the voltmeter should not have one of its terminal grounded.

For the circuit shown in Fig. 4.7 to function properly, the lower end of R_{B_1} must be at zero volt with respect to $+V_{CC}$ and $-V_{EE}$. The positive and negative supply voltages may be derived from two batteries, as shown in Fig. 4.13(*a*) or from two dc power supply circuits, as shown in Fig. 4.13(*b*).

In both cases, the negative terminal of the positive supply is connected to the positive terminal of the negative supply. The point at which the two supplies are connected is called the *common terminal* or *zero voltage terminal*. It is identified by a symbol of an inverted triangle as shown in Figs. 4.13(*a*), 4.13(*b*). In many electronic circuits, this common terminal is grounded. In electronic voltmeter circuits, this terminal is not grounded, simply to avoid the problem discussed above. When the common terminal is not grounded, the voltmeter supply voltages are said to be *floating*. This means that the common terminal assume the absolute voltage (with respect to ground) of any terminal to which it may be connected.

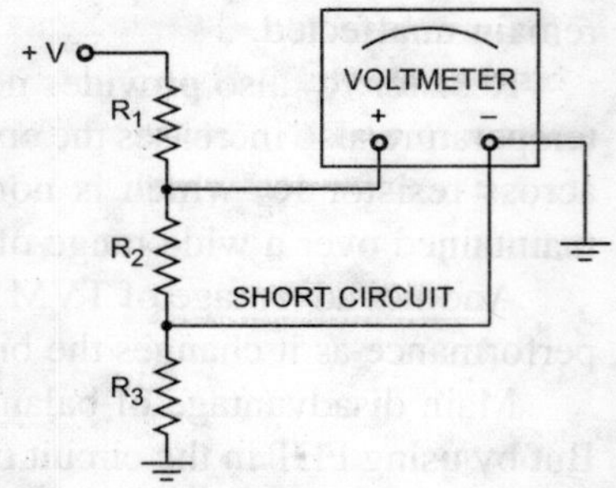

Fig. 4.12

Although the electronic voltmeter supply voltages are allowed to *float*, some instruments have their common terminal connected to ground via a capacitor (usually of 0.1 μF). If the batteries are used as supply, the capacitor is connected to the chassis. When power supply is used in the voltmeter, the chassis and capacitor are grounded. Thus, while measuring unknown voltage in a transistor circuit, the common

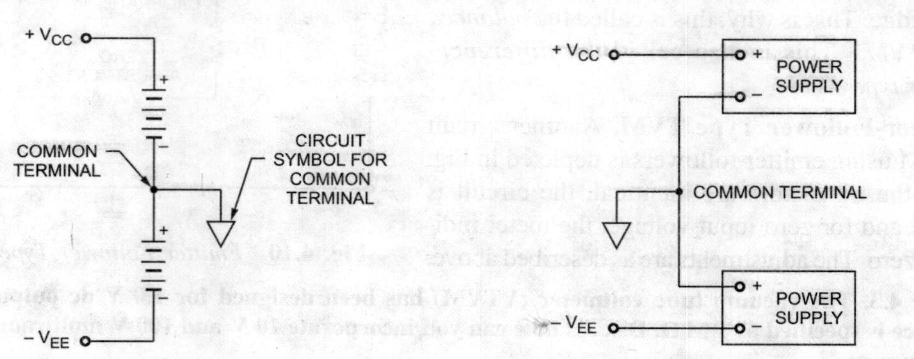

(*a*) ± *Supply Using Batteries* (*b*) ± *Supply Using Power Supplies*

Fig. 4.13

terminal introduces a capacitance to ground whenever it is connected in the circuit. The voltmeter common terminal should always be connected to the transistor circuit ground or zero voltage terminal so that any effect on conditions within the circuit (oscillations or phase shifts) is avoided. All voltages are then measured with respect to ground or zero terminal voltage.

4.3 OPERATIONAL AMPLIFIER VOLTMETER CIRCUITS

4.3.1. Op-Amp Voltage Follower Voltmeter. In an op-amp voltage follower shown in Fig. 4.14, the output voltage exactly tracks the input voltage both in sign and magnitude. Because of three unique characteristics of extremely high input resistance (of the order of 10^5 MΩ), unity transmission gain, and extremely low output resistance (of the order of few mΩ), the voltage follower circuit is an ideal circuit for use in voltmeter circuits.

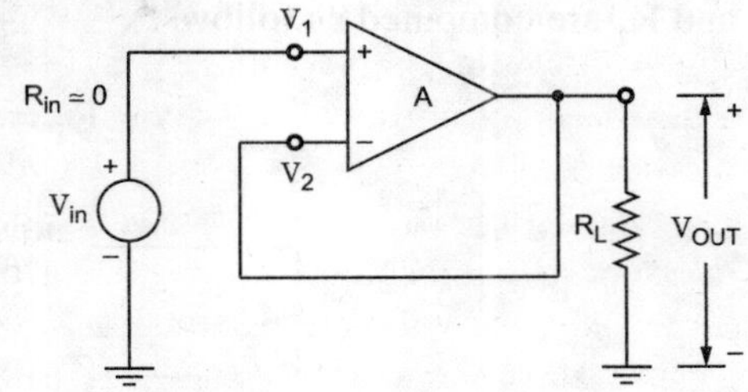

Fig. 4.14 *Voltage or Unity Follower*

The op-amp voltage follower voltmeter circuit is shown in Fig. 4.15. This circuit has a much higher input resistance and lower output resistance than the emitter follower. In this circuit, unlike the emitter follower, there is no base-emitter voltage drop from input to output. As discussed, earlier, the attenuator select the voltmeter range.

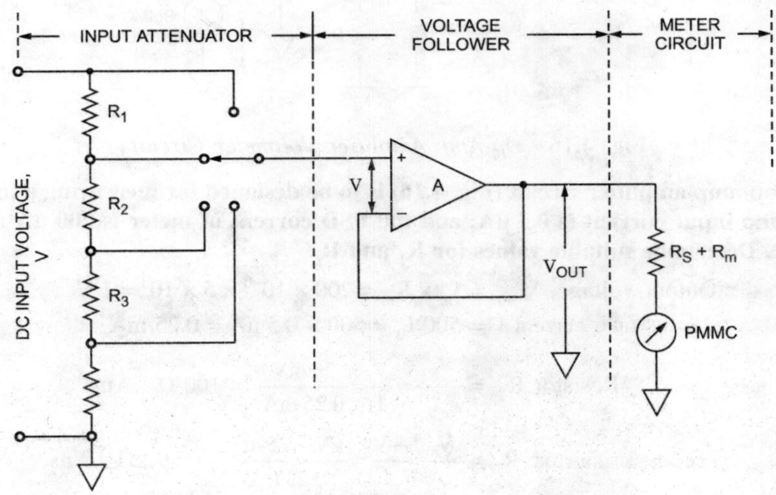

Fig. 4.15 *An IC Op-Amp Voltage Follower Voltmeter Circuit*

4.3.2. Op-Amp Amplifier Voltmeter. Like a transistor amplifier, an IC operational amplifier circuit can be employed for amplification of low voltages to levels measurable by a deflection instrument. In this circuit, inverting (–) input terminal is grounded through R_1 and output is applied to the inverting input terminal through feedback circuit composed of two resistors R_1 and R_f. Input voltage V is applied to the noninverting input terminal of op-amp. The output voltage is given by the equation

$$V_{out} = \left(1 + \frac{R_f}{R_1}\right) V$$

The circuit to known as a *noninverting amplifier* because it uses a feedback and input signal is applied to the noninverting input terminal of the op-amp. This circuit has a very high input resistance, very low output resistance and a voltage gain of $A_f = \left(1 + \frac{R_f}{R_1}\right)$.

An op-amp noninverting amplifier voltmeter circuit can be very easily designed. Current I through R_f and R_1 is first selected very much larger than the op-amp input bias current I_B. Then the resistor R_f and R_1 are computed as follows:

$$R_1 = \frac{V}{I} \quad \text{and} \quad R_f = \frac{V_{out} - V}{I}$$

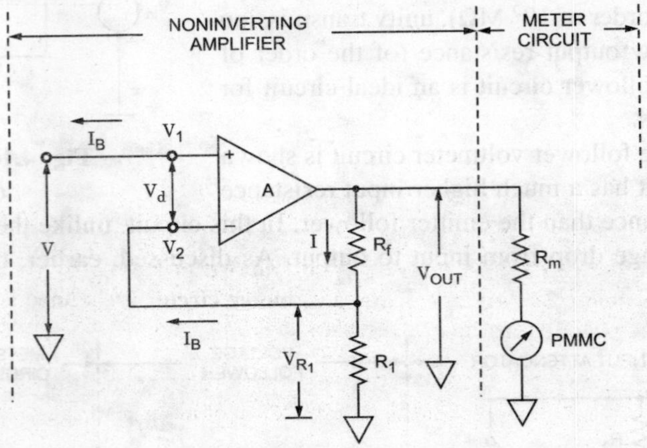

Fig. 4.16 *Op-Amp Amplifier Voltmeter Circuit*

Example 4.4. An op-amp amplifier circuit (Fig. 4.16) is to be designed for measuring a maximum input of 25 mV. The op-amp input current is 0.5 µA, and the FSD current of meter is 200 µA and meter circuit resistance is 5 kΩ. Determine suitable values for R_1 and R_f.

Solution: Output voltage, $V_{out} = I_m \times R_m = 200 \times 10^{-6} \times 5 \times 10^3 = 1\text{ V}$

Let us take current $I = 500\, I_B = 500 \times 0.5\ \mu A = 0.25\text{ mA}$ $\quad \because I >> I_B$

$$\text{Resistor } R_1 = \frac{V_{R_1}}{I} = \frac{V}{I} = \frac{25\text{ mV}}{0.25\text{ mA}} = 100\ \Omega \quad \textbf{Ans.}$$

$$\text{Feedback resistor, } R_f = \frac{V_{out} - V}{I} = \frac{1\text{ V} - 25\text{ mV}}{0.25\text{ mV}} = 3.9\text{ k}\Omega \quad \textbf{Ans.}$$

4.3.3. Op-Amp Voltage-To-Current Converter Voltmeter Circuit. Op-amp voltage-to-current converter (Fig. 4.17) provides a current proportional to input voltage.

$$\text{Output current, } I_L = I = \frac{V_{in}}{R_1}$$

From above equation it is obvious that the output current I_L is independent of load resistance R_L and is proportional to the input voltage V_{in}. This is because of virtual ground at the inverting input terminal of the op-amp.

If R_L is substituted by meter circuit in Fig. 4.17 we can have voltage-to-current converter voltmeter circuit, as shown in Fig. 4.18.

$$\text{Meter current, } I_m = I + I_B \simeq I \qquad \because I >> I_B$$

and voltage drop across R_1, V_{R1} = Input voltage, V

$$\text{So } I_m \simeq I \simeq \frac{V_{R_1}}{R_1} \simeq \frac{V}{R_1}$$

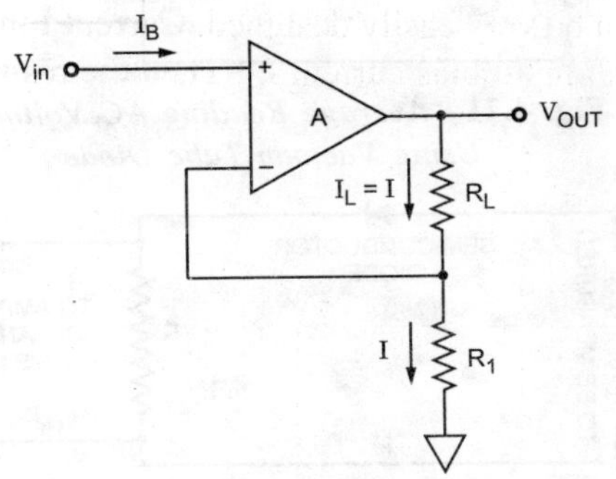

Fig. 4.17 *Voltage-To-Current Converter*

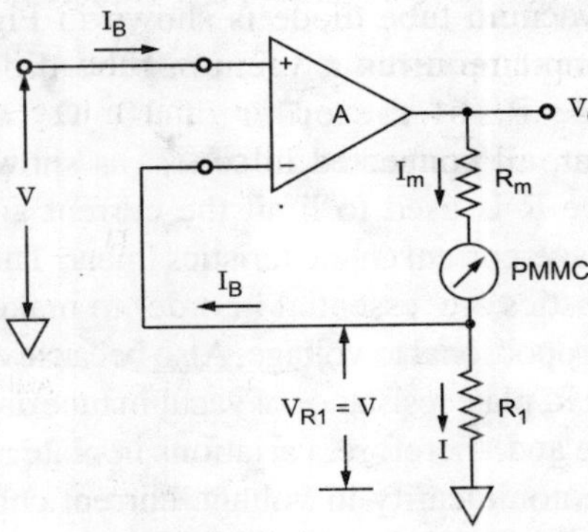

Fig. 4.18 *Op-Amp Voltage-To-Current Converter Voltmeter Circuit*

Example 4.5. Determine the value of R_1 for the circuit shown in Fig. 4.18. Maximum input voltage is 1.2 V. The FSD current of the meter circuit is 1 mA and meter circuit resistance is 10 Ω. Also determine the maximum voltage at the output terminal of op-amp.

Solution: Resistance, $R_1 = \frac{\text{Input voltage}}{\text{Full-scale deflection current of meter}} = \frac{V}{I_{m(FSD)}} = \frac{1.2 \text{ V}}{1 \text{ mA}} = 1.2 \text{ k}\Omega$ **Ans.**

Output voltage, $V_{out} = I(R_m + R_1) = 1 \text{ mA} \times (100\ \Omega + 1.2 \text{ k}\Omega) = 1.3 \text{ V}$ **Ans.**

4.4 AC ELECTRONIC VOLTMETERS

As already mentioned in previous section, in electronic ac voltmeters a rectifier is used either before (Fig. 4.19) or after the amplifier (Fig. 4.20) depending on whether the amplifier is used is dc or ac. In the former case (Fig. 4.19) the advantage is of economical amplifiers and the arrangement is usually employed in low-priced voltmeters.

Broadly ac voltmeters can be divided into three categories.

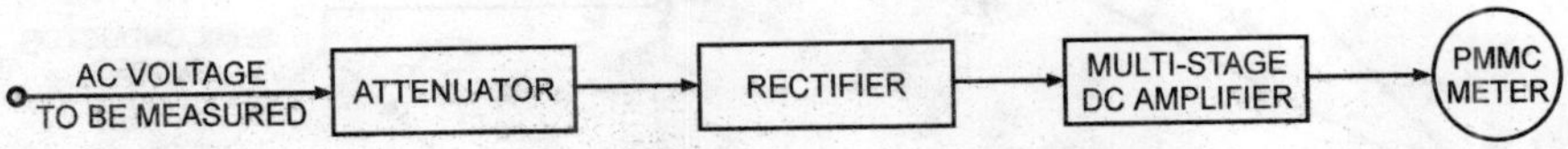

Fig. 4.19 *Block Diagram of AC Voltmeter*

Fig. 4.20 *Block Diagram of AC Voltmeter*

4.4.1. Average Reading AC Voltmeters. Normally ac voltmeters are average responding type and the meter is calibrated in terms of the rms values for a sine wave. Since most of the voltage measurements involve sinusoidal waveform, this method of measuring rms value of ac voltages works satisfactorily and is less expensive than true rms responding voltmeters. However, in case of measurement of non-sinusoidal waveform voltage, this meter will give high or low reading depending on the form factor of the waveform of the voltage to be measured.

The circuit diagram for an average reading voltmeter using a vacuum tube diode is shown in Fig. 4.21. The arrangement requires a vacuum tube diode, an high resistance R (of the order of 10^5 Ω) and PMMC instrument, all connected in series, as shown in figure. Resistance R is used to limit the current and make the plate voltage-current characteristics linear. The linear plate characteristics are essential in order to make the current directly proportional to voltage. Also because of high series resistance R, plate resistance of vacuum tube diode becomes negligible and, therefore, variations in plate resistance do not cause nonlinearity in voltage-current characteristics. In this way we get the scale of PMMC instrument uniform and independent of variations of tube plate resistance. Voltage across the high resistance is fed to dc amplifier and output of the amplifier is fed to PMMC instrument. Circuit diagram of an average reading ac voltmeter using semiconductor diode is shown in Fig. 4.22.

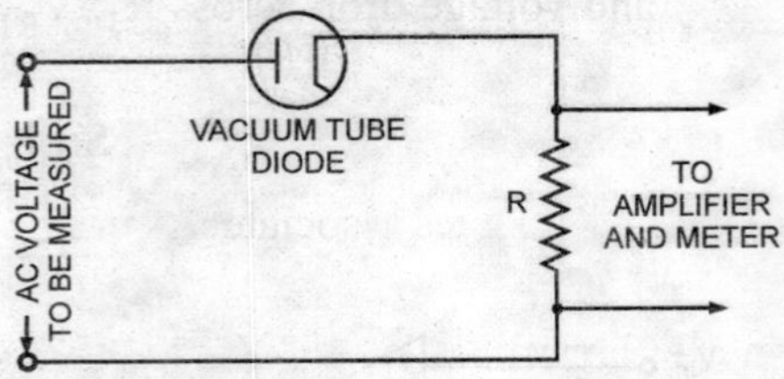

Fig. 4.21 *Average Reading AC Voltmeter Using Vacuum Tube Diode*

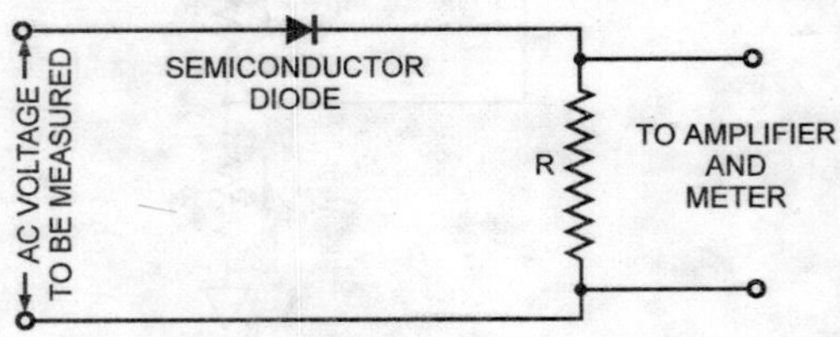

Fig. 4.22 *Average Reading AC Voltmeter Using Semiconductor Diode As Half-Wave Rectifier*

The diode conducts during the + ve half cycle and does not conduct during the –ve half cycle, as shown in Fig. 4.23.

The average current through the meter will be given by the expression

$$I_{av} = \frac{V_{av}}{2R} = \frac{V_{rms}}{2 \times 1.11 \times R} = 0.45 \frac{V_{rms}}{R} \qquad \text{... (4.1)}$$

V_{rms} is the effective or rms value of applied voltage and 1.11 is the form factor of sinusoidal wave. R is multiplied by 2 because the voltmeter operates on half-wave rectification.

It is worthnoting that this instrument can be used to indicate dc voltages but in such a case the instrument readings will have to be multiplied by 2 × 1.11 *i.e.* 2.22 as the diode conducts all the time.

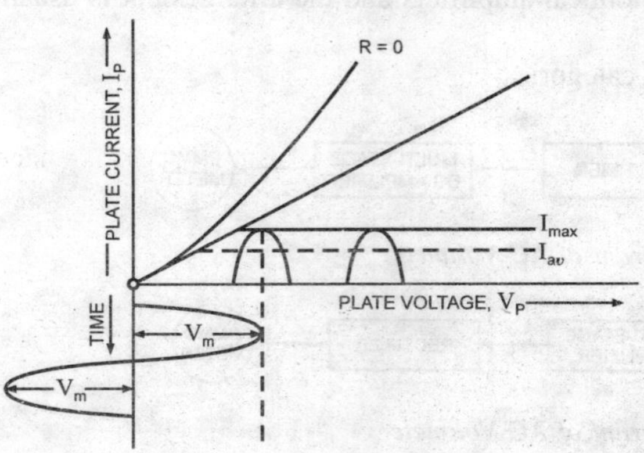

Fig. 4.23

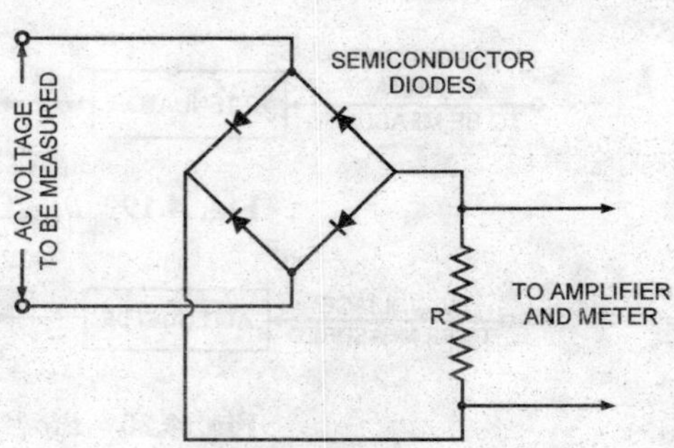

Fig. 4.24 *Average Reading AC Voltmeter Using Semiconductor Diodes As Full-Wave Rectifier*

Circuit diagram of an average reading ac voltmeter using semiconductor diodes as a full–wave rectifier is shown in Fig. 4.24. In this case average current through the meter will be $\frac{V_{rms}}{1.11\,R}$.

Main advantages associated with these voltmeters are that they are simple in construction, have high input impedance, low power consumption and uniform scale.

Main disadvantage of these voltmeters is that these operate in audio-frequency range. In radio-frequency range, distributed capacitance of the high resistance R introduces error in the readings.

Another disadvantage of such a voltmeter is that due to nonlinear volt-ampere characteristic for lower voltage the readings of the voltmeter at lower voltage are not correct.

The resistance of the circuit under test must be much small, otherwise it will load the circuit.

4.4.2. Peak Reading AC Voltmeters. The circuit diagrams for peak reading voltmeter using vacuum tube and semiconductor diode are shown in Figs. 4.25 and 4.26 respectively.

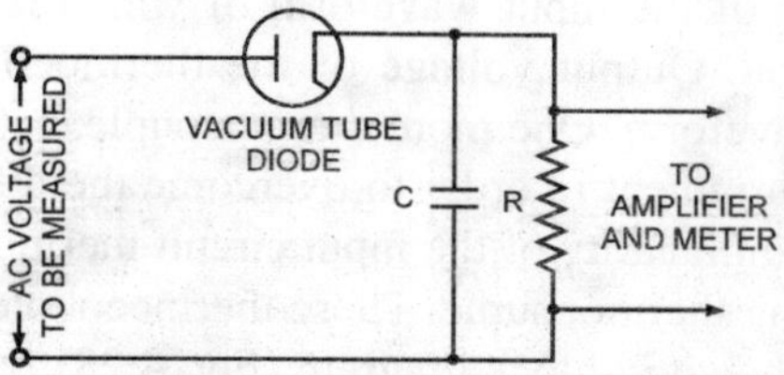

Fig. 4.25 *Peak Reading AC Voltmeter Using Vacuum Tube Diode*

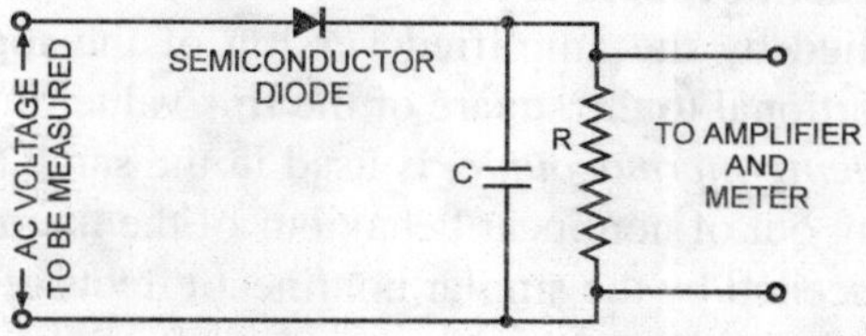

Fig. 4.26 *Peak Reading AC Voltmeter Using Semiconductor Diode*

In this type of voltmeters capacitor C is charged to the peak value of the applied voltage and capacitor is discharged through the high resistance R between two peaks of the wave which results in a small fall in capacitor voltage. But this voltage is again built up during next peak of the wave, as shown in Fig. 4.27. So voltage across capacitor C and resistance R remains almost constant and equal to the peak value of the applied voltage.

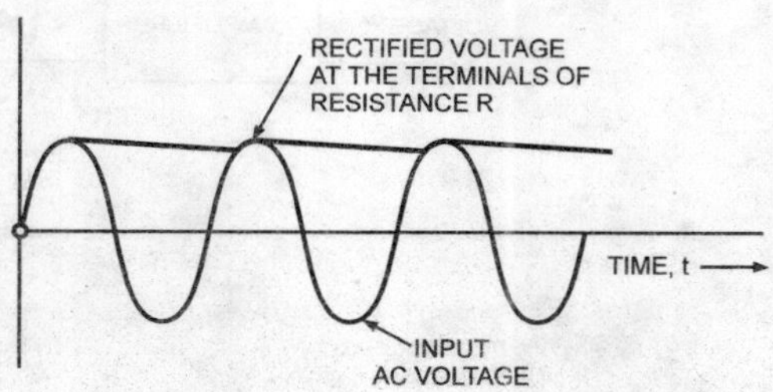

Fig. 4.27

Either the average voltage across R or the average current through R, can be used to indicate the peak value of applied voltage.

In case the vacuum tube diode (or semiconductor diode), series resistance R shunted by capacitance C and PMMC are connected in series across the source of unknown voltage, the current through the PMMC will indicate the peak value of applied voltage.

In case, the circuit shown in Fig. 4.25 or in Fig. 4.26 making use of rectifying diode, series resistance R, dc amplifier and PMMC is employed, the average voltage across R will indicate the peak value of applied voltage. This alternative is preferred, as explained earlier, the power consumption can be reduced by making series resistance R high. By making series resistance R high a less sensitive type of PMMC instrument can also be used. The high value input resistance also gives more linear relationship between peak applied voltage and the instrument indication. Also the performance of the diode with inputs consisting of pulses and modulated waves is improved.

The dc amplifier associated with the diode rectifier should be provided with stabilizing means in order to prevent drift in the indication of the output meter. Usually a voltage regulated supply combined with a compensating circuit is used.

The use of high series resistance R, associated with dc amplifier, no doubt results in a high input resistance but at the same time it implies that an applied voltage of sufficient amplitude is required so that the system acts as a peak voltage device.

The main disadvantage of this system is with regards to measurement of low voltage. If the applied voltage is too small, then there is a flow of some current throughout the cycle of the voltage because of high velocity of emission of electrons, and the input resistance may be a few hundred ohms and it defeats the very purpose with which electronic instruments are used.

4.4.3. True RMS Reading Voltmeters. RMS value of the sinusoidal waveform is measured by the average reading voltmeter of which scale is calibrated in terms of rms value. This method is quite simple and less expensive. But sometimes rms value of the non-sinusoidal waveform is required to be measured. For such a measurement a true rms reading voltmeter is required.

True rms reading voltmeter gives a meter indication by sensing heating power of waveform which is proportional to the square of the rms value of the voltage.

Thermocouple is used to measure the heating power of the input waveform of which heater is supplied by the amplified version of the input waveform. Output voltage of the thermocouple is proportional to the square of the rms value of the input waveform. One more thermocouple, called the *balancing thermocouple*, is used in the same thermal environment in order to overcome the difficulty arising out of nonlinear behaviour of the thermocouple. Nonlinearity of the input circuit thermocouple is cancelled by the similar nonlinear effects of the balancing thermocouple. These thermocouples form part of a bridge in the input circuit of a dc amplifier, as shown in block diagram (Fig. 4.28).

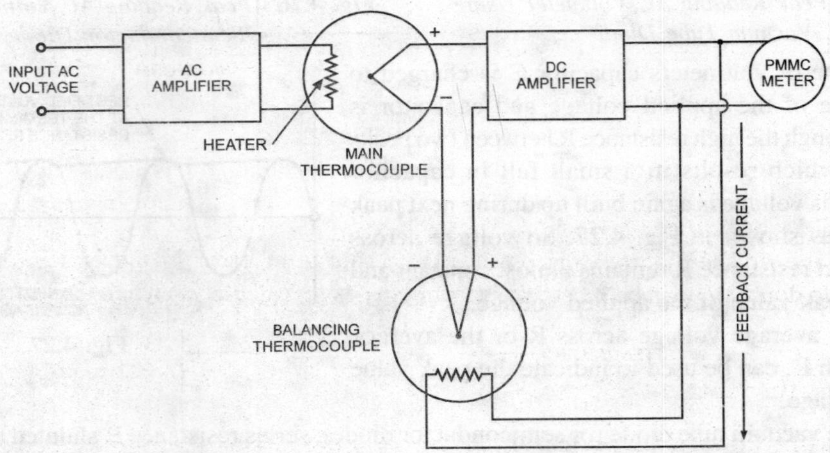

Fig. 4.28 *Block Diagram of True RMS Reading Voltmeter*

AC waveform to be measured is applied to the heating element of the main thermocouple through an ac amplifier. Under absence of any input waveform, output of both thermocouples are equal so error signal, which is input to dc amplifier is zero and, therefore, indicating meter connected to the output of dc amplifier reads zero. But on the application of input waveform, output of main thermocouple upsets the balance and an error signal is produced, which gets amplified by the dc amplifier and feedback to the heating element of the balancing thermocouple. This feedback current reduces the value of error signal and ultimately makes it zero to obtain the balanced bridge condition. In this balanced condition, feed-back current supplied by the dc amplifier to the heating element of the balance thermocouple is equal to the ac current flowing in the heating element of main thermocouple. Hence this direct current is directly proportional to the rms value of the input ac voltage and is indicated by the meter connected in the output of the dc amplifier. The PMMC meter may be calibrated to read the rms voltage directly.

By this method, rms value of any voltage waveform can be measured provided that the peak excursions of the waveform do not exceed the dynamic range of the ac amplifier.

4.4.4. Op-Amp AC Voltmeter Circuits. An IC op-amp voltage follower voltmeter circuit described in Art. 4.3.1 is for measuring dc voltages. If we connect a rectifier in series with the meter circuit, as shown in Fig. 4.29, the circuit is converted into a half-wave rectifier voltmeter. As already mentioned in Art. 4.3.1, the output from the voltage follower is exactly the same as the input. Thus, the voltage supplied to the meter circuit is simply a half-wave rectified version of the input voltage V from the attenuator. The operation of such a circuit is just explained in Art. 4.4.1. Coupling capacitor C_1 is usually provided at the input of an ac voltmeter to block unwanted dc voltages.

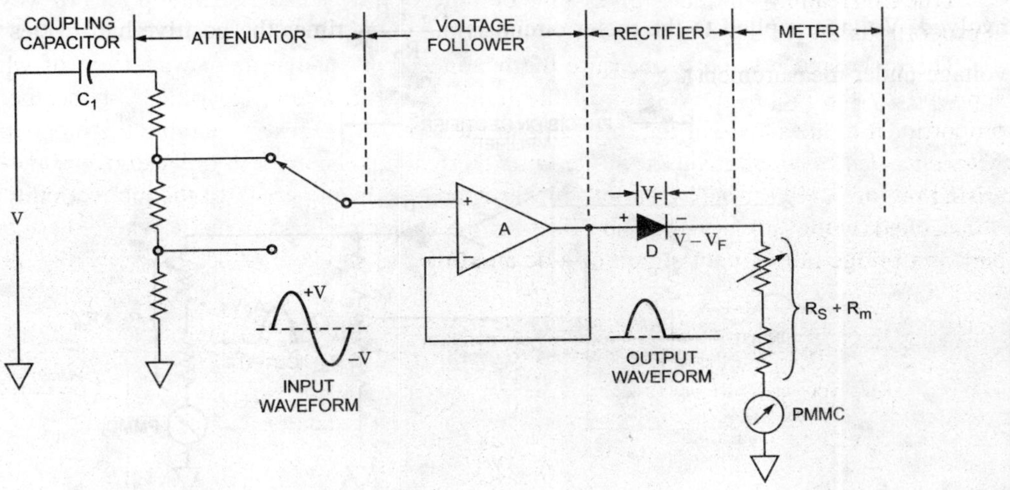

Fig. 4.29

The voltage drop V_F across the rectifier is a source of error in the circuit described above (Fig. 4.29). No doubt, V_F can be taken into account in design calculation when the instrument is used to

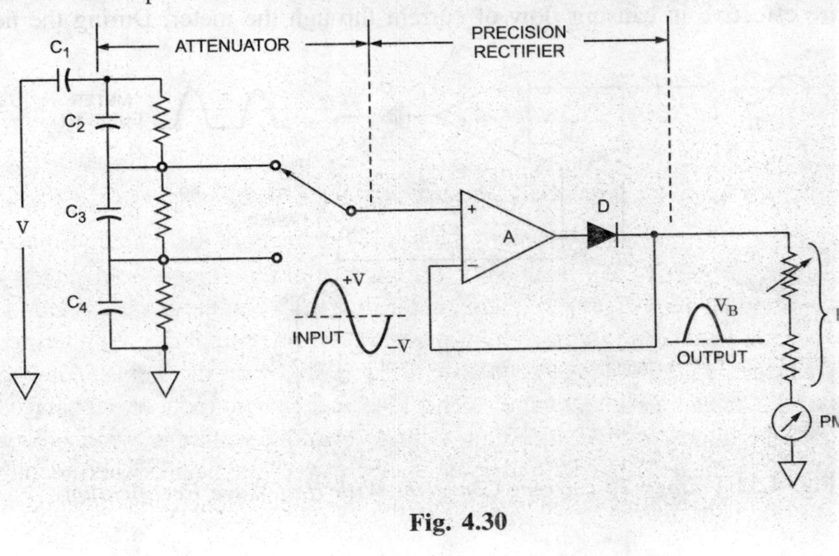

Fig. 4.30

indicate full scale. However, for other readings, an error will occur due to V_F. Also, the rectifier voltage drop is not always exactly 0.7 V, as usually assumed for silicon diode, and it varies with variations in temperature. However, such errors can be avoided if the voltage follower feedback connection to the op-amp inverting terminal is taken from the cathode of rectifier D instead of from the amplifier output (Fig. 4.30). Consequently, the half-wave rectified output precisely follows the positive half cycle of the input voltage. There is no rectifier voltage drop from input to output. This circuit is called the *precision rectifier* capacitors. C_2, C_3 and C_4 are connected across the attenuator resistors in order to compensate for the input capacitance of the amplifier.

For measurement of voltages of very small magnitudes, the input voltage is required to be amplified before being rectified and applied to the meter circuit. Amplification is combined with half-wave rectification as shown in Fig. 4.31. The amplification is precise and there is no rectifier voltage drop involved. Voltage applied to the meter circuit is $\frac{R_2 + R_3}{R_3}$ times the positive half-cycles of the input (voltage under measurement).

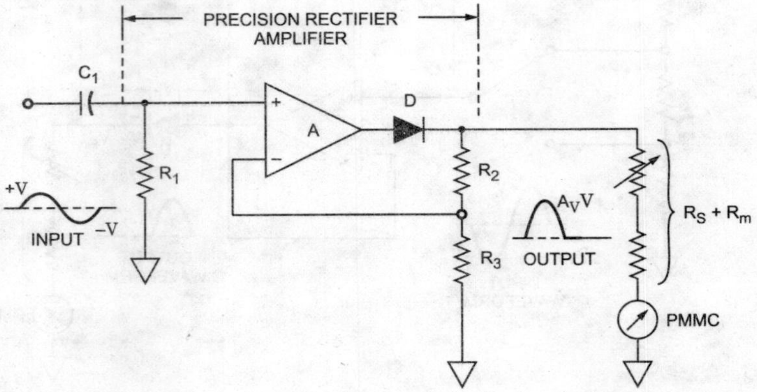

Fig. 4.31

The circuit depicted in Fig. 4.32 is a *voltage-to-current converter* with half-wave rectification. The circuit operates exactly as explained in Art. 4.3.3, with the exception that only the positive half cycles of the ac input are effective in causing flow of current through the meter. During the negative half

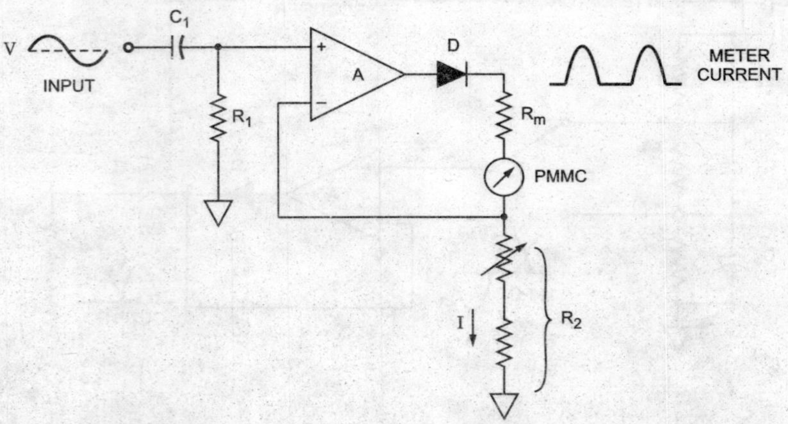

Fig. 4.32 *Voltage-To-Current Converter With Half-Wave Rectification*

cycle, the diode is reverse biased and no current flows through the meter or through resistor R_2. The meter peak current is $I_{max} = \frac{V_{max}}{R_2}$ and the average meter current is $I_{av} = 0.318\ I_{max}$.

A full-wave bridge rectifier is shown in Fig. 4.33. When the input voltage is positive, the op-amp output is positive, diodes D_1 and D_4 are forward biased and the current flows through the meter downward. When the input is negative, diodes D_2 and D_3 are forward biased and the current again passes through the meter downward. The meter peak current is limited to $I_{max} = \frac{V_{max}}{R_2}$ and the average meter current is $0.637\ I_{max}$.

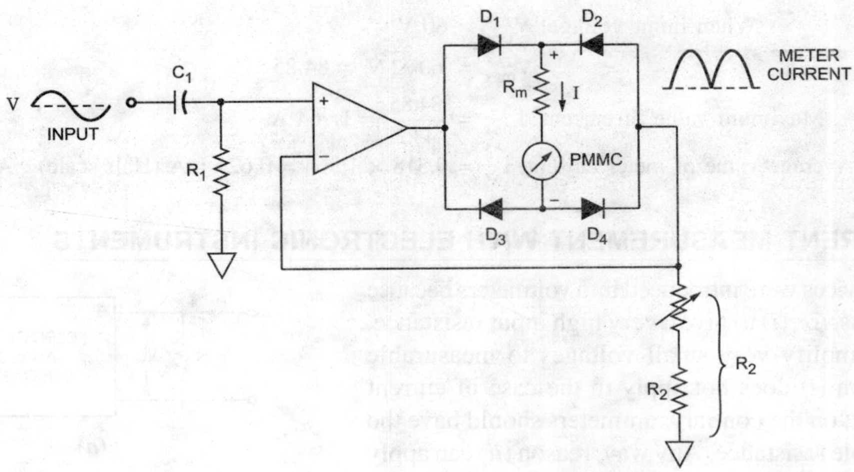

Fig. 4.33 *Voltage-To-Current Converter With Full-Wave Rectification*

Instead of a full-wave bridge rectifier, some electronic instruments make use of the half-bridge full-wave rectifier. A half-bridge rectifier could be useful in electronic voltmeter that uses a transistor amplifier. Half-bridge rectifier circuit connected to the output stage of a transistor amplifier is shown in Fig. 4.34. Output transistor Q passes current through resistor R_3, developing a dc voltage at the collector of the transistor which might be above or below the zero voltage level. Capacitors C_1 and C_2 block the direct current from the meter circuit and pass the ac when ac voltage appears at the amplifier output.

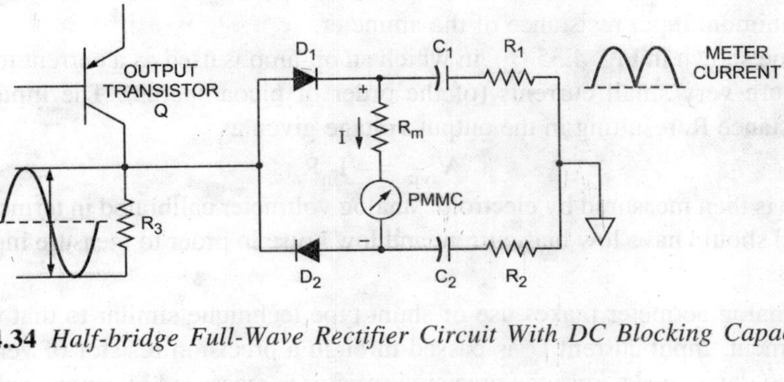

Fig. 4.34 *Half-bridge Full-Wave Rectifier Circuit With DC Blocking Capacitors*

Example 4.6. An half-wave rectifier electronic voltmeter circuit is shown in Fig. 4.32. The meter has FSD current of 1.25 mA and coil of resistance 1.2 kΩ. Determine the value of R_2 that will provide full-scale deflection when the ac input voltage is 120 mV (rms). Also determine the meter deflection when the input is 60 mV.

Solution: For FSD, the average meter current, I_{av} = 1.25 mA

$$\text{Peak value of meter current, } I_{max} = \frac{1}{0.318} I_{av} = \frac{1.25}{0.318} = 3.93 \text{ mA} \quad \because I_{av} = 0.318\, I_{max}$$

$$\text{Input voltage peak value, } V_{max} = \sqrt{2}\, V_{rms} = \sqrt{2} \times 120 = 169.7 \text{ mV}$$

$$R_2 = \frac{V_{max}}{I_{max}} = \frac{169.7 \text{ mV}}{3.93 \text{ mA}} \; 43.2\ \Omega \quad \textbf{Ans.}$$

$$\text{When input voltage, } V_{rms} = 60 \text{ V}$$

$$V_{max} = 60\sqrt{2} \text{ V} = 84.85$$

$$\text{Maximum value of current, } I_{max} = \frac{84.85}{43.2} = 1.964 \text{ A}$$

$$\text{Average value of meter current, } I_{av} = 0.318 \times 1.964 = 0.625 \text{ mA (Half scale)} \quad \textbf{Ans.}$$

4.5 CURRENT MEASUREMENT WITH ELECTRONIC INSTRUMENTS

Electronic devices were introduced into voltmeters because of two reasons viz. (*i*) to give a very high input resistance, and (*ii*) to amplify very small voltages to measurable levels. Reason (*i*) does not apply in the case of current measurement; on the contrary, ammeters should have the lowest possible resistance. Any way, reason (*ii*) can apply in the case of very low current levels.

An electronic analog dc ammeter can be constructed either making use of shunt type technique [Fig. 4.35 (*a*)] or by making use of an op-amp [Fig. 4.35 (*b*)].

In the shunt type technique, shown in Fig. 4.35 (*a*), the input current is converted into voltage by passing it through a precision resistor, R and this voltage, which is proportional to input current, is then measured by the electronic analog voltmeter. The scale of the voltmeter is calibrated in micro-amperes or milliamperes. The value of precision resistor R is kept smallest possible so as to offer minimum input resistance of the ammeter.

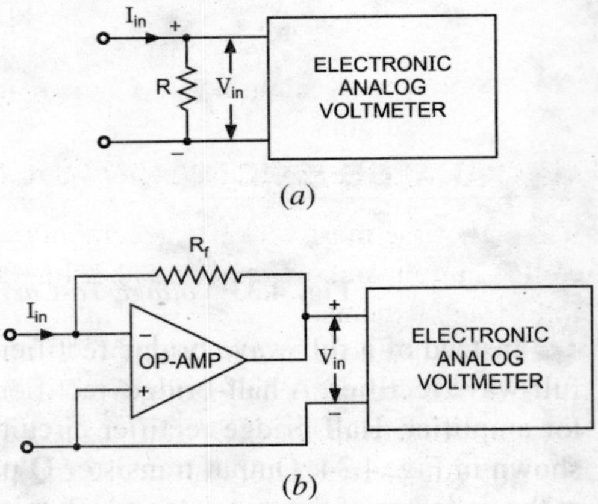

Fig. 4.35 *Electronic Analog DC Ammeter*

The technique shown in Fig. 4.35 (*b*), in which an op-amp is used as a current to voltage converter, is used to measure very small currents (of the order of picoamperes). The input current I_{in} flows through the resistance R_f resulting in the output voltage given as

$$V_{out} = -I_{in}R_f$$

This voltage is then measured by electronic analog voltmeter calibrated in terms of units of current. The op-amp used should have low bias current and low noise in order to measure input currents of very small magnitude.

Electronic analog ac meter makes use of shunt type technique similar to that used in case of dc current measurement. Input current I_{in} is passed through a precision resistor of very small magnitude and the voltage so developed across the precision resistor is measured by an ac analog voltmeter.

The basic circuit of an analog electronic ammeter to measure currents of very small magnitudes is depicted in Fig. 4.36. The small voltage drop across shunt resistor R_S is amplified before being applied to the PMMC instrument. This approach is just as applicable to the measurement of small alternating currents as it is to dc measurements. For ac measurements, an ac electronic voltmeter using rectifier (discussed in Art. 4.4) is employed instead of a dc instrument.

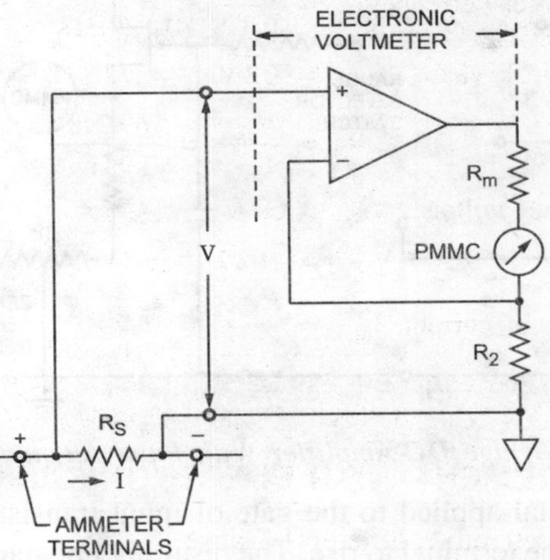

Fig. 4.36 *Current Measurement With Electronic Instruments*

4.6 ANALOG ELECTRONIC MULTIMETER

It is one of the most versatile general purpose instruments capable of measuring dc and ac voltages as well as current and resistances. The solid-state electronic multimeter (or VOM) generally consists of the following elements:

1. A balanced bridge dc amplifier and a PMMC meter.
2. An attenuator in input stage to select the proper voltage range.
3. A rectifier for converting of an ac input voltage to proportionate dc value.
4. An internal battery and additional circuitry for providing the capability of resistance measurement.
5. A *function switch* for selecting various measurement functions of the meter such as voltage, current or resistance.

In addition, the instrument is usually provided with a built-in power supply for operation on ac mains and, in most cases, one or more batteries for operation as a portable test instrument.

The schematic diagram of a balanced-bridge dc amplifier using two field effect transistors (FETs) is given in Fig. 4.37. It is to be noted that two FETs used in circuit should be reasonably well matched for current gain to ensure circuit thermal stability. The two FETs and the source resistors R_1 and R_2, together with zero adjust control resistor R_3, constitute a bridge circuit. The PMMC meter is connected between the source terminals of the FETs, representing two opposite corners of the bridge.

In the absence of input signal, the gate terminals of the FETs are at ground potential and the transistors operate under identical quiescent conditions. Ideally no current should flow through the PMMC movement but in practice, on account of some mismatch between the two FETs and slight tolerance differences in the values of various resistors a current does flow and causes the meter movement to deflect from zero position. This current is reduced to zero by the adjust control resistor R_3. Now the bridge is balanced.

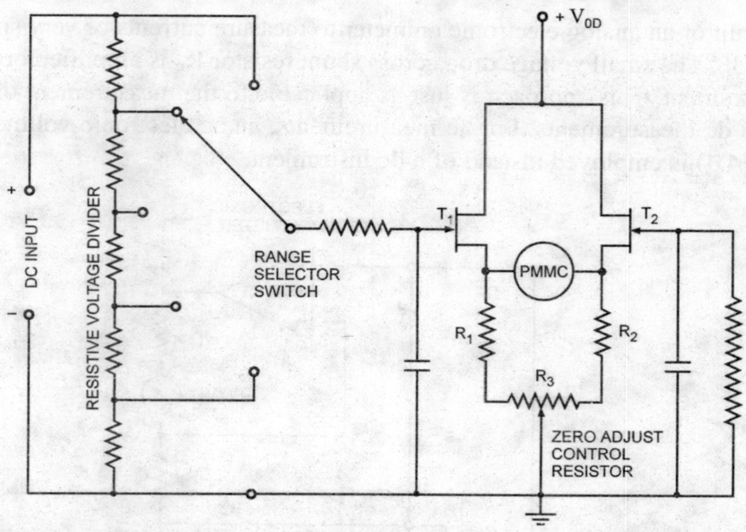

Fig. 4.37 *Balanced-Bridge DC Amplifier With Input Attenuator and Indicating Meter*

With a positive input signal applied to the gate of input transistor T_1, its drain current increases causing the voltage at the source terminal to rise. The resulting unbalance between the two transistors T_1 and T_2 source voltages is shown by the meter movement, whose scale is calibrated in terms of the magnitude of the applied input voltage.

The maximum voltage that can be applied to the gate of input transistor T_1 is determined by its operating range, which is usually of the order of a few volts. The range of the bridge can easily be extended by employing an input attenuator or a range switch, as illustrated in Fig. 4.38.

The unknown dc input voltage is applied through a large resistor in the probe body to a resistive voltage divider. Thus, with the range switch in the 3 V position as illustrated, the voltage at the gate of the input transistor T_1 is developed across 8 MΩ resistor of the total resistance of 11.3 MΩ and the circuit is so arranged that the PMMC meter gives full-scale deflection with 3 V applied to the tip of the probe. With the range switch in 1,200 V position, the gate voltage is developed across 20 kΩ of the total divider resistance of 11.3 MΩ and an input voltage of 1,200 V will be required to cause the full-scale meter deflection.

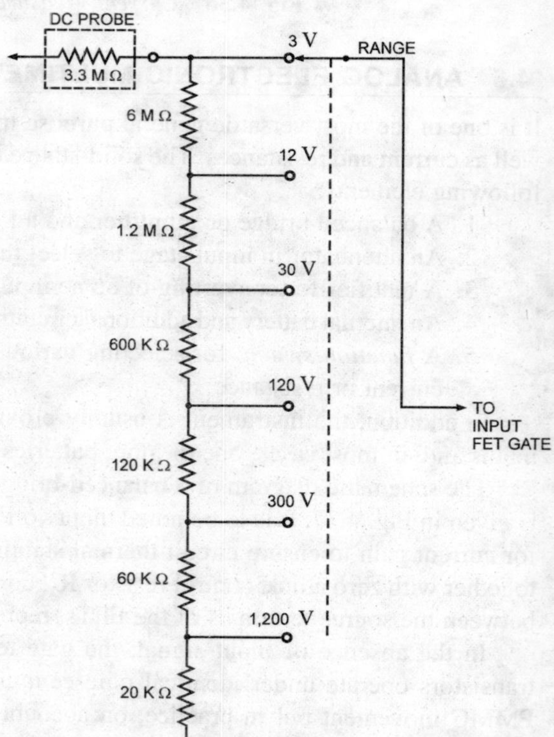

Fig. 4.38 *Typical Input Voltage Attenuator For a Multimeter*

Measurement of Resistance by Multimeter or VOM. When the multimeter's function switch is placed in the OHM position, the unknown resistor is connected in series with an internal battery, and the PMMC meter simply measures the voltage drop across it (unknown resistor, R_X).

A typical circuit is given in Fig. 4.39. In the given circuit, a separate divider network, employed only for the measurement of resistance, provides a number of different resistance ranges. With the unknown resistor R_X connected to the OHM terminals of the multimeter, the battery supplies current through one of the range resistors and the unknown resistor to ground. Voltage drop across unknown resistor R_X, V_X is applied to the input of the bridge amplifier and the PMMC movement deflects. Since the voltage drop across R_X is directly proportional to its resistance, the meter scale can be calibrated in terms of kΩ.

The worthnoting point is that this instrument indicates increasing resistance from left to right whereas the conventional meter indicates increasing resistance from right to left. This is because the electronic multimeter reads a large resistance as a higher voltage, whereas the conventional multimeter indicates a higher resistance as a smaller current.

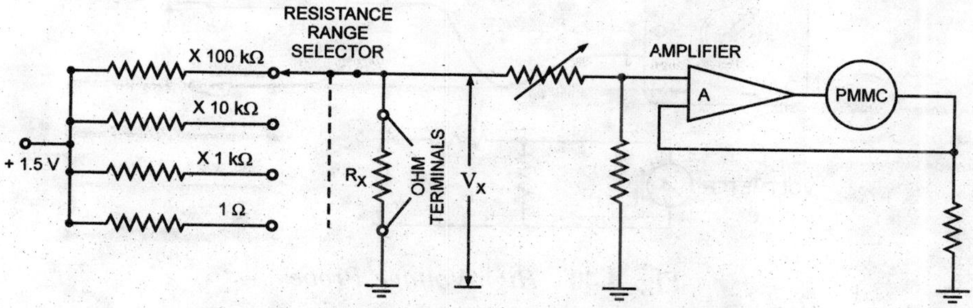

Fig. 4.39 *Measurement of Resistance By a Multimeter*

4.7 MULTIMETER PROBES

A probe can be any conductor used to establish a connection between the circuit under test and the measuring instrument. The conductor could be a piece of bare wire, a multimeter lead, or a piece of unterminated coaxial cable. These simple probes, however, do not fulfil the essential purpose of a probe, that is, to extract minimal energy from the circuit under test and transfer it to a measuring instrument with maximum fidelity. The bare wire can load the input amplifier with its high capacitance and inductance or even cause a short circuit; multimeter leads are unshielded and are often susceptible to stray pickup; and the unterminated coaxial cable will severely capacitively load the circuit under test (100 pF per metre typically). Also, the unterminated coaxial cable is usually resonant at certain frequencies and does not permit faithful transfer of the signal to the test instrument due to reflections. Emphasis should be on reducing the reflections and other effects associated with unterminated coaxial probe cables to the minimum and on reducing the effect of coaxial probe cable's capacitance in high-speed probing.

The ideal probe offers the following key attributes:

1. Ease of connection 2. Absolute signal fidelity 3. Zero signal source loading and 4. Complete noise immunity.

But in practice it is almost impossible to achieve these properties.

There are many probes and adapters available for use with multimeters and analog and digital electronic instruments that can extend the ranges of measurement, or adapt the instrument for measurement of temperature or other quantities. Some of these are described below.

High Voltage Probe. This is basically a high resistance passive divider probe with high probe tip resistance (typical value is 500 MΩ). So the attenuation factor is very high. Several high voltage probes are available to provide 100X or 1000X compensated dividers. Because of the larger attenuation factors required for high voltage applications, the input capacitance is reduced to approximately 3 pF). These probes are used for input voltage exceeding 500 V. These probes have the following features.

1. Very high dynamic range; So very high voltages can be measured with this probe, and it is the safest at voltages over 500 V.
2. Physically larger in size.

A high-voltage probe using a resistive divider to reduce the voltage to a measurable level is depicted in Fig. 4.40.

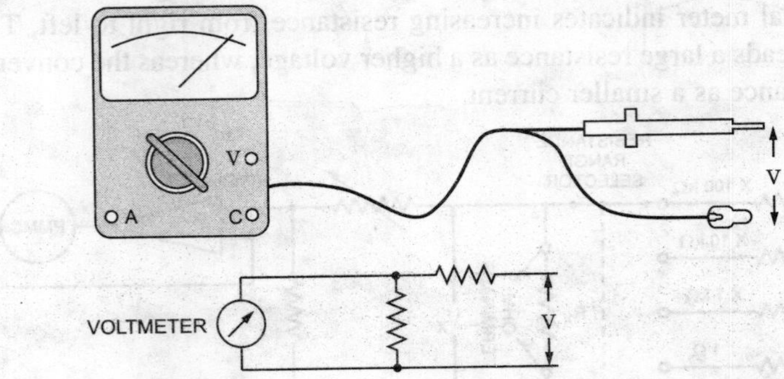

Fig. 4.40 *High-Voltage Probe*

Fluke high voltage probe for digital multimeters is shown in Fig. 4.41. These are available in ratings of 15 KVDC/10 KVAC; 40 KVDC/28 KVAC; 6 KVDC/AC. These probes can work with any multimeter which has input impedance of 10 MΩ (minimum). Their bandwidth is 60 Hz and attenuation factor × 1000.

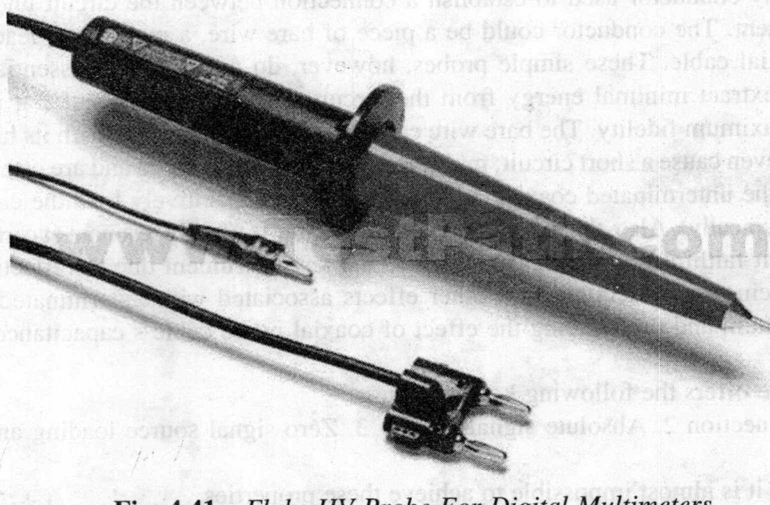

Fig. 4.41 *Fluke HV Probe For Digital Multimeters*
[Courtesy Tektronix]

High Current Probes. Alternating currents induce voltage in a transformer resulting from the buildup and collapse of flux fields as the direction of flow of current reverses. The transformer action is the basis for ac current probes. The ac current probe head is actually a coil (primary) that has been wound to precise specifications on a magnetic core. When this probe head is held within a specified orientation and proximity to an ac current carrying conductor, voltage is induced in the secondary coil.

This probe voltage is directly proportional to ac current in the primary coil. This current-related voltage can be measured by a voltmeter or multimeter.

A high-current probe using a current transformer for reducing an alternating current to a measurable level is shown in Fig. 4.42. The transformer core opens, as shown, to close around a conductor carrying current to be measured. The conductor can be treated as a single-turn primary on the transformer, and the secondary winding then determines the measurable current level. Typically, the 1 mA ac scale on the multimeter is converted into a 1 A scale by the use of current probe. The circuit needs not to be broken or altered in any way, as the current probe measures whatever current is passing through its closed jaws.

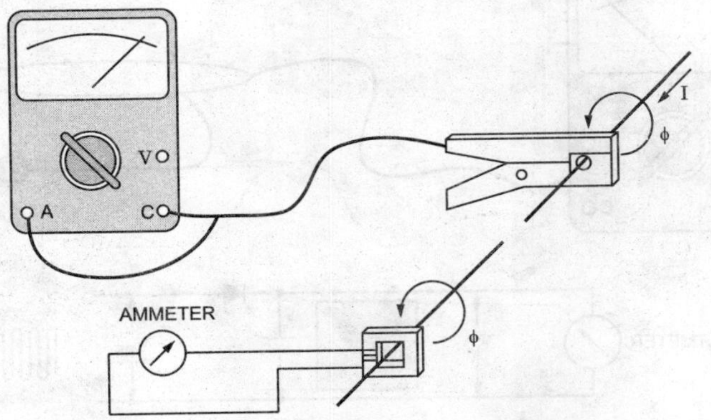

Fig. 4.42 *High-Current Probe Using a CT*

Alternating current probes are passive in that they do not require external power. A steady-state direct current, however, will not induce a current in a transformer, so this probe cannot be used for measuring direct currents.

AC/DC Current Probes. The other type of current probe uses the principle of the Hall effect. The Hall effect produces voltage output when direct current is applied to a semiconductor device present in a dc generated magnetic field. This technique requires an external power supply, but does measure both ac and dc. Consequently, a direct current probe is an active device, in that it needs external power.

The Hall effect probe is shown in Fig. 4.43. The Hall effect transducer contained in the probe operates on the principle that a small voltage is produced at the edges of a flat current carrying conductor

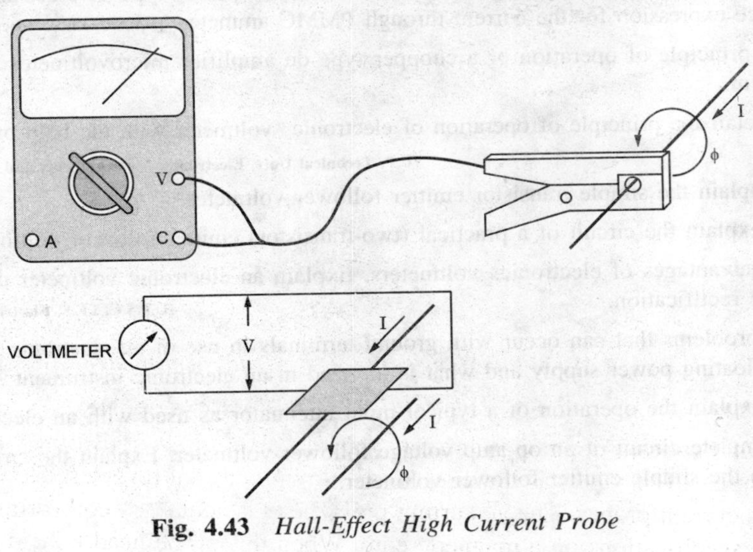

Fig. 4.43 *Hall-Effect High Current Probe*

in the presence of a magnetic field. The output is connected to the meter voltage terminals, typically resulting in the multivolt scale being read as a current scale with 1 mV representing 1 mA.

Radio-Frequency (RF) Probe. The magnitude of RF voltage cannot be measured directly on a multimeter because the upper-frequency limit of the instrument is not high enough. In an RF probe, depicted in Fig. 4.44, the RF wave is rectified and converted to a dc voltage equal to its peak value. With proper attenuation, the voltage can be measured on the meter and read as an rms quantity.

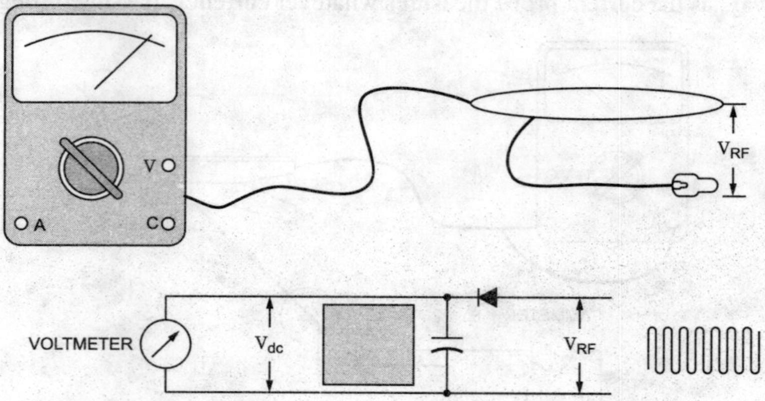

Fig. 4.44 *RF Probe*

EXERCISES

1. What are the advantages of TVM's over conventional meters ? **[D.E.I. Agra. Elec. Measurements May 1989]**
2. Discuss the advantages of electronic voltmeter as compared to conventional type of voltmeters. Describe the working of basic dc voltmeter using FET at the input. Why are zero and calibration controls provided in these instruments ?
3. Draw and describe a schematic diagram of a difference amplifier type of electronic voltmeter using FETs. Deduce expression for the current through PMMC ammeter. **[U.P.S.C. I.E.S. Electrical Engineering I, 1998]**
4. What is the principle of operation of a chopper type dc amplifier microvoltmeter ? Explain using its block diagram.
5. Discuss in detail the principle of operation of electronic voltmeter with the help of a circuit diagram. **[U.P. Technical Univ. Electronics Measurements and Instrumentation 2004-05]**
6. Draw and explain the simple transistor emitter follower voltmeter.
7. Sketch and explain the circuit of a practical (two-transistor) emitter follower voltmeter.
8. Discuss the advantages of electronics voltmeters. Explain an electronic voltmeter using bridge circuit for full-wave rectification. **[U.P.S.C. I.E.S. Electrical Engineering - I, 1996]**
9. Explain the problems that can occur with ground terminals in use of an electronic voltmeter. What is meant by a floating power supply and what is its need in an electronic instrument ?
10. Sketch and explain the operation of a typical input attenuator as used with an electronic voltmeter.
11. Draw the complete circuit of an op-amp voltage follower voltmeter. Explain the circuit operation, and compare it to the simple emitter follower voltmeter.

12. Draw and explain the circuit of an op-amp noninverting amplifier voltmeter.
13. Draw and explain the operation of an electronic voltmeter using an op-amp voltage-to-current converter.
14. Explain the construction and working of an average reading VTVM. Describe its advantages and disadvantages. Explain how this meter can be converted to a rms reading voltmeter.
15. Explain the construction and principle of operation of a thermocouple type true rms ac ammeter.
16. What are the differences amongst average, peak and rms responding electronic voltmeters ? Explain the working of an rms responding electronic voltmeter with the help of its diagram.
17. Give the salient features of average, peak and rms responding electronic voltmeters. Explain how dc and ac currents are measured with the help of electronic meters.
18. Explain the working principle and operation of a peak-to-peak responding electronic voltmeter with the help of its circuit diagram. What may be the sources of errors in this meter ?
19. Draw and explain the operation of circuits of half-wave rectifier ac electronic voltmeters using (*a*) a voltage follower and (*b*) a precision rectifier.
20. Draw the circuit diagram of an ac electronic voltmeter using a half-wave precision rectifier with amplification. Explain the circuit operation.
21. Draw the circuit of an ac electronic voltmeter using a voltage-to-current converter with half-wave rectification. Explain the operation of the circuit.
22. Draw the circuit of an ac electronic voltmeter using a voltage-to-current converter with full-wave rectification. Explain the circuit operation.
23. Draw and explain the circuit of a half-bridge full-wave rectifier as used with an ac electronic voltmeter.
24. Explain the principle of operation of electronic analog dc ammeter. Draw the circuit for measurement of very small currents using electronic analog dc ammeter.
25. Explain how an electronic analog voltmeter can be used to measure alternating currents.
26. What is the principle of electronic multimeters ?
27. Discuss the construction and working of a multimeter.

[Pb. Technical Univ. Electronics Measurements and Instrumentation December 2005]

28. Explain the operation of high-voltage probe and an ac high current probe.
29. Explain the operation of a Hall effect high current probe and RF probe.
30. Write short notes on the following:
    (*i*) Emitter follower voltmeter. (*ii*) Attenuator.
    (*iii*) Op-amp voltage follower voltmeter. (*iv*) Op-amp ac voltmeter circuits.
    (*v*) Current measurement with electronic instruments. (*vi*) Multimeter probes.

SHORT ANSWER TYPE QUESTIONS WITH ANSWERS

Q. 1. Why ordinary voltmeter cannot be used in electronic circuits ?

Ans. A moving coil voltmeter draws a large current, hence a large power from the test circuit during measurement. Generally current drawn by the voltmeter does not make much difference in electrical circuit but when measurement is involved in electronic circuits, current drawn by the voltmeter load the test circuit, and therefore, voltmeter shows an error in the reading.

Q. 2. What is VTVM ? **[PTU December 2005]**

Ans. VTVM is a vacuum tube voltmeter. It makes the use of vacuum tube(s) as active device.

Q. 3. Enlist the types of VTVMs.

Ans. Average reading, peak reading, slide back, logarithmic, balanced bridge and electrometer VTVMs are the different types of VTVM.

Q. 4. Enlist TVMs.

Ans. 1. TVM with cascaded transistor.

2. TVM with FET (field effect transistor).

3. Balanced bridge TVM.

4. Chopper amplifier.

Q. 5. Where is VTVM used ? **[PTU December 2004]**

Ans. VTVMs are used for the measurement of the dc and ac voltages.

Q. 6. What is difference between VTVM and TVM ?

Ans. Electronic voltmeter using vacuum tube is called a VTVM while that using a transistor is called a TVM.

Q. 7. In what respect TVM is inferior to VTVM?

Ans. The only disadvantage with TVM is its low input impedance which it offers in comparison to VTVM. But this shortcoming is overcome by using FET in input stage of voltmeter.

Q. 8. What are the advantages of TVM over VTVM ? **[PTU December 2005]**

Ans. The advantages of TVMs over VTVMs are given below :

1. A TVM does not require any warm up time because of absence of heating element.
2. Use of transistor makes the instrument portable because of its light weight.
3. A transistorized instrument can be operated on low voltage *i.e.* on battery so TVM is well suited for the field work where power source is not available.
4. VTVM cannot measure current directly because of its high resistance whereas TVM can do.
5. TVM consumes very less power because of absence of heating element.

Q. 9. What is the difference between dc voltmeter and ac voltmeter ?

Ans. Fundamentally, dc and ac voltmeters are not much different, dc voltmeter consists of a dc amplifier succeeded by a conventional dc meter while an ac voltmeter consists of an additional rectifier either before or after the amplifier.

Q. 10. Why is direct-coupled amplifier dc voltmeter commonly used ?

Ans. Because of its low cost.

Q. 11. What are the shortcomings of a direct-coupled dc voltmeter ?

Ans. The drawback of a direct-coupled dc voltmeter is that it has to work under specified ambient temperature to have the required accuracy otherwise excessive drift problem occurs during operation. It can be used only for measuring voltage of the order of millivolts owing to limited amplifier gain. The drift problem in direct-coupled dc amplifiers restricts the use of a number of stages required for a high gain.

Q. 12. What are the advantages of emitter follower voltmeter ?

Ans. The main advantages of emitter follower voltmeter are high input resistance and low output resistance.

Q. 13. What is the main drawback of basic emitter follower voltmeter circuit ? How this drawback is avoided in practical emitter follower voltmeter circuit ?

Ans. The main drawback of basic emitter follower voltmeter circuit is that the transistor base-emitter voltage V_{BE} introduces an error. However, this error is eliminated by using a potential divider and an additional emitter follower as shown in Fig. 4.7.

Q. 14. How is op-amp voltage follower voltmeter superior than the transistor emitter follower voltmeter ?

Ans. The op-amp voltage follower voltmeter has the advantages of much higher input resistance and lower output resistance than those of transistor emitter follower voltmeter.

Q. 15. What are the advantages of average reading ac voltmeters ?

Ans. Main advantages associated with average reading ac voltmeters are that they are simple in construction, have high input impedance, low power consumption and uniform scale.

Q. 16. How can an average reading VTVM be converted to an RMS reading voltmeter ?

Ans. The average reading voltmeter can be converted into an RMS reading voltmeter by calibrating its meter scale in terms of the rms value of a sinusoidal wave.

Q. 17. What are the disadvantages of average reading voltmeters ?

Ans. Main disadvantage of average reading voltmeters is that they operate in audio-frequency range. In radio-frequency range, distributed capacitance of the high resistance R introduces error in the readings. Another disadvantage of such a voltmeter is that due to nonlinear volt-ampere characteristic for lower voltage the readings of the voltmeter at lower voltage are not correct.

Q. 18. Why is dc amplifier associated with diode rectifier in peak reading ac voltmeters provided with stabilising means ?

Ans. The dc amplifier associated with diode rectifier should be provided with the stabilising means in order to prevent drift in the indication of the output meter.

Q. 19. For what type of measurement is true rms reading voltmeter required ?

Ans. For measuring rms value of non-sinusoidal waveform a true rms reading voltmeter is required. It gives a meter indication by sensing heating power of waveform which is proportional to the square of the rms value of the voltage.

Q. 20. What is main source of error in an IC op-amp voltage follower voltmeter circuit shown in Fig. 4.29 ? How is it can be avoided ?

Ans. The voltage drop across the rectifier, V_F is a source of error in the circuit shown in Fig. 4.29. This error can be eliminated by taking the voltage follower feedback connection to the inverting terminal from the cathode of rectifier instead of from the amplifier output (Fig. 4.30).

Q. 21. What is VOM ?

Ans. The volt-ohm-milliammeter (VOM) is another name of multimeter. It is also called AVO meter.

Q. 22. What is a multimeter ?

Ans. The instrument, which contains a function switch to connect the appropriate circuits to the d'Arsonval movement, is often called a multimeter or ampere-volt-ohm (AVO) meter or volt-ohm-milliammeter (VOM).

Q. 23. Indicate the various quantities that can be measured with a multimeter.

Ans. Multimeter is used for measurement of current (dc as well as ac), voltage (dc as well as ac) and resistance. With the external source, high resistance (exceeding 1 MΩ), inductance and capacitance can be measured with this instrument.

Q. 24. What are the shortcomings of multimeter.

Ans. It may be noted that the ac sensitivity of a half-wave ac meter is only 45 percent of the dc sensitivity, therefore, the loading effect is greater while measuring ac voltages than that while measuring corresponding dc voltages.

The rectifying action depends upon the frequency to some extent due to capacitance effect and rectifier instruments show lower readings. Instrument indications may be in error by as much as 0.5% decrease for every 1 kHz rise in frequency.

Q. 25. What is the principle of operation of ac current probe ?

Ans. High levels of alternating current can be reduced by using a current transformer, and this principle is used in the ac current probe.

PROBLEMS

1. The simple emitter follower circuit shown in Fig. 4.6 has V_{CC} = 20 V, meter circuit resistance of 9.3 kΩ, PMMC meter FSD current is 1 mA and transistor h_{FE} = 100. Determine the meter current when voltage under measurement is 10 V. Also determine the voltmeter input resistance. **[Ans.** 1 mA, 1 MΩ**]**
2. An emitter-follower voltmeter circuit shown in Fig. 4.7 has emitter resistances of 3.9 kΩ each and V_{CC} is ± 12 V. Determine emitter currents when input voltage is zero volt. Determine also, the meter circuit voltage when input voltage is (*i*) 1.5 V (*ii*) 0.5 V. **[Ans.** 2.9 mA, 1.5 V and 0.5 V**]**
3. An op-amp amplifier circuit (Fig. 4.16) is to be designed for measuring a maximum input of 20 mV. The op-amp input current is 0.25 μA, and FSD current of meter is 100 μA and meter circuit resistance is 10 kΩ. Determine the suitable values of R_1 and R_f. **[Ans.** 100 Ω, 4.9 kΩ**]**
4. Determine the value of R_1 for the circuit shown in Fig. 4.18. Maximum input voltage is 1 V. The FSD current of the meter is 1 mA and the meter circuit resistance is 250 Ω. Also determine the maximum voltage at the output terminal of the op-amp. **[Ans.** 1 kΩ, 1.25 V**]**

❑❑❑

5

Digital Meters

5.1 INTRODUCTION

Digital instruments are becoming more and more popular because of their inherent advantages over analog instruments, such as, greater speed, increased accuracy, better resolution, reduction in operator errors and the ability to provide automatic measurements in system application.

Digital instruments use logic circuits and techniques for carrying out measurements of quantities. Any digital instrument may be viewed as an arrangement of logic gates that change states at very high speed in the process of carrying out a measurement.

The major advantage of digital instruments is the readability of the measurement result because of the digital read-out. In analog instruments, the user performs the function of an analog-to-digital converter (ADC). The user must read the analog scale properly, possess some skill in interpolation, be able to use mirrored scales and in general, have a good eye.

The second major advantage of digital instruments over analog instruments is the accuracy. In general, digital instruments are more accurate than analog ones. For example, a low-cost digital multimeter is 10 times more accurate than a comparably priced analog multimeter.

Digital instruments provide better resolution than analog instruments. A typical value of resolution of a digital instrument may be one part in 10^6 *i.e.* 1 μV can be read on the 1 V input range.

The digital instruments also have greater speed. The maximum speed of reading and period of sampling is interrelated. The digital voltmeters may have sampling rate from a few samples per second to 1,000 samples per second. However, input capacitance and filter limits the sampling speed to 100 or below. Also, it is impossible to read out at high reading speed.

A digital instrument is quite different from an analog instrument with digital read-out. The circuit required for measurement in a digital instrument is of digital design, while the measuring circuit of a digital read-out analog instrument is of analog design and only the indicating device is of digital design. A digital read-out analog instrument, in general, is no more accurate than the same analog instrument with analog read-out, except that digital display is unambiguous and can be read more quickly.

Virtually all digital instruments provide a digital display of the measurand. A fixed number of pulses proportional to the measurand are entered in a digital counter and the counter actuates the digital display. So the counting circuits are vital parts of the digital instruments.

A digital voltmeter (DVM) essentially consists of an analog-to-digital converter (ADC), a set of seven-segment numerical displays, and the necessary BCD-to-seven segment drivers. Digital multimeters are electronic volt-ohm-milliammeters (VOMs) with digital displays.

If a pulse waveform is used to toggle a digital counter for a time period of exactly 1 second, the counter registers the number of pulses that occur per second; *i.e.*, it registers the frequency of the input waveform. If the count is 500, the frequency is 500 pulses per second. A digital frequency meter is a digital counter combined with an accurate timing system.

Digital instruments, particularly digital voltmeters or multimeters are employed for measurement of analog quantities. Conversion of an analog signal into an equivalent digital signal is, therefore, essential. Thus analog-to-digital converters (ADCs) are most important parts of digital instruments, and will, therefore, be discussed in the following Arts. separately.

5.2 DIGITAL-TO-ANALOG AND ANALOG-TO-DIGITAL CONVERSION

Because most sensors have analog output while much data processing is accomplished with digital computers, analog-to-digital and digital-to-analog conversion obviously play an important role. The process of changing an analog signal to an equivalent digital signal is accomplished with the help of an *analog-to-digital converter* (ADC). For example, an ADC is used to convert an analog signal from a transducer, (measuring some physical quantity such as temperature, pressure, position, rotational speed or flow rate) into an equivalent digital signal. An analog-to-digital converter (ADC) is often referred to as an *encoding device*, as it is employed for encoding signals for entry into a digital system.

Digital-to-analog conversion involves translation of digital information into equivalent analog information and this is accomplished by the use of *digital-to-analog converter* (DAC). DACs are used whenever the output of a digital circuit has to provide an analog voltage or current to drive an analog device. As an example, the output from a digital system might be converted into an analog control signal for adjusting the motor speed or the furnace temperature, or for controlling almost any physical variable. Computers can be programmed to generate the analog signals (through a DAC) required for testing analog circuitry. A digital-to-analog converter (DAC) is sometimes considered a *decoding device.*

Digital-to-analog (D/A) conversion is a straight forward process and is considerably easier than analog-to-digital (A/D) conversion. In fact, DACs are used as components in some ADCs. So we will consider D/A conversion first.

5.3 DIGITAL-TO-ANALOG (D/A) CONVERSION

Basically, D/A conversion is the process of taking a value represented in digital code (such as simple binary or BCD) and converting it into a voltage or current which is proportional to the digital value. As already mentioned, D/A conversion is accomplished by the use of digital-to-analog converter abbreviated as D/A converter or DAC.

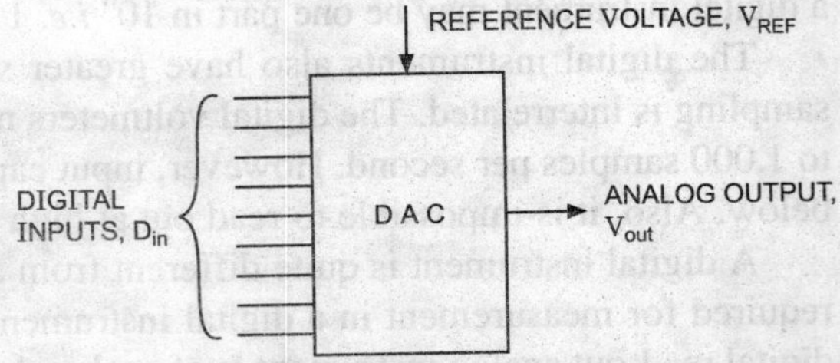

Fig. 5.1 *Basic Function of a DAC*

The basic function of a DAC is shown in Fig. 5.1

Digital input D_{in} is defined as an *n*-bit digital word such that

$$D_{in} = D_0 \times 2^{-n} + D_1 \times 2^{-(n-1)} + D_2 \times 2^{-(n-2)} + \dots D_{n-1} 2^{-1}$$

where D_0 is a binary digit and equals a 1 or a 0. D_0 is also defined as the least significant bit (LSB) and D_{n-1} the most significant bit (MSB). The analog output signal, V_{out} is related to the digital input signal, D_{in} through an analog voltage reference, V_{REF}. The relationship between these three signals is given by

$$V_{out} = V_{REF}(D_0 \times 2^{-n} + D_1 \times 2^{-(n-1)} + D_2 \times 2^{-(n-2)} + \ldots D_{n-1} 2^{-1}) = V_{REF} \times D_{in}$$

V_{LSB} may be defined as to be the voltage change when one LSB changes or mathematically

$$V_{LSB} = \frac{V_{REF}}{2^n}$$

The transfer characteristics of an ideal 3-bit DAC are shown in Fig. 5.2. The worthnoting point is that the maximum value of output voltage V_{out} is not reference voltage V_{REF} but rather $V_{REF}(1 - 2^{-n})$, or equivalent to $V_{REF} - V_{LSB}$. In practical circuits, the transfer characteristics are nonlinear and non-monotonic.

There are various techniques which can be used to realize a DAC. The most commonly used techniques are discussed below :

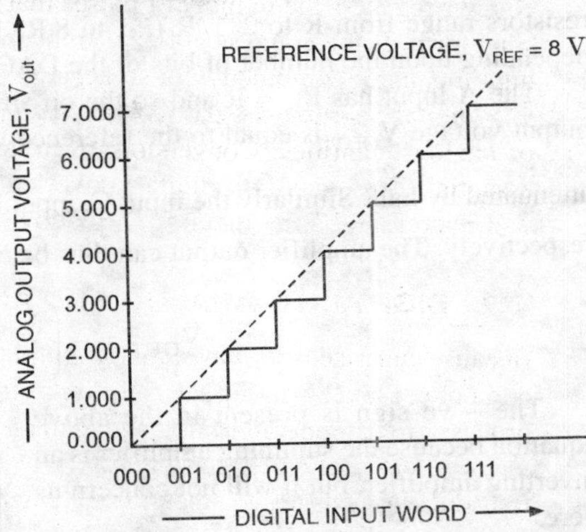

Fig. 5.2 *Transfer Characteristics of an Ideal 3-bit DAC*

5.3.1. Weighted-Resistor DAC. The basic configuration of a simple DAC is shown in Fig. 5.3. It consists of a precision resistor ladder network, a reference precision voltage supply, logic inputs, semiconductor switches and an operational amplifier (op-amp). The inputs A, B, C, D, ...H are binary inputs which are assumed to have values of either 0 V (LOW) or 8 V (HIGH). When the input is HIGH, the switch closes and connects a precision reference supply to the input resistor and when the input is LOW the switch is open. The reference supply produces a very stable, precise voltage required for generating an accurate analog output. The *op-amp* is used as a summing amplifier, which produces the weight sum of the binary inputs.

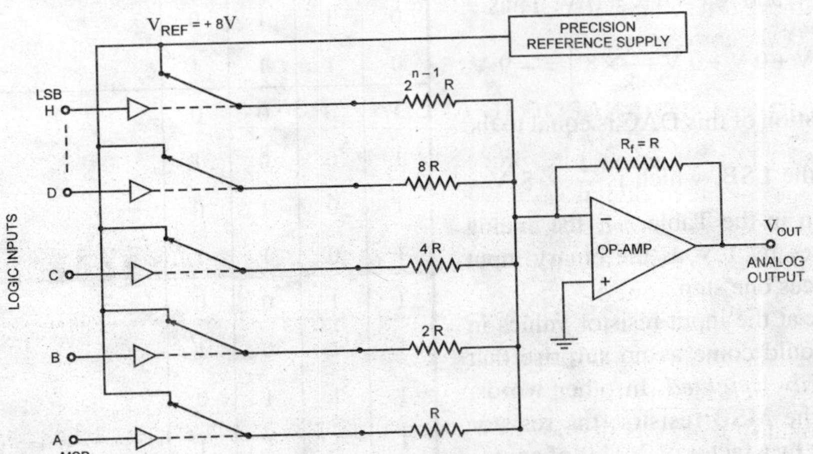

Fig. 5.3 *DAC Circuitry*

In an 8-bit code input the switch A is actuated by most significant bit (MSB) and the switch H is actuated by least significant bit (LSB). If the input binary number is 10,000,000 then switch A is closed and others are open. The output voltage which depends upon feedback resistor, is equal to the reference voltage.

It is known that the summing amplifier multiplies each input voltage by the ratio of feedback resistor R_f to the corresponding input resistor R_{in}. In this circuit R_f = R (say of 1 kΩ) and the input resistors range from R to 2^{n-1} R (*i.e.* to 8 R, 128 R, 2048 R in case of 4-bit, 8-bit and 12-bit DAC) depending upon the number of bits of the DAC.

The A input has R_{in} = R and so the op-amp possess the voltage at A with no attenuation *i.e.* the output voltage V_{OUT} is equal to the reference voltage, V_{REF}. The B input has R_{in} = 2 R, so it will be attenuated by half. Similarly the input C, input D and input H will be attenuated by 1/4, 1/8 and $\frac{1}{2^{n-1}}$ respectively. The amplifier output can thus be expressed as

$$V_{OUT} = -\left(V_A + \frac{1}{2}V_B + \frac{1}{4}V_C + \frac{1}{8}V_D + \ldots \frac{1}{2^{n-1}}V_H\right) \quad \ldots(5.1)$$

The – ve sign is present in the above equation because the summing amplifier is an inverting amplifier, but it will not concern us here.

Clearly, the summing amplifier output is an analog voltage which represents a weighted sum of the digital inputs, as shown by the Table 5.1, for a 4-bit DAC. This Table lists all the possible input conditions and the resultant amplifier output voltage. The output is evaluated for any input condition by setting the appropriate inputs to either 0 V or 8 V. For example, if the digital input 1001, then $V_A = V_D$ = 8 V and $V_B = V_C$ = 0 V. Thus

$$V_{OUT} = -\left(8\text{ V} + 0\text{ V} + 0\text{ V} + \frac{1}{8}\times 8\right) = -9\text{ V}.$$

The resolution of this DAC is equal to the weighting of the LSB, which is $\frac{1}{8} \times 8$ V = 1 V. As shown in the Table 5.1, the analog output increases by 1 V as the binary input number advances one step.

Table 5.1

Input Code				*Output Voltage*	
A	*B*	*C*	*D*		
0	0	0	0	0	
0	0	0	1	–1.0	LSB
0	0	1	0	–2.0	
0	0	1	1	–3.0	
0	1	0	0	–4.0	
0	1	0	1	–5.0	
0	1	1	0	–6.0	
0	1	1	1	–7.0	
1	0	0	0	–8.0	
1	0	0	1	–9.0	
1	0	1	0	–10.0	
1	0	1	1	–11.0	
1	1	0	0	–12.0	
1	1	0	1	–13.0	
1	1	1	0	–14.0	MSB
1	1	1	1	FULL	SCALE

If we look at the input resistor values in Fig. 5.3. It should come as no surprise that they are *binarily weighted.* In other words, starting with the MSB resistor, the resistor values increase by a factor of 2. This of course provides the desired weighting in the output voltage.

There are two serious drawbacks of the above D/A conversion system. The biggest problem is the large difference in resistor values between the LSB and MSB, especially in high-resolution (or many bit) DAC. For example if the MSB resistor is of 1 kΩ in a 10-bit DAC, the LSB resistor will be 512 kΩ.

With the current IC fabrication technology, it is very difficult to produce resistors of values over a wide resistance range that maintain an accurate ratio especially with variations in temperature. Moreover, the MSB resistor is required to handle a much greater current than that used for LSB resistor. For example, in a 10-bit system, the current through the MSB resistor will be more than 500 times as large as the current through the LSB resistor. For these reasons it is preferable to employ a circuit that uses resistances of fairly close values. One of the most widely used DAC circuit that satisfies the above requirement is the *R-2R ladder network* (binary ladder), where the resistors used have only two values and that in the ratio of only 2 to 1.

5.3.2. Binary Ladder or R-2R Ladder DAC. A DAC using R-2R ladder network with four input voltages, representing 4-bits of digital data and dc voltage output is illustrated in Fig. 5.4.

The output current, I_{OUT} depends on the positions of the four switches and the digital inputs D_0, D_1, D_2, D_3 control the states of the switches. The current is allowed to flow through an op-amp current-to-voltage converter to give V_{OUT}.

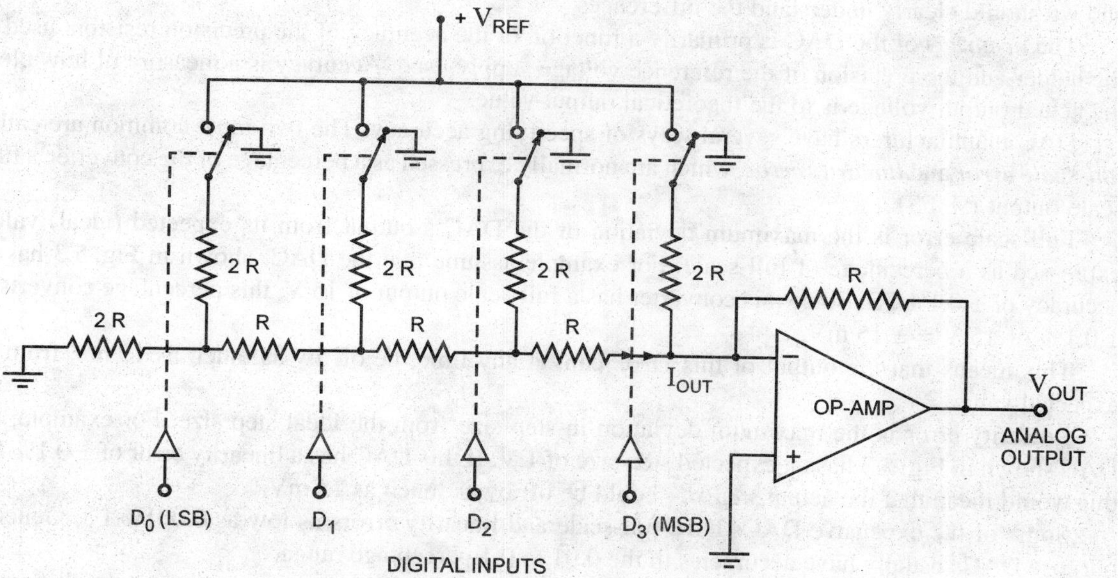

Fig. 5.4 *DAC Using R-2R Ladder Network With Four Input Voltage and DC Voltage Output*

The output voltage (analog), V_{OUT} is proportional to the digital input and is given by the equation

$$V_{OUT} = \frac{D_0 \times 2^0 + D_1 \times 2^1 + D_2 \times 2^2 + D_3 \times 2^3}{2^4} V_{REF} \qquad ...(5.2)$$

For example, if the digital input is 1010 then output voltage V_{OUT} will be given by the equation

$$V_{OUT} = \frac{0 \times 1 + 1 \times 2 + 0 \times 4 + 1 \times 8}{16} V_{REF} = \frac{10}{16} V_{REF}$$

The function of the ladder network is to convert the 16 possible binary values (from 0000 to 1111) into one of 16 voltage levels in steps of $\frac{V_{REF}}{16}$.

Equation (5.2) can be rewritten in general form as follows

$$V_{OUT} = \frac{D_0 \times 2^0 + D_1 \times 2^1 + D_2 \times 2^2 + D_3 \times 2^3 + D_4 \times 2^4 + ... D_{n-1} \times 2^{n-1}}{2^n} \times V_{REF} \quad ...(5.3)$$

Thus we can have more digital or binary inputs and greater quantization for each step by using more sections of ladder network. For example, an 8-stage ladder network could extend the number of voltage steps or voltage resolution to $\frac{V_{REF}}{2^8}$ or $\frac{V_{REF}}{256}$.

In general, the voltage resolution for an *n*-stage ladder network is given as $\frac{V_{REF}}{2^n}$.

5.3.3. DAC Accuracy and Resolution. Two very important aspects of the digital-analog-converter (DAC) are the resolution and accuracy of the conversion. There is a definite distinction between the two and we should clearly understand the differences.

The *accuracy* of the DAC is primarily a function of the accuracy of the precision resistors used in the ladder and the precision of the reference voltage supply used. Accuracy is a measure of how close the actual output voltage is to the theoretical output value.

DAC manufacturers have several ways of specifying accuracy. The two most common are called *full-scale error* and *linearity error*, which are normally expressed as a percentage of the converter's full-scale output (% FS).

Full-scale error is the maximum deviation of the DAC's output from its expected (ideal) value, expressed as a percentage of full-scale. For example assume that the DAC shown in Fig. 5.3 has an accuracy of ± 0.1% FS. Since this converter has a full-scale output of 15 V, this percentage converts to ± 0.1% of 15 V = ± 15 mV.

This means that the output of this DAC can, at any time, be off by as much as 15 mV from its expected value.

Linearity error is the maximum deviation in step size from the ideal step size. For example, the DAC shown in Fig. 5.3 has an expected step size of 1 V. If this DAC has a linearity error of ± 0.1% FS, this would mean that the actual *step size* could be off by as much as 15 mV.

Some of the expensive DACs have full-scale and linearity errors as low as 0.001% FS. General-purpose DACs usually have accuracies in the 0.01 to 0.1 percentage range.

Resolution, on the other hand, defines the smallest increment in the voltage that can be discerned. Resolution is primarily a function of the number of bits in the digital input signal; that is, the smallest increment in output voltage is determined by the LSB.

Referring to Table 5.1 we can see the resolution is 1 V, since V_{OUT} can change by no less than 1 V when the digital input value is changed. The resolution is always equal to the weight of LSB and is also referred as to *step size,* since it is the amount that V_{OUT} will change as the digital input value is changed from one step to the next.

For above reason, manufacturers usually specify a DAC resolution as the number of bits. An 8-bit DAC has a finer (smaller) resolution than a 6-bit DAC.

The system designer must decide what resolution is required on the basis of the required system performance. The resolution limits how close the DAC output can come to a given analog value. Generally, the cost of DACs increases with the number of bits and so the designer will use as few bits as necessary.

Example 5.1. Determine the (*i*) resolution, (*ii*) full-scale output and weight of each input bit for the DAC shown in Fig. 5.3. Assume $V_{REF} = 10$ V.

Determine also the full-scale output when the feedback resistor R_f is made one-fourth of R.

Solution : The MSB passes with unity gain, so its weight in the output is equal to V_{REF} *i.e.* 10 V **Ans.**

$$\text{The second MSB weight} = \frac{V_{REF}}{2} = \frac{10}{2} = 5 \text{ V } \textbf{Ans.}$$

$$\text{The third MSB weight} = \frac{V_{REF}}{4} = \frac{10}{4} = 2.5 \text{ V } \textbf{Ans.}$$

$$\text{The fourth MSB (or LSB) weight} = \frac{V_{REF}}{8} = \frac{10}{8} = 1.25 \text{ V } \textbf{Ans.}$$

(*i*) The resolution of the DAC is equal to the weight of the LSB *i.e.* 1.25 V **Ans.**

(*ii*) Full scale output = 10 + 5 + 2.5 + 1.25 = 18.75 V **Ans.**

If R_f is reduced to one-fourth, each input will be 4 times smaller than the values above. Thus the full-scale output will be reduced in the same ratio and becomes $\frac{18.75}{4} = 4.6875$ V. **Ans.**

Example 5.2. A simple digital-to-analog conversion circuit is shown in Fig. 5.5. The '1' signal indicates the 'ON' position while '0' the 'OFF' position. If the input signal ($D_3 D_2 D_1 D_0$) is given by 1011, then find the analog output V_{out}.

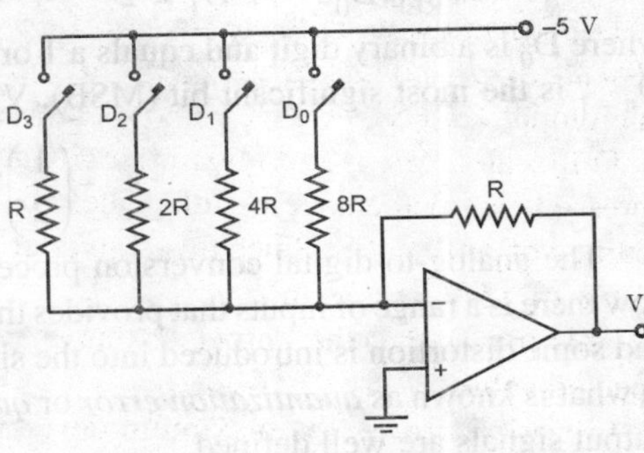

Fig. 5.5

Solution : Reference voltage, $V_{REF} = -5$ V

The output voltage is given as

$$V_{out} = -\left[V_A + \frac{1}{2}V_B + \frac{1}{4}V_C + \frac{1}{8}V_D\right]$$

Here $V_A = V_C = V_D = 5$ V and $V_B = 0$

$$\text{So output voltage, } V_{out} = -\left[-5\text{ V} + 0 + \frac{-5\text{ V}}{4} + \frac{-5\text{ V}}{8}\right]$$

$$= 5\text{ V} + 0 + 1.25\text{ V} + 0.625\text{ V} = 6.875 \text{ V } \textbf{Ans.}$$

Example 5.3. Determine the output voltages caused by each bit in a 6-bit ladder if the input levels are 0 = 0 V and 1 = + 16 V. Determine the resolution and full-scale output of this circuit.

Find out the voltage from the above ladder for a digital input of 101011.

Solution :

$$\text{First MSB output} = \frac{D_{n-1} \times 2^{n-1}}{2^n} = \frac{D_{n-1}}{2} = \frac{16}{2} = 8 \text{ V } \textbf{Ans.}$$

$$\text{Second MSB output} = \frac{D_{n-2} \times 2^{n-2}}{2^n} = \frac{D_{n-2}}{2^2} = \frac{16}{4} = 4 \text{ V } \textbf{Ans.}$$

$$\text{Third MSB output} = \frac{D_{n-3} \times 2^{n-3}}{2^n} = \frac{D_{n-3}}{2^3} = \frac{16}{8} = 2 \text{ V } \textbf{Ans.}$$

$$\text{Fourth MSB output} = \frac{16}{16} = 1 \text{ V } \textbf{Ans.}$$

$$\text{Fifth MSB output} = \frac{16}{32} = 0.5 \text{ V } \textbf{Ans.}$$

$$\text{Sixth MSB (LSB) output} = \frac{16}{64} = 0.25 \text{ V } \textbf{Ans.}$$

The resolution is equal to the weight of the LSB *i.e.* 0.25 V **Ans.**

The full scale output occurs for digital input of 111111

$$i.e.\ 8 + 4 + 2 + 1 + 0.5 + 0.25 = 15.75 \text{ V } \textbf{Ans.}$$

The voltage output for a digital input of 101011

$$V_{OUT} = \frac{D_0 \times 2^0 + D_1 \times 2^1 + D_2 \times 2^2 + D_3 \times 2^3 + D_4 \times 2^4 + D_5 \times 2^5}{2^6}$$

$$= \frac{16 \times 1 + 16 \times 2 + 0 \times 4 + 16 \times 8 + 0 \times 16 + 16 \times 32}{64} = 10.75 \text{ V } \textbf{Ans.}$$

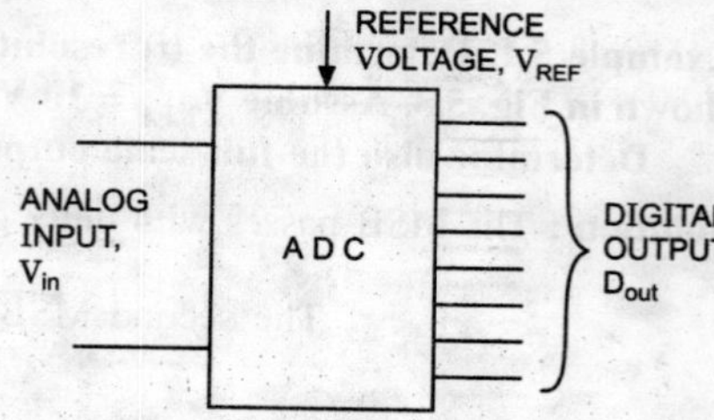

Fig. 5.6 *Basic Function of an ADC*

5.4 ANALOG-TO-DIGITAL (A/D) CONVERSION

The basic function of an ADC is shown in Fig. 5.6. The digital output, D_{out} is the *n*-bit digital output word while V_{in} and V_{REF} are the analog input and reference signals.

For an ADC, these signals are related by equation

$$V_{REF}(D_0 2^{-n} + D_1 \times 2^{-(n-1)} + D_2 \times 2^{-(n-2)} + \ldots + D_{n-1} 2^{-1}) = V_{in} \pm V_x$$

where D_0 is a binary digit and equals a 1 or a 0. D_0 is also defined as the least significant bit (LSB) and D_{n-1} is the most significant bit (MSB). V_x is given by the following equation

$$-\left(\frac{1}{2}\right)V_{LSB} \le V_x < \left(\frac{1}{2}\right)V_{LSB}$$

The analog-to-digital conversion process is known as *quantization*. The worthnoting point is that now there is a range of inputs that provides the same digital output word. There is loss of some information and some distortion is introduced into the signal during the given process. This signal ambiguity results in what is known as *quantization error* or *quantization noise*. DAC does not have this effect because the output signals are well defined.

The analog-to-digital (A/D) conversion is the process of converting an analog input voltage into an equivalent digital signal. The operation is somewhat more complex and time-consuming than the D/A conversion. A number of different methods have been developed and used for A/D conversion. Few of these will be described here.

5.4.1. Ramp A/D Conversion. This is the simplest and popular method of A/D conversion and employs a digital counter as the register and allows the clock to increment the counter one step at a time until the reference ramp voltage, V_{REF} becomes equal to or exceeds the analog input voltage, V_A. It is called a *digital-ramp ADC* because the waveform of DAC's output, V_{REF} is a step-by-step ramp (actually a staircase), like the one shown in Fig. 5.7 (*b*). It is also referred to as a *counter-type* ADC. Such an ADC quantizes an analog signal through conversion of the signal to a time duration pulse.

A ramp ADC is shown in Fig. 5.7. It consists of a digital counter, a DAC, an analog comparator and a control AND gate. The digital counter advances from a zero count while the reference voltage V_{REF}, output of the DAC driven by the counter, increases one voltage increment for each count step. A comparator circuit, receiving both DAC's output (reference ramp voltage V_{REF}) and analog input voltage V_A, provides a signal to stop the count when V_{REF} rises above V_A. The counter value at this time is the digital output.

For explaining the operation let us assume that analog input voltage V_A is positive. When a start pulse is applied, it resets the counter to zero and initiates the operation of the DAC which produces a linear ramp output signal $V_{REF} = K\,t$ where K is a constant and t is the time. A comparator then

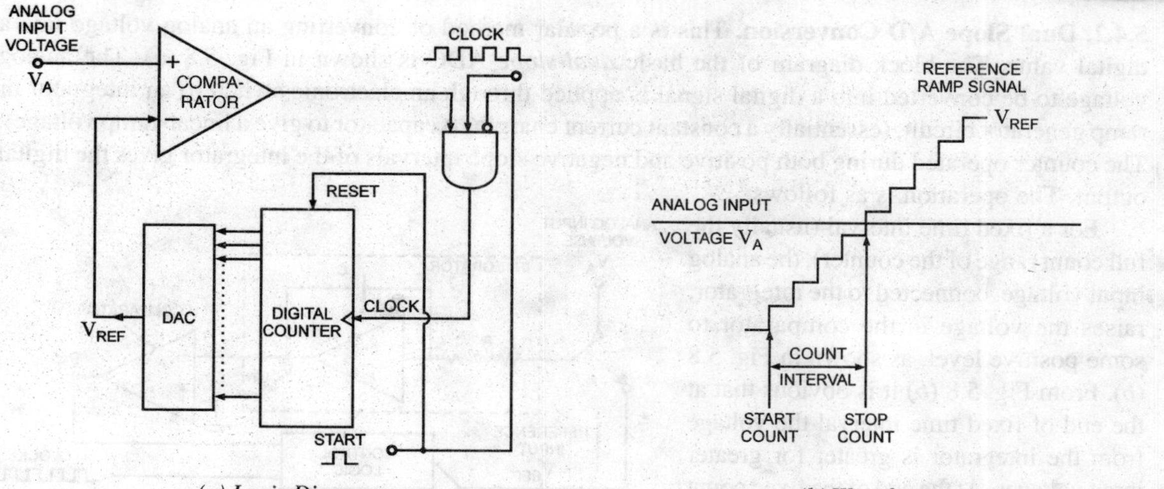

(a) Logic Diagram (b) Waveform

Fig. 5.7 *Ramp ADC*

continuously compares the reference ramp signal, V_{REF} with analog input voltage V_A. The resulting output of the comparator represents a binary 0 when V_{REF} is greater than V_A and a binary 1 when V_{REF} is less than V_A. While the comparator output is 1, pulses from a clock pulse generator are counted by the digital counter. When V_{REF} becomes equal to or exceeds V_A, the comparator output changes to a binary zero. Also, by the action of AND gate, clock pulses are prevented from entering the counter. The time during which the comparator output remains in the 1 state is proportional to the magnitude of the input signal V_A. Also, the count in the counter at the instant the comparator changes state is proportional to the time interval that the comparator output is 1. Thus, the count in the counter is a digital representation of the input signal.

The amount of voltage change stepped by the reference ramp signal depends on the reference voltage applied and on the number of count bits employed. A 10-stage counter operating a 10-stage DAC using a reference voltage of 5 V would step each count by a voltage of $\frac{V_{REF}}{2^{10}} = \frac{5}{1,024}$ V = 4.9 mV.

This would result in a conversion resolution of 4.9 mV. The clock rate of counter would affect the time required to carry out a conversion. A clock rate of 1 MHz operating a 10-stage counter would require a maximum conversion time of 1,024 × 1 μs = 1.025 ms.

The minimum number of conversions that could be carried out each second would then be

$$\text{Number of conversions} = 1000/1.025 \cong 976 \text{ conversions per second.}$$

Since on an average, with some conversions requiring little count time and other near maximum count time a conversion time of $\frac{1.025}{2}$ *i.e.* 0.513 ms would be required, the average number of conversions would be 976 × 2 = 1,952 conversions per second. A slower clock rate would result in few conversions per second. A converter employing fewer count stages (and less conversion resolution) would carry out more conversions per second. The conversion accuracy depends on the accuracy of the comparator.

This type of a ADC may be used for conversion of less than 10 or 12 bits at speeds that do not exceed several thousand conversions per second. Higher speeds can be obtained at lower resolution. Added to simplicity is the advantage of excellent differential linearity, accounting for the very wide use of this type of ADC, particularly in such applications as the generation of histograms as encountered in the field of nuclear experimentation.

5.4.2. Dual Slope A/D Conversion. This is a popular method of converting an analog voltage into a digital value. The block diagram of the basic *dual-slope* ADC is shown in Fig. 5.8 (*a*). The analog voltage to be converted into a digital signal is applied through an electronic switch to an integrator or ramp generator circuit, (essentially a constant current charging a capacitor to give a linear-ramp voltage). The counter operated during both positive and negative slope intervals of the integrator gives the digital output. The operation is as follows.

For a fixed time interval (usually the full count range of the counter), the analog input voltage, connected to the integrator, raises the voltage in the comparator to some positive level, as shown in Fig. 5.8 (*b*). From Fig. 5.8 (*b*) it is obvious that at the end of fixed time interval the voltage from the integrator is greater for greater input voltages. At the end of the fixed count interval, the count is set zero and the electronic switch connects the integrator to a reference or fixed input. The integrator output or input to capacitor then decreases at a fixed rate, as shown in Fig. 5.8 (*b*). The counter advances during this time. The integrator output voltage decreases at a fixed rate until it drops below the comparator reference voltage, at which the control logic receives a signal (the comparator output) to stop the count. The count shown by the counter at this time represents the digital output of the ADC.

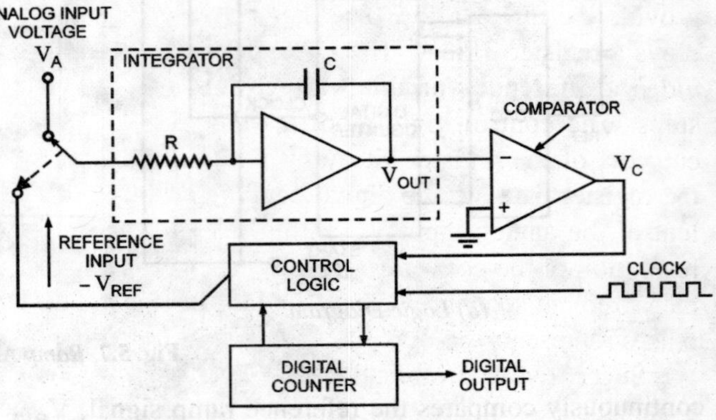

(a) Logic Diagram

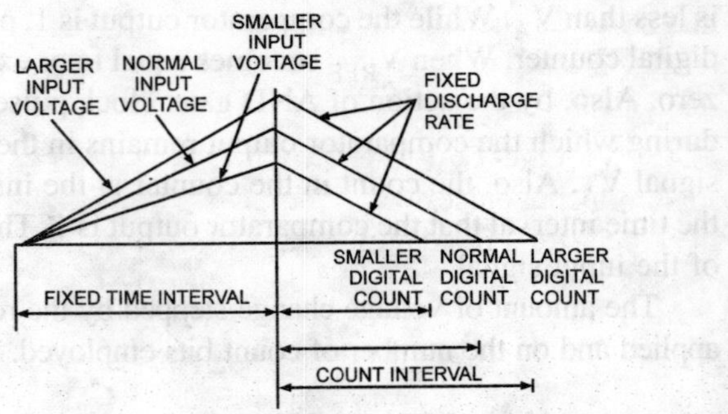

(b) Waveform

Fig. 5.8 *Dual-Slope ADC*

Typically, a dual-slope ADC is limited to conversion speeds of from 1,000 to 2,000 samples per second at a resolution of 10 or 12 bits as a result of limitations on the counter counting speed. The conversion rate can be increased to approximate 30,000 samples per second at a 14-bit resolution without markedly increasing the logic requirements by using a two-step integration during the second integration period.

The averaging characteristics and cancellation of errors that usually limit the performance of a ramp-type ADC are the main advantages of this ADC. The integration characteristic provides the average value of the input signal during the period of first integration. Consequently, disturbances, such as spurious noise pulses, are minimized. Performance analysis of these ADCs also indicates that some other types of errors cancel out as well. Long-term drifts in the time constant, as may result from temperature variations or aging, do not affect conversion accuracy. Also long-term alteration in clock frequency have no effect.

5.4.3. Successive-Approximation A/D Conversion. This is one of the most widely used method of A/D conversion. Though it employs more complex circuitry than that used by ramp A/D conversion but it has much shorter conversion time. In addition, it has a fixed value of conversion time that does not depend upon the value of the analog input.

This type of ADC makes direct comparison between an unknown input signal and a reference signal. The basic arrangement of a successive-approximation ADC shown in Fig. 5.9 is similar to the digital ramp ADC. This type of ADC, however, does not employ a counter to provide the input to the DAC but employs a register instead. The DAC provides a reference variable voltage in steps. The control logic modifies the contents of the register bit by bit until the register data are the digital equivalent of the analog input V_A within the resolution of the converter. Usually the measurement sequence selects the largest step of the DAC output voltage first. The number of clock pulses represents the digital output of the DAC.

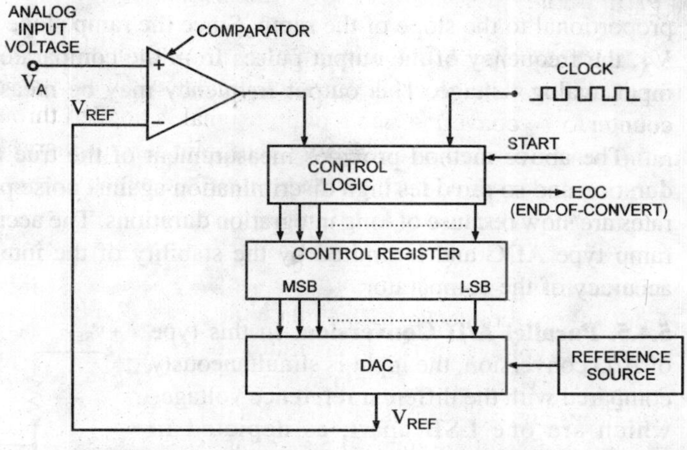

Fig. 5.9 *Successive-Approximation ADC*

Successive-approximation ADC can be employed at conversion speeds of up to about 100,000 samples per second at resolutions of up to 16 bits (not including sign). At lower resolutions, speeds of over 250,000 samples per second are practical. Factors to be considered in the design and applications of these ADCs include stability and regulation of the reference voltage source, overload and recovery characteristics of the comparator, characteristics of the analog switches and speed and response of the ladder network.

5.4.4. Voltage-To-Frequency A/D Conversion. An analog voltage can be converted into digital form, by producing pulses whose frequency is proportional to the analog voltage. These pulses are counted by a counter for a fixed duration and the reading of the counter will be proportional to the frequency of the pulses, and hence, to the analog voltage.

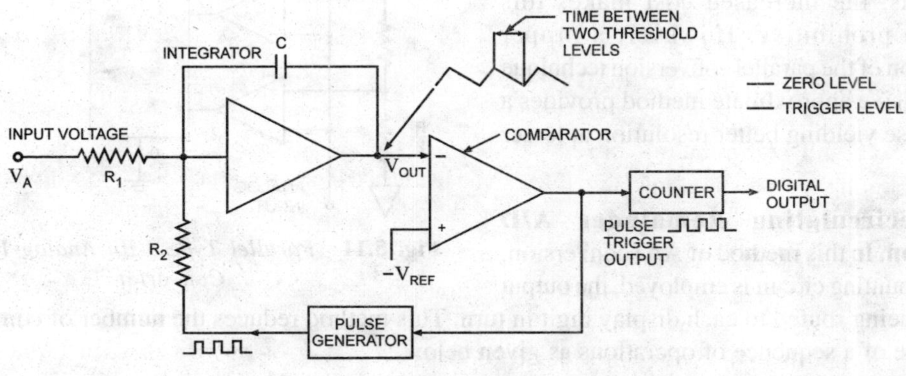

Fig. 5.10 *Voltage-To-Frequency A/D Conversion*

A block diagram of a voltage-to-frequency ADC is given in Fig. 5.10. The analog input voltage V_A is applied to an integrator which in turn produces a ramp signal whose slope is proportional to the input voltage. When the output voltage V_{OUT} attains a certain value (a preset threshold level), a trigger pulse is produced and also a current pulse is generated which is used to discharge the integrator capacitor C. Now a new ramp is initiated. The time between successive threshold level crossings is inversely proportional to the slope of the ramp. Since the ramp slope is proportional to the input analog voltage V_A, the frequency of the output pulses from the comparator is, therefore, directly proportional to the input analog voltage. This output frequency may be measured with the help of a digital frequency counter.

The above method provides measurement of the true average of the input signal over the ramp duration and so provides high discrimination against noise present at the input. However, the digitizing rates are slow because of long integration durations. The accuracy of this method is comparable with the ramp type ADC and is limited by the stability of the integrator time constant and the stability and accuracy of the comparator.

5.4.5. Parallel A/D Conversion. In this type of A/D conversion, the input is simultaneously compared with the different reference voltages which are one LSB apart, as depicted in Fig. 5.11. The outputs of the comparators are then logically combined to provide the output digital code by the priority encoder that puts out the digital code for the highest-order input received by it. Its main advantage is very high speed of conversion, as the speed is restricted only by the switching time of the comparators and gates. Its main drawback is its complexity. For 3-bit resolution, 7 comparators and one 8-bit priority encoder are required, as shown in Fig. 5.11, for 4-bit resolution, 15 comparators and two 8-bit priority encoders are required while for 5-bit resolution, 31 comparators and four 8-bit encoders are required. For higher resolutions, the increased cost makes this technique prohibitive. However, a proper combination of the parallel conversion technique and successive approximate method provides a compromise yielding better resolution at lesser cost.

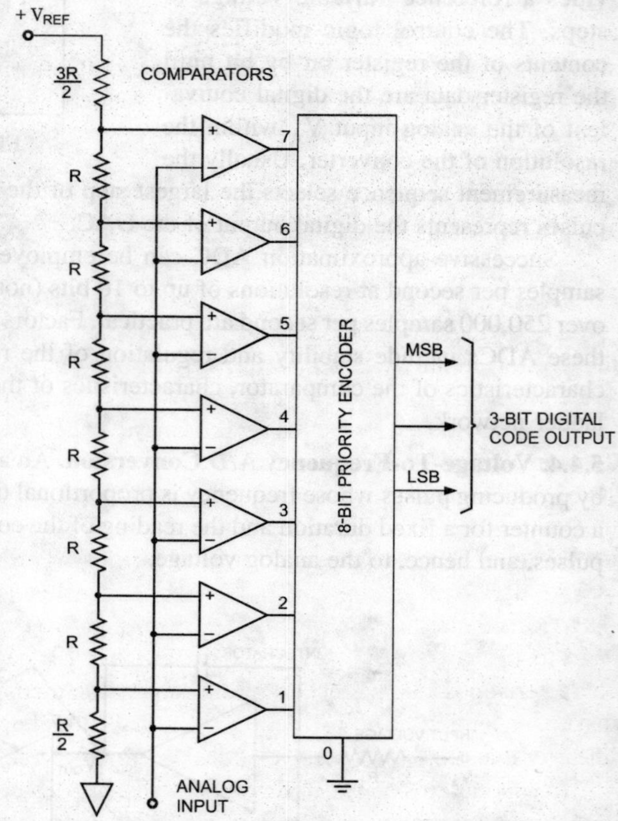

Fig. 5.11 *Parallel Type 3-Bit Analog-To-Digital Converter*

5.4.6. Recirculating Remainder A/D Conversion. In this method of A/D conversion, only one counting circuit is employed, the output from this being routed to each display digit in turn. This method reduces the number of components by making use of a sequence of operations as given below.

(*i*) Sampling of analog input, (*ii*) Determination of most significant digit of the input, (*iii*) Storing of difference of the first digit worth and the input in a capacitor, (*iv*) Amplification of this remainder by a factor of 10, (*v*) Determination of next most significant digit and (*vi*) Storage of the remainder. The steps (*iv*), (*v*) and (*vi*) are cycled until the magnitude of the analog input is displayed with the full resolution available.

Digital voltmeter (DVM) making use of this technique will be described later on [Art. 5.7(6)].

5.5 ADC RESOLUTION AND ACCURACY

Since the analog-to-digital converter (ADC) is a closed-loop system involving both analog and digital systems, the overall accuracy must include errors from both the analog and digital positions. In determination of the overall accuracy it is easiest to separate the two sources of error.

If we assume that all components are operating properly, the source of the digital error is simply determined by the resolution of the system. In digitizing an analog voltage, we are trying to represent a continuous analog voltage by an equivalent set of digital numbers. When the digital levels are converted back into analog form by the ladder, the output is the familiar staircase waveform. This waveform is a representation of the input voltage but is certainly not a continuous signal. It is, in fact, a continuous signal composed of a number of discrete steps. In efforts of reproducing the analog input signal, the best that can be done is to get on the step which most nearly equals the input voltage in amplitude.

The simple fact that the ladder voltage has steps in it leads to the digital error in the system. The smallest digital step, or quantum, is due to the LSB and can be made smaller only by increasing the number of bits in the counter. This inherent error is often called the *quantization error* and is commonly ± 1 bit. By centering the comparator, as with the continuous converter, the quantization error can be made $\pm\frac{1}{2}$ LSB.

The main source of analog error in the ADC is probably the comparator. Other sources of error are the resistors in the ladder, the reference voltage supply ripple and noise. These can, however, usually be made secondary to the sources of error in the comparator.

The sources of error in the comparator are centred around variations in the dc switching point. The dc switching point is the difference between the input voltage levels that make the output to change state. Variations in switching are due primarily to offset, gain and linearity of the amplifier employed in the comparator. These parameters, usually vary slightly with input voltage levels and quite often with temperature. These changes give rise to the analog error in the system.

An important measure of converter performance is given by the *differential linearity*. *Differential linearity* is a measure of the variation in voltage-step size that makes the converter to change from one state to the next. It is usually expressed as a per cent of the average step size. This performance characteristic is also a function of the conversion method and is best for the converters having counters that count continuously. The counter-type and continuous-type converters usually have better differential linearity than do the successive-approximation type converters.

In general, it is considered good practice to construct converters having analog and digital errors of approximately the same magnitudes. There are many arguments for and against this, and any final argument would depend on the situation. For example, an 8-bit converter would have a quantization error of $\frac{1}{2^8} = \frac{1}{256} \cong 0.4$ percent. It would then seem reasonable to construct this converter to an accuracy of 0.5% in an effort to achieve an overall accuracy of 1.0 percent. This might mean constructing the ladder to an accuracy of 0.1%, the comparator to an accuracy of 0.2%, and so on, as these errors are all cumulative.

Example 5.4. The reference voltage of a dual slope ADC is 5 V. The integrator has a capacitor of 1 μF and resistance of 100 kΩ. The time taken to read an unknown voltage V_x is 0.2 s. Find V_x.

Solution : Time constant for the integrator, $T = RC = 100 \times 10^3 \times 1 \times 10^{-6} = 0.1$ s

Reference voltage, $V_{REF} = 5$ V

$$\text{Unknown voltage, } V_x = \frac{\text{Time constant of the circuit}}{\text{Time taken to read unknown voltage}} \times V_{REF} = \frac{0.1}{0.2} \times 5 = 2.5 \text{ V} \quad \textbf{Ans.}$$

Example 5.5. What are the methods of conversion of analog to digital (A to D) signals ? Compare the methods. Show how 180 V will be converted into 8-bit digital form using successive approximation technique. If the clock frequency is 75 MHz, what will be the conversion time ?

Solution : For an 8-bit converter reference voltage V_{REF} be taken as 100 V.

$$\text{So MSB} = 100 \text{ V}$$

For setting $D_7 = 1$; Output voltage $= V_{REF} \times \frac{2^7}{2^8} = \frac{100}{2^1} = 50$ V

Since 180 – 100 = 80 > 50; Set $D_7 = 1$

For setting $D_6 = 1$; Output voltage $= V_{REF} \times \frac{2^6}{2^8} = \frac{100}{2^2} = 25$ V

Hence for setting $D_7 = 1$ and $D_6 = 1$

$$\text{Output voltage} = 50 + 25 = 75 \text{ V}$$

Since 80 > 75; Set $D_6 = 1$

For setting $D_5 = 1$, $D_6 = 1$ and $D_7 = 1$;

$$\text{Output voltage} = V_{REF} \times \frac{2^5}{2^8} + 25 + 50 = \frac{100}{8} + 25 + 50 = 87.5 \text{ V}$$

Since 80 < 87.5 hence set $D_5 = 0$

All other digits will be set to zero or 1. Output will be accordingly indicated as a result of successive approximation. The converted 8-bit digital form will be 1110010 **Ans.**

$$\text{Conversion time, } T = \frac{1}{f} = \frac{1}{75 \times 10^6} = 13.33 \times 10^{-9} \text{ s or } 13.3 \text{ ns} \quad \textbf{Ans.}$$

Example 5.6. An 8-bit successive approximation type of ADC uses a clock frequency of 1 MHz. Calculate the time of conversion. *[U.P.S.C. I.E.S. Elec. Engineering I 2003]*

Solution : Time of conversion, t_c = Number of bits × conversion time, $T_C = 8 \times \frac{1}{10^6} = 8\ \mu s$ **Ans.**

Example 5.7. A dual slope integrating type of A/D converter has an integrating capacitor of 0.1 μF and a resistance of 100 kΩ. If the reference voltage is 2 V and the output of the integrator is not to exceed 10 V, what is the maximum time up to which the reference voltage can be integrated ?

Solution : The circuit diagram of an integrator is given in Fig. 5.12 (*a*).

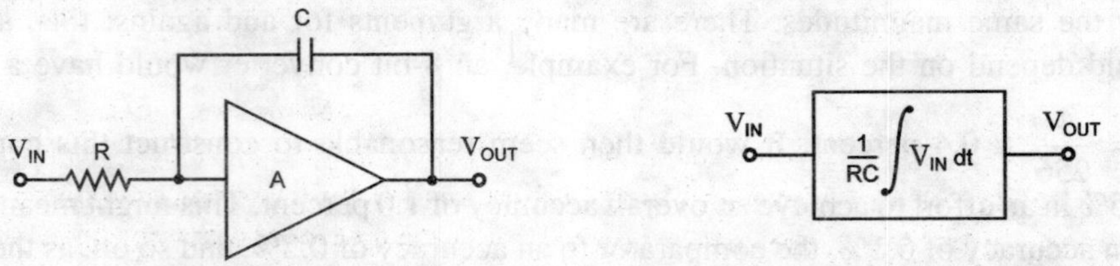

(a) Circuit Diagram *(b) Symbolic Representation*

Fig. 5.12

$$\text{Output voltage, } V_{out} = \frac{-1}{RC}\int_0^t V_{IN}\,dt = -\frac{V_{IN}t}{RC}$$

Given that $V_{in} = 2$ V; $V_{out} = 10$ V;

$R = 100$ kΩ; $C = 0.1\ \mu F$.

$$\text{So } 10 = \frac{2 \times t}{100 \times 1{,}000 \times 0.1 \times 10^{-6}}$$

or the maximum time up to which the reference voltage can be integrated,

$$t = 0.05 \text{ s} \quad \text{or} \quad 50 \text{ ms} \quad \textbf{Ans.}$$

5.6 SEVEN SEGMENT DISPLAYS

A very common requirement in modern electronics is that of displaying alphanumeric characters: digital watches, pocket calculators, and digital instruments are all examples of devices that use such displays. The best known display of this type is the so-called *seven-segment display*. It comprises seven independently-accessible photoelectrical segments (*e.g.*, LED, LCD, gas-discharge, fluorescent, or filament-type segments) arranged in the form shown in Fig. 5.13(*a*). The segments are conventionally notated from *a* to *g* in the manner shown in Fig. 5.13(*b*), and it is possible to make them display any numeral from 0 to 9 or any alphabetic character from A to F (in a mixture of upper and lower case letters) by activating these segments in different combinations, as illustrated in Truth Table (Table 5.2)

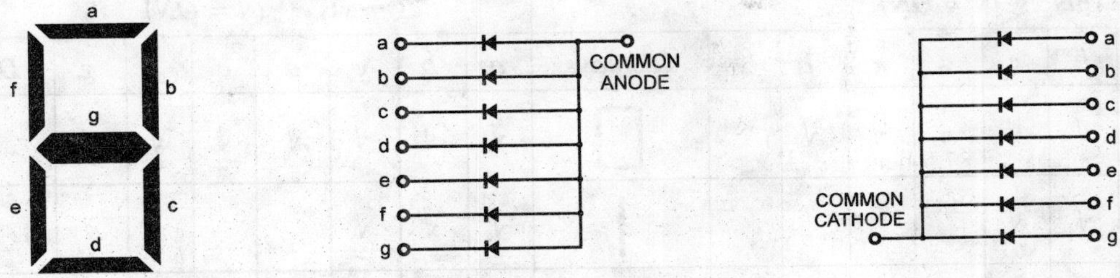

(a) Standard Form and Notations of a Seven Segment Display *(b) Common Anode Configuration* *(c) Common Cathode Configuration*

Fig. 5.13

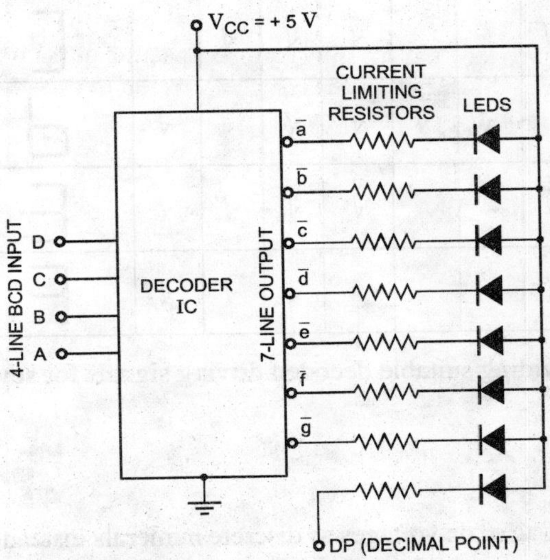

Fig. 5.14 *LED Seven Segment Display Driver*

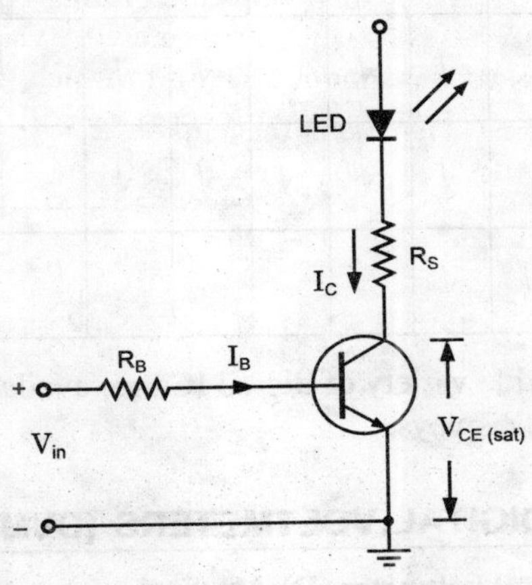

Fig. 5.15 *Transistor Circuit For Switching ON and OFF a LED*

The wiring pattern of LED *seven segment displays* is simplified by having one terminal as common for all seven segments. The "common anode" and "common cathode" internal connections for LED displays are illustrated in Figs. 5.13 (*b*) and 5.13 (*c*) respectively. The common-cathode LEDs have to be driven by "active-high" decoder ICs and the common-anode LEDs have to be driven by "active-low" decoders. Such a decoder connected to a seven-segment display digit is shown in Fig. 5.14. It may be noted that series resistors are included for each of the LEDs corresponding to the seven segments. Further an additional LED corresponding to the decimal point is also provided in the display, which again has a current limiting series resistor. The typical value of the series resistor is 150 Ω for a segment current of about 20 mA.

To illustrate the operation of the circuit, let us assume that the ABCD input is D = 0, C = 1, B = 1, A = 0, which is BCD for 6. With these inputs the decoder/driver outputs will be driven LOW (connected to the ground), allowing current to flow through the LED segments thereby displaying the digit 6. The and outputs will be HIGH (open), so the LED segments *a* and *b* cannot conduct.

LEDs are usually switched on and off by transistor circuit shown in Fig. 5.15.

Table 5.2

Segments (√ = ON)								*Segments (√ = ON)*							
a	*b*	*c*	*d*	*e*	*f*	*g*	*Displays*	*a*	*b*	*c*	*d*	*e*	*f*	*g*	*Displays*
√	√	√	√	√	√		0	√	√	√	√	√	√	√	8
	√	√					1	√	√	√			√	√	9
√	√		√	√		√	2	√	√	√		√	√	√	A
√	√	√	√			√	3			√	√	√	√	√	b
	√	√			√	√	4	√			√	√	√		C
√		√	√		√	√	5		√	√	√	√		√	d
√		√	√	√	√	√	6	√			√	√	√	√	E
√	√	√					7	√				√	√	√	F

A wide variety of digital ICs are available for providing suitable decoded driving signals for seven segment displays.

5.7 DIGITAL VOLTMETERS (DVMs)

The digital voltmeter (DVM) displays measurements of ac or dc voltages as discrete numerals instead of a pointer deflection on a continuous scale as in analog instruments. It is a versatile and accurate instrument that is employed in many laboratory measurement applications. Because of development and perfection of IC modules, the size, power requirements and cost of the digital voltmeter has been drastically reduced and, therefore, DVMs can actively compete with conventional analog instruments, both in price and portability.

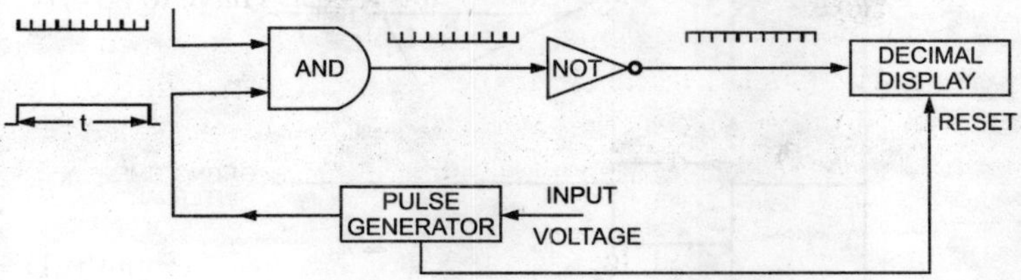

Fig. 5.16 *Block Diagram of DVM*

The block diagram of a simple digital voltmeter is shown in Fig. 5.16. The unknown voltage signal is fed to the pulse generator which generates a pulse whose width is directly proportional to the input unknown voltage. The output of the pulse generator is applied to one leg of an AND gate. The input signal to the other leg of the AND gate is a train of pulses. The output of the AND gate is, thus, a positive trigger train of duration *t* seconds and the inverter converts it into a negative trigger train. The counter counts the number of triggers in *t* seconds which is proportional to the voltage under measurement. Thus, the counter can be calibrated to indicate voltage in volts directly.

Thus, we see that the DVM described above is an ADC which converts an analog signal into a train of pulses, the number of which is proportional to the input voltage. So a digital voltmeter can be made by using any one of the A/D conversion methods discussed in Art. 5.4 and can be represented by a block diagram shown in Fig. 5.17. Thus the DVMs can be classified on the basis of ADCs used.

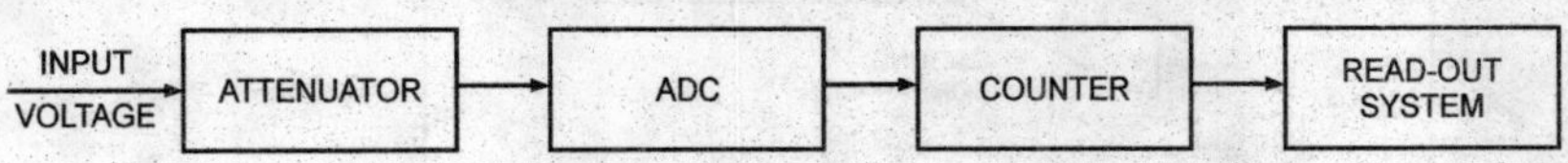

Fig. 5.17

The input range of the DVM may vary form ± 1.00000 V to ± 1,000.00 V and its limiting accuracy is as high as ± 0.005 per cent of the reading. Its resolution may be 1 part in 10^6, giving 1 μV reading of the 1 V input range. It has a high input resistance of the order of 10 MΩ and input capacitance of the order of 40 pF.

Digital voltmeters (DVMs) employing different A/D conversion methods are discussed below :

1. *Ramp Type DVM*. The operation of a ramp type DVM is based on the measurement of the time that a linear ramp voltage takes to change from the level of the input voltage to zero voltage or vice-versa. This time interval is measured with an electronic time interval counter and the count is displayed as a number of digits on the electronic indicating tubes of the voltmeter output read-out.

The operating principle and block diagram of a *ramp type DVM* are given in Figs. 5.18 (*a*) and 5.18 (*b*) respectively.

At the start of measurement a ramp voltage is initiated, this voltage can be positive-going or negative going. In the Fig. 5.18 (*a*) negative going voltage ramp is illustrated. The ramp voltage is continuously compared with the voltage under measurement (or unknown voltage). At the instant the value of ramp voltage becomes equal to the voltage under measurement, a coincidence circuit, called the *input comparator,* generates a pulse which opens a gate, as shown in Fig. 5.18 (*b*). The ramp voltage continues to fall till it reaches zero value (or ground level). At this instant another comparator, called the *ground comparator*, generates a stop pulse. The output pulse from this ground comparator closes the gate. The time duration of the gate opening is proportional to the value of the dc input voltage.

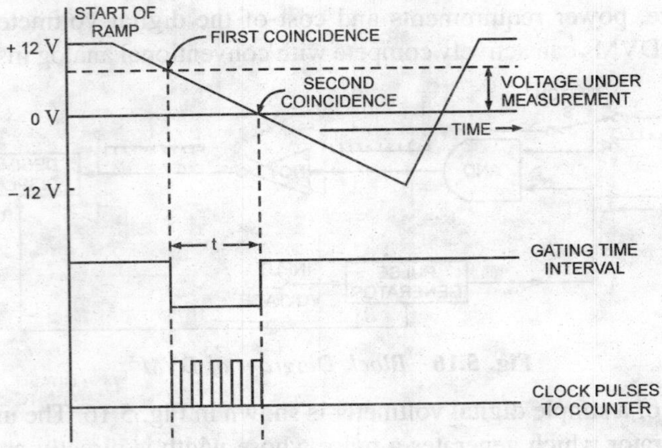

(a) *Voltage-To-Time Conversion*

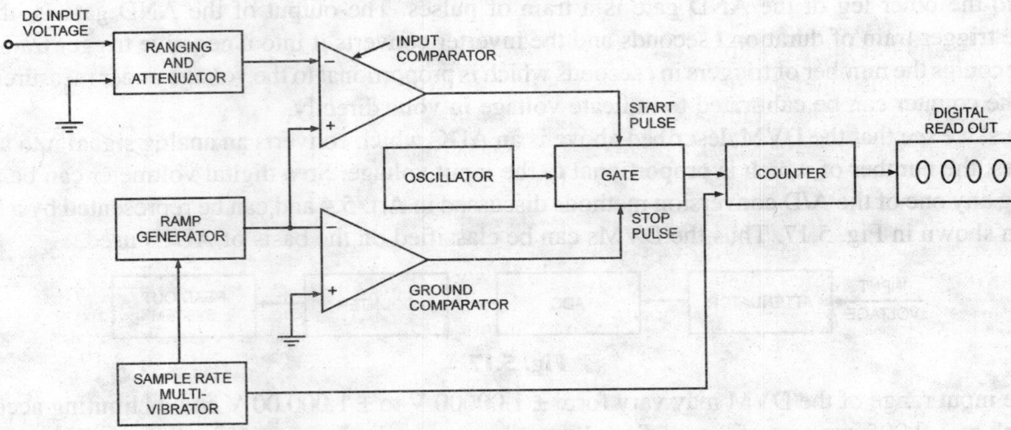

(b) *Block Diagram of a Ramp Type DVM*

Fig. 5.18

The time elapsed between opening and closing of gate is t, as illustrated in Fig. 5.18 (*a*). During this time interval pulses from a clock pulse oscillator pass through the gate and are counted and displayed. The decimal number indicated by the read-out is a measure of the value of the input voltage.

The sample rate multivibrator determines the rate at which the measurement cycles are initiated. The sample rate circuit provides an initiating pulse for the ramp generator to start its next ramp voltage. At the same time a reset pulse is generated, which resets the counter to zero state.

2. Dual Slope Integrating Type DVM. The block diagram of a dual slope integrating type DVM is given in Fig. 5.19 (*a*). The dual-slope ADC consists of five blocks viz. an op-amp employed as an integrator, a level comparator, a basic clock for generating time pulses, a set of decimal counter and a block of logic circuitry.

For a fixed time interval (usually the full count range of the counter), the analog input voltage (voltage to be measured) is applied through the switch S to the integrator which raises the voltage in the comparator to some positive value, as illustrated in Fig. 5.19 (*b*). Obviously at the end of fixed time interval the voltage from the integrator will be greater for larger input voltage *i.e.* rate of increase of voltage or slope is proportional to the analog input voltage. At the end of the fixed count interval, the count is set at zero and the switch S is shifted to the reference voltage V_{REF} of opposite polarity. The

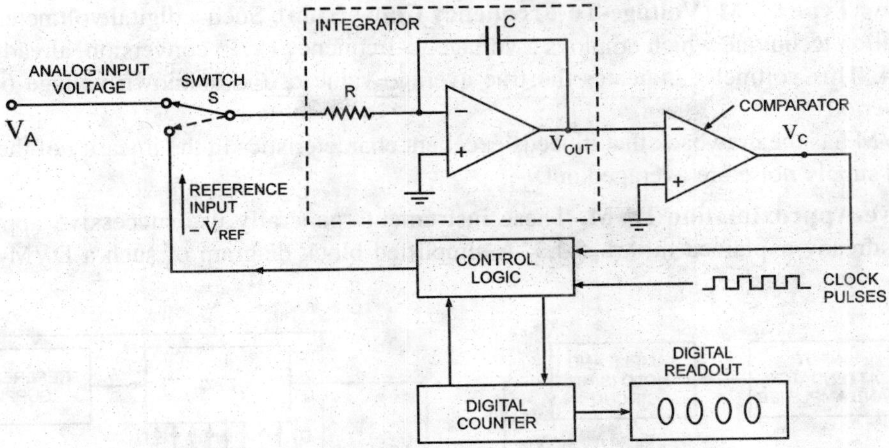

(a) Block Diagram of a Dual Slope Integrating Type DVM

integrator output or input to capacitor C then starts decreasing at a fixed rate, as illustrated in Fig. 5.19 (*b*). The counter advances during this time. The integrator output voltage decreases at a fixed rate until it drops below the comparator reference voltage, at which the control logic receives a signal (the comparator output) to stop the count. The pulse counted by the counter thus has a direct relation with the input voltage V_A.

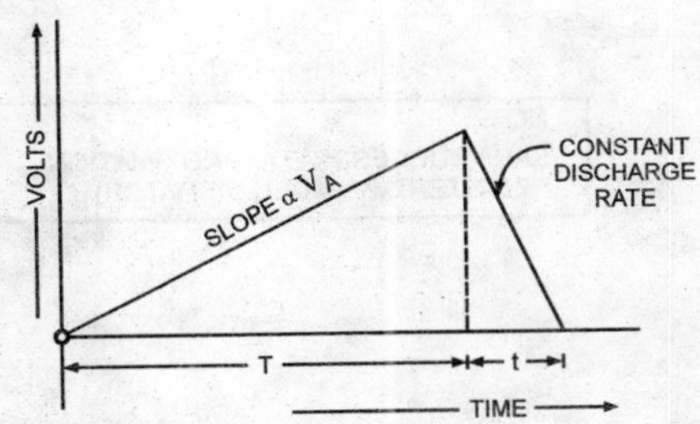

(b)

Fig. 5.19

During charging, the output voltage V_{OUT} is given as

$$V_{OUT} = \frac{-1}{RC}\int_0^T V_A \, dt$$

$$= \frac{-V_A T}{RC} \qquad ...(5.4)$$

During discharging, the output voltage V_{OUT} is given as

$$V_{OUT} = \frac{1}{RC}\int_0^t -V_{REF} \, dt = \frac{-V_{REF}\, t}{RC} \qquad ...(5.5)$$

From Eqs. (5.4) and (5.5) we have

$$\frac{-V_A T}{RC} = \frac{-V_{REF}\, t}{RC}$$

$$\text{or } V_A = \frac{V_{REF}\, t}{T}$$

Thus the count shown by the counter at time *t*, which is proportional to the input voltage (voltage under measurement) V_A, is displayed as the measured voltage.

The averaging characteristics and cancellation of errors that usually limit the performance of ramp type DVM are the main advantages of such DVM. The integration characteristics provide the average value of the input signal during the period of first integration. Consequently, disturbances, such as spurious noise pulses, are minimized. Long-term drifts in the time constant, as may result from temperature variations or aging, do not affect conversion accuracy. Also long-term alterations in clock frequency have no effect.

3. Integrating Type DVM (Voltage-To-Frequency Conversion). Such a digital voltmeter makes use of an integration technique which employs a voltage-to-frequency (V/F) conversion, already explained in Art. 5.5.4. This voltmeter indicates the true average value of the unknown voltage over a fixed measuring period.

This DVM has the drawback that it needs excellent characteristics in the linearity of the ramp. The ac noise and supply noise are averaged out.

4. Successive-Approximation DVM. These instruments generally use successive approximation conversion, already explained in Art. 5.4.3. A simplified block diagram of such a DVM is given in Fig. 5.20.

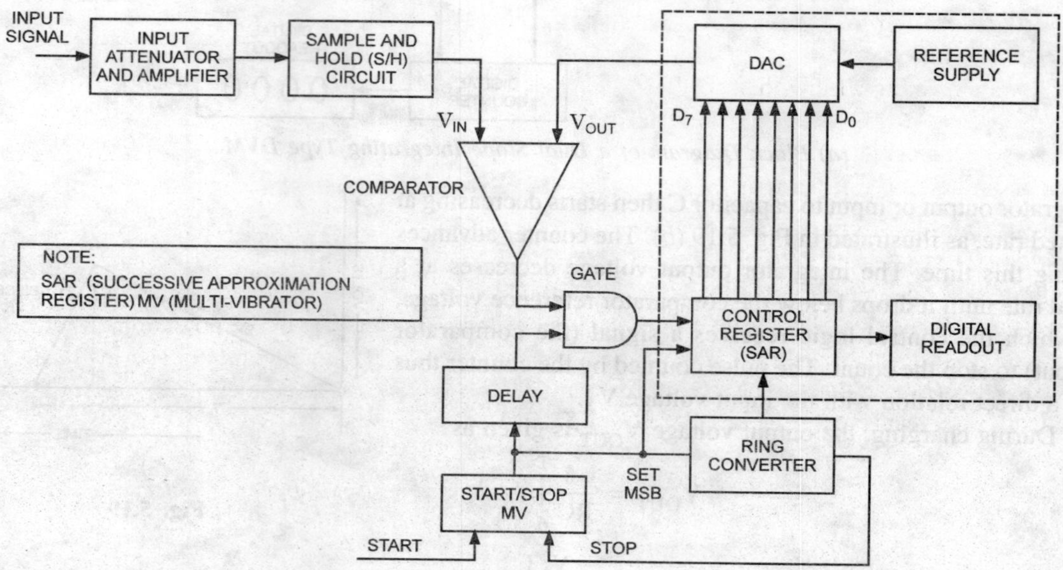

Fig. 5.20 *Simplified Block Diagram of Successive Approximation DVM*

At the start of the measurement cycle, a start pulse is applied to the start/stop multivibrator (MV). This sets a 1 in the most significant digit (MSB) of the control register and a 0 in all the bits of less significance. Assuming an 8-bit control register, its reading would then be 10000000. This initial setting of the control register causes the output of DAC to be one-half the reference voltage *i.e.* $\frac{1}{2}$ V_{REF}. This converter output is compared with the unknown input by the comparator. In case the input voltage exceeds the converter reference voltage, the comparator produces an output that causes the control register to retain the 1 setting in its MSB, and the converter continues to supply its reference output voltage of $\frac{1}{2}$ V_{REF}.

The ring counter then advances one count, shifting a 1 in the second MSB of the control register and its reading becomes 11000000. This causes DAC to increase its reference output by 1 increment to $\frac{1}{4}$ V_{REF} *i.e.* $\frac{1}{2}$ V_{REF} + $\frac{1}{4}$ V_{REF}, and another comparison with the unknown input takes place. If in this case the total reference voltage exceeds the unknown voltage, the comparator produces an output that makes the control register to reset its second MSB to 0. The converter output then returns to its previous value and awaits another input from the control register (SAR) for the next approximation. When the ring counter advances by 1, the third MSB is set to 1 and the converter output rises by next increment of $\frac{1}{2}V_{REF} + \frac{1}{8}V_{REF}$. The measurement cycle thus proceeds through a series of successive approximations as illustrated in Fig. 5.21, retaining or rejecting the converter output in the manner explained. Finally, when the ring counter reaches its last count, the measuring cycle stops, and the digital output of the control register represents the final approximation of the unknown voltage.

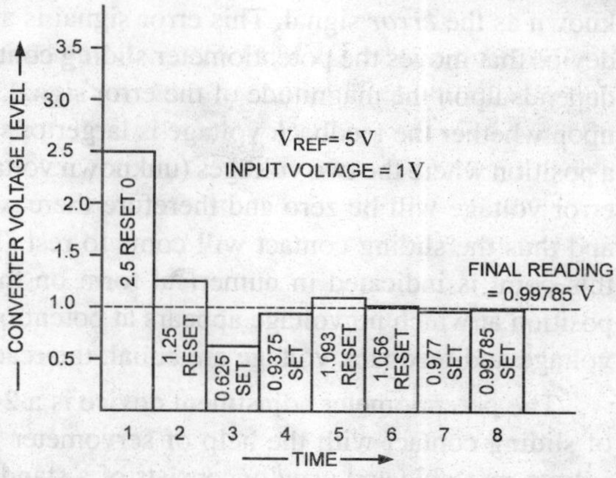

Fig. 5.21 *Successive Approximation Illustration*

With input voltages other than dc, the input level changes during digitization and decisions made during conversion are not consistent. However, this can be avoided by using a sample and holding (S/H) circuit and placing it in the input, directly following the input attenuator and amplifier.

5. Potentiometric Type DVM. A potentiometric type DVM makes use of voltage comparison technique—by comparing the unknown voltage with a reference voltage, V_{REF} whose value is fixed by the setting of the calibrated potentiometer. Balance or null condition is obtained by making adjustment of potentiometer setting. Under null or balance condition, the dial setting of the potentiometer indicates the unknown voltage. In this type of DVM, the balance is obtained automatically (not manually). In fact, this DVM is a *self-balancing potentiometer*. The only exception is that a potentiometric DVM is provided with a read-out to display the voltage under measurement.

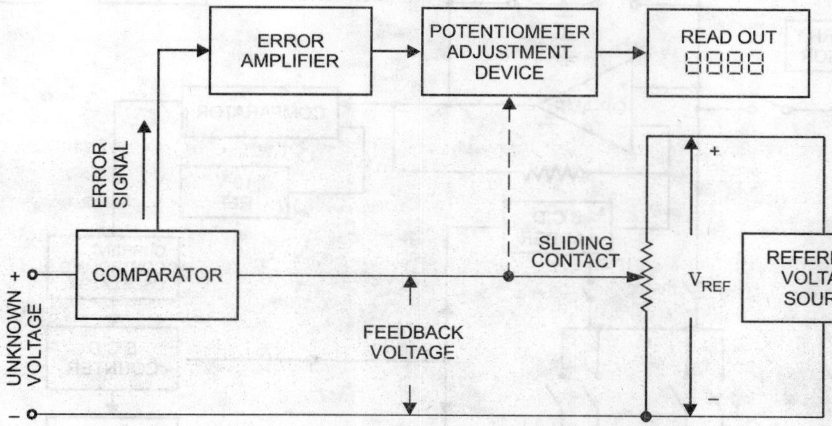

Fig. 5.22 *Block Diagram of a Potentiometric Type DVM*

Figure 5.22 shows the block diagram of basic circuit of potentiometric type DVM. The unknown voltage (voltage under measurement) is filtered and attenuated to a suitable level. This input voltage is applied to a comparator, also called the *error detector*, which may be a chopper. The fixed voltage source provides the reference voltage V_{REF}. V_{REF} is applied to the potentiometer. Sliding contact controls the magnitude of feedback voltage. This feedback voltage is applied to the comparator. The unknown voltage and feedback voltage are compared in the comparator. The output voltage of the comparator is the difference of the above two voltages (unknown voltage and feedback voltage), and is known as the *error* signal. This error signal is amplified and is supplied to a potentiometer adjustment device that moves the potentiometer sliding contact. The magnitude by which the sliding contact moves depends upon the magnitude of the error signal. The direction of movement of sliding contact depends upon whether the feedback voltage is larger or smaller than the input voltage. The sliding contact takes a position where the two voltages (unknown voltage and feedback voltage) are equal. In such a condition, error voltage will be zero and therefore there will be no input to the potentiometer adjustment device and thus the sliding contact will come to rest. The position of the potentiometer adjustment device at this point is indicated in numerical form on the digital read-out device associated with it. Since the position at which no voltage appears at potentiometer adjustment device is the one where the unknown voltage and feedback voltage are equal, the read-out device indicates the value of unknown voltage.

The potentiometer adjustment device is a 2-phase servomotor. The automatic action of adjustment of sliding contact with the help of servometer is known. The reference voltage V_{REF} source must be extremely stable and usually consists of a standard cell or zener diode source.

6. Recirculating Remainder Type DVM. A schematic diagram of this type of DVM is depicted in Fig. 5.23. This makes proper utilization of components, and has a strobed display, resulting in lesser power consumption, facilitating battery operation. As the same capacitors are employed for storing all the digits, each is determined with the same precision. The sampling time is small (4 ms) and the conversion rate is fast, providing a reading rate of 40 per second. The accuracy is limited due to the following factors :

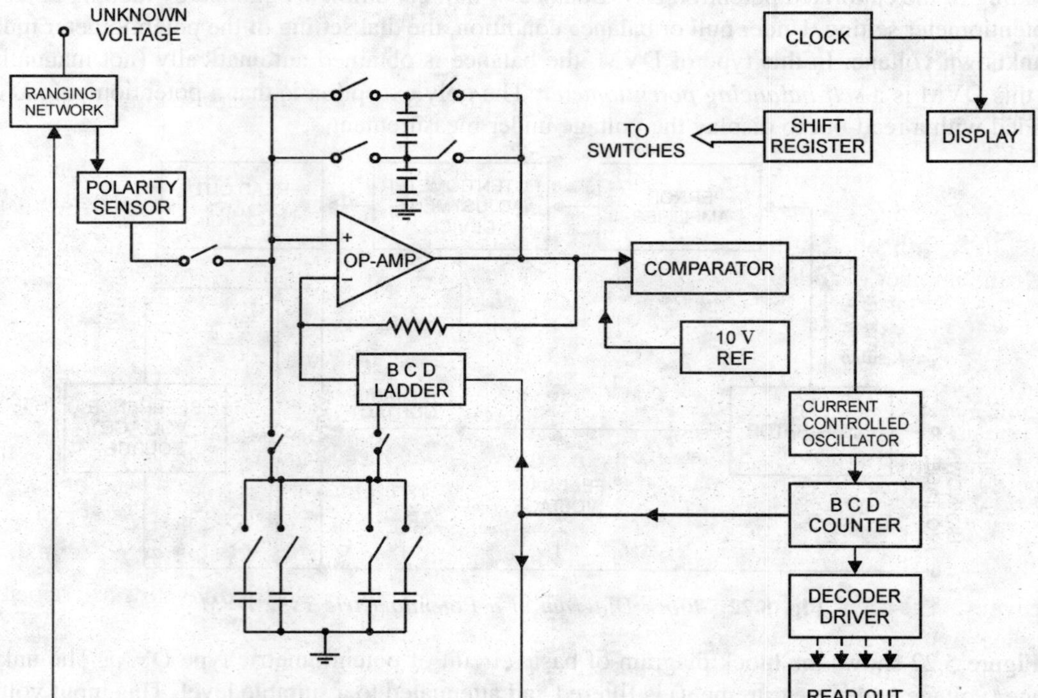

Fig. 5.23

(*i*) Reference voltage (*ii*) stability of op-amp (*iii*) leakage and dielectric storage of the remainder storage capacitor which is much less of a problem in this conversion technique in comparison to that in dual slope method. This is because only the remainder is stored.

5.8 RANGE SELECTION

Switching arrangement for DVM range selection is depicted in Fig. 5.24. The attenuator functions exactly as explained for an analog electronic voltmeter. An input of 1.999 V or less is applied directly to the DVM comparator. In this case, the ganged rotary switch selects the decimal point after the first numeral in the display. The decimal point may be a LCD or LED that has one terminal energized, so that it switches *on* when the other terminal is grounded, as shown in the figure. When the voltage range is switched on to 19.99 V, the decimal point is selected after the second numeral. When the range is 199.9 V, the third decimal point is energized.

Many DVMs have automatic ranging circuits (*auto-ranging*) which evaluate the input voltage, select the appropriate range, and illuminate the correct decimal point. In manual range selection, greatest accuracy is achieved by using the range that provides the largest possible numerical display.

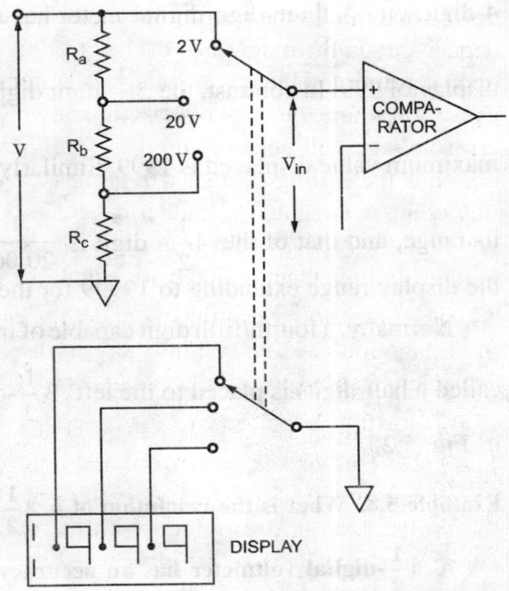

Fig. 5.24 *Range Selection For DVM*

5.9 CHARACTERISTICS OF DIGITAL METERS

Resolution, sensitivity and accuracy are the few specifications which characterise digital meters. These are explained as below.

Resolution. It is defined as the number of digit positions or simply the number of digits used in a meter. Hence a three-digit display on the digital meter for 0–1 V range will be able to indicate from 000 to 999 mV, with smallest increment (resolution) of 1 mV.

If the number of full digits is *n*, then resolution

$$R = \frac{1}{10^n}$$

For $n = 4$,
$$R = \frac{1}{10^4} = 0.0001 \quad \text{or} \quad 0.01\%$$

Sensitivity. It is the smallest change in input which a digital meter is able to detect. Thus, it is the full-scale value of the lowest voltage range multiplied by the resolution of the meter. Sensitivity is given as

$$S = (fs)_{min} \times R$$

where $(fs)_{min}$ is the lowest full-scale value of digital meter, and R is the resolution in decimal.

Accuracy. The accuracy specifications of digital meters normally include two quantities : (*i*) a percentage of range, and (*ii*) a percentage of reading.

As a result, the ambiguity of a measurement is highest in any range at the lowest input that cannot be measured in a lower range. For example, a $4\frac{1}{2}$-digit DVM having an accuracy specification of ± 0.05% of reading ± 0.01% range may indicate anywhere from 2.018 to 2.022 V in the 10 V range for a standard input of 2.020 V. (The total error, here, is equivalent to about ± 0.1% of the applied input).

Depending on the resolution possible, a digital meter may be referred to as 3-digit, $3\frac{1}{2}$-digit,

4-digit, etc. A three-digit digital meter has a resolution of $\frac{1}{1,000}$th of the range and has a maximum display of 999. In contrast, the $3\frac{1}{2}$ digit digital meter has a resolution of $\frac{1}{2,000}$th of the range, and the maximum value displayed is 1999. Similarly, the resolution of a four-digit digital meter is $\frac{1}{10,000}$th of its range, and that of the $4\frac{1}{2}$ – digit is $\frac{1}{20,000}$ of the range, with the display range extending to 19999 for the latter.

Normally, a fourth/fifth digit capable of indicating 0 or 1 (hence called a half digit) is placed to the left. $3\frac{1}{2}$-digit display is shown in Fig. 5.25.

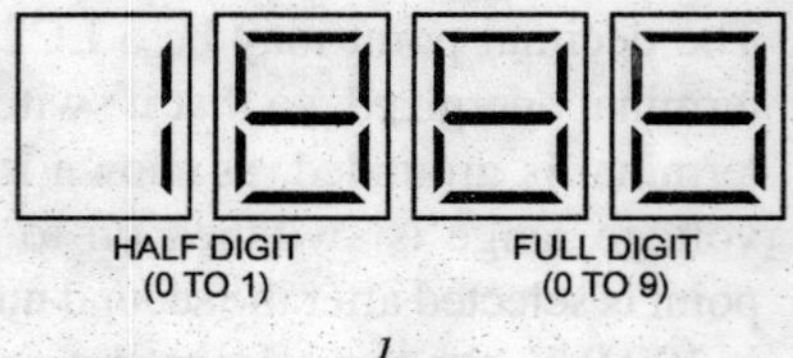

Fig. 5.25 *$3\frac{1}{2}$-Digit Display*

Example 5.8. What is the resolution of a $3\frac{1}{2}$-digit display on 1 V and 10 V ranges ?

A $3\frac{1}{2}$-digital voltmeter has an accuracy specification of ± 0.5% of reading ± 1 digit. What is the possible error in volt, when the instrument is reading 2.00 V on 5 V scale ?

Solution : Number of full digits on display, $n = 3$

$$\text{So resolution, R} = \frac{1}{10^n} = \frac{1}{10^3} = 0.001$$

i.e. the meter cannot distinguish between values that differs from each other by less than 0.001 of full-scale.

For full-scale range of 1 V, the resolution is (fs) × R *i.e.* 1 × 0.001= 0.001 V **Ans.**

For full-scale range of 10 V, the resolution is 10 × 0.001 = 0.01 V **Ans.**

The display for 2 V reading on 10 V scale of $3\frac{1}{2}$ digital meter would be 02.00

The digit in the least significant digit has a value of fs × R *i.e.* $5 \times \frac{1}{10^3} = 0.005$ V

$$\text{Total possible error} = \frac{0.5}{100} \times \text{reading} + \text{value of digit in LSD} = \frac{0.5}{100} \times 2 + 0.005$$

$$= 0.01 + 0.005 = 0.015 \text{ V} \quad \textbf{Ans.}$$

Example 5.9. A 4½ Digit Voltmeter is used to measure voltage. Find:

1. Resolution

2. How would 16.58 would be displayed on a 10 V range?

3. How would 0.7254 be displayed on 1 V and 10 V range?

[U.P. Technical Univ. Electronic Measurements and Instrumentation 2006-07]

Solution: Resolution, $R = \frac{1}{10^4} = 0.0001$

There are 5 digit places in 4½ display, so 16.58 would be displayed as 16.580 V on a 10 V range **Ans.**

Resolution on 1 V range is 0.0001 = 0.0001 V

Any reading up to 4th decimal can be displayed.

Hence 0.7254 will be displaced as 0.7254 **Ans.**

Resolution on 10 V range is 0.00001 × 10 = 0.001 V

Hence decimals upto third place can be displaced.

Hence 0.7254 will be displaced as 0.725 instead of 0.7254 **Ans.**

Example 5.10. A $3\frac{1}{2}$ digital voltmeter is used for measuring voltage

(*i*) Find resolution of the instrument.

(*ii*) How would a voltage of 14.53 V be displayed on 10 V scale ?

(*iii*) How would a reading of 14.53 V be displayed on 100 V scale ?

[M.D. Univ. Electronic Measurements and Instrumentation December 2006]

Solution : Number of full digits on display, $n = 3$

$$\text{So resolution, } R = \frac{1}{10^3} = 0.001$$

i.e. the meter cannot distinguish between values that differs from each other by less than 0.001 of full scale.

For full-scale range of 10 V, the resolution is $10 \times 0.001 = 0.01$ V **Ans.**

For full-scale range of 100 V, the resolution is $100 \times 0.001 = 0.1$ V **Ans.**

The display of 14.53 V reading on 10 V scale would be 14.530 **Ans.**

The display of 14.53 V reading on 100 V scale would be 0145.3 **Ans.**

Example 5.11. Explain the operational principle of a successive approximation type DVM supported by necessary diagrams. If the DVM has the maximum range of 255 volts, determine the time it will take to read the unknown voltage V_x = 180 V, using a clock oscillator of 10 kHz.

Solution : Maximum range of DVM, $V_{max} = 255$ V

Unknown voltage, $V_x = 180$ V

Frequency of clock oscillator, $f = 10 \text{ kHz} = 1 \times 10^4$ Hz

$$\text{Time taken to read the unknown voltage} = \frac{V_{max} - V_x}{2\pi f V_{max}} = \frac{255 - 180}{2\pi \times 10^4 \times 255}$$

$$= 4.68 \times 10^{-6} \text{ second} = 4.68 \text{ µs} \quad \textbf{Ans.}$$

5.10 DIGITAL MULTIMETERS

Digital multimeter (DMM) is basically a digital voltmeter and may be used for the measurement of voltage, current (dc or ac) and resistance. All quantities other than dc voltage are first converted into an equivalent dc voltage by some device.

The block diagram of a basic digital multimeter is given in Fig. 5.26.

For measurement of ac voltage the input voltage is converted into a dc voltage by means of a rectifier. A compensated attenuator is employed. Many manufacturers provide the same attenuator for both dc and ac measurements.

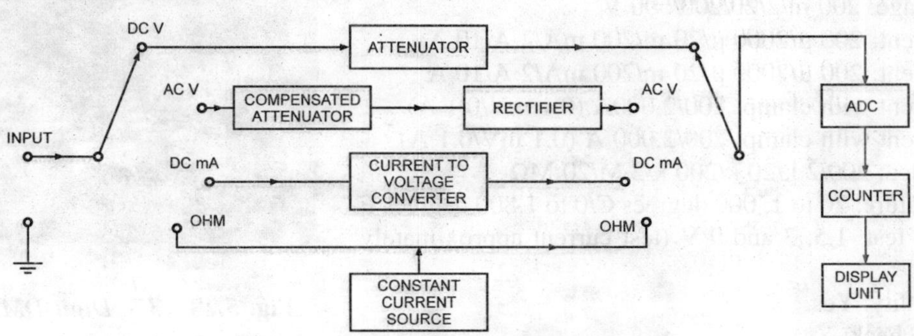

Fig. 5.26 *Block Diagram of Digital Multimeter*

For measurement of resistance a constant current, depending on the range, supplied from a battery or a constant current source is passed through the resistance under measurement and the voltage developed across it is measured. The resistance value is displayed in ohms.

For measurement of current, the unknown current is passed through a precision resistor in many commercial digital multimeters and the voltage developed across the precision resistor is measured. The current value is displayed in mA.

For measurement of current, a current-to-voltage converter may also be used, as illustrated by block diagram in Fig. 5.27.

The current under measurement is applied to the summing junction at the input of the op-amp. The current in the feedback resistor I_R is equal to the input current I_{IN} because of very high input impedance of the op-amp. The current I_R causes a voltage drop across one of the resistors, which is proportional to the input current I_{IN}. Different resistors are employed for different ranges.

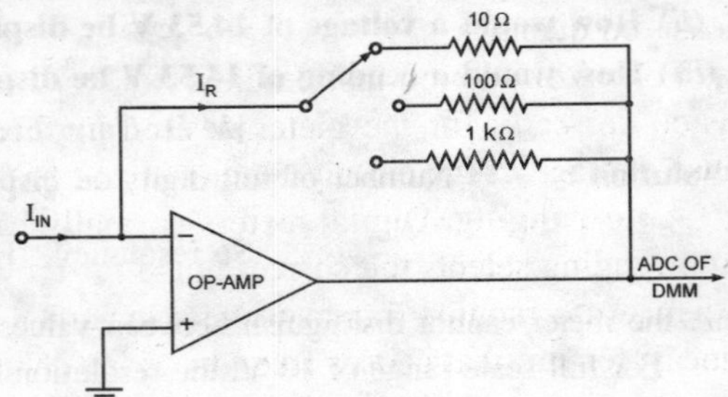

Fig. 5.27 *Block Diagram of Current-To-Voltage Converter*

The performance of a digital multimeter largely depends on the A/D converter. Low-cost hand held multimeters use a single chip $3\frac{1}{2}$ digit ADC whereas high precision $5\frac{1}{2}$ and $6\frac{1}{2}$ digit multimeters make use of microprocessor controlled discrete ADCs. Digital multimeters give very high input impedance, better accuracy and resolution. Such multimeters have autoranging, autopolarity and auto zero facilities.

The hand-held (or pocket type), $3\frac{1}{2}$ digit digital multimeter DMM with auto-range and data hold function is shown in Fig. 5.28. The main features and technical specifications for the DMM shown in Fig. 5.28 are as follows:

$3\frac{1}{2}$ digits LCD maximum reading 1999 and backlight
12 functions for DCV, ACV, DCA, ACA and OHM
Diode check
Auto power off
Display: 1999 counts and digit is 16 mm high
DC voltage: 200 m/2/20/200/600 V
AC voltage: 200 m/2/20/200/600 V
DC current: 200 μ/2000 μ/20 m/200 mA/2 A/10 A
AC current: 200 μ/2000 μ/20 m/200 mA/2 A/10 A
DC current with clamp: 200/2,000A (0.1 mV/0.1 A)
AC current with clamp: 200/2,000 A (0.1 mV/0.1 A)
Resistance: 200/2 k/20 k/200 k/2 M/20 MΩ
Temperature: 20 to 1, 000 degrees C/0 to 1,800 degrees F
Battery test: 1.5, 3 and 9 V (test current approximately 60 mA)
Continuity: Yes
Diode check: Yes

Fig. 5.28 $3\frac{1}{2}$ *Digit DMM With Autorange and Data-hold Function*

The central knob (function switch) has many positions and can be switched to any position which is required for the measurement to be carried out. For example, if the meter is switched to 200 V DC, then 200 V is the maximum voltage that can be measured. This is sometimes known as 200 V FSD (full-scale deflection).

A minus sign also appears in front of the numerical display, when required, to indicate a negative measured quantity.

For multifunction instruments, it is very important to use correct terminals and to set the function switch correctly for the meter desired application. Incorrect connection and function selection may cause damage to the circuit under test and to the instrument.

Auto-ranging. Digital meters normally have auto-ranging, so that range selection is not required. Auto-ranging selects the correct range for the quantity under test so that the most significant digits are shown. Auto-ranging meters usually include a facility to freeze the meter to a particular range, because a measurement that causes frequent range changes is distracting to the user.

Auto-polarity for direct current readings, indicates if the applied voltage is positive (agrees with meter lead labels) or negative (opposite polarity to meter leads).

Sample and hold latches the most recent reading for examination after the meter is removed from the circuit under test.

Burden voltage. On both dc and ac current ranges, there is an instrument terminal voltage drop. This voltage drop is known as the *burden voltage* and is specified in volts per ampere. This value can change depending on the range the meter selects, because different ranges usually use different shunt resistors. Since this voltage drop is inserted into the circuit under test, its effect on the circuit must be taken into account. For instance, a burden voltage of 5 mV/mA represents an ammeter resistance of 5 Ω and for such a meter, a current of 250 mA would result in a voltage drop of 250 mA × 5 mV/mA = 1,250 mV = 1.25 V. Most DMMs have the current measuring circuits provided with fuses for the protection against flow of excessive currents.

Accuracy. Digital multimeters generally takes measurements with accuracy superior to their analog counterparts. Standard portable digital multimeters claim to be capable of carrying out measurements with accuracy of 0.5% on the dc voltage ranges. A multimeters quoted accuracy is specified as being that of the lower (mV) dc range, and is known as the "basic dc volts accuracy" figure. However dc voltage ranges, current, resistance, ac and other ranges will usually have a lower accuracy than the basic dc volts figure. The accuracy of digital instruments, as already stated in Art. 5.9, is usually stated as percentage of range + percentage of reading.

Sensitivity and Input Impedance. The current load, or how much current is drawn from the circuit under test may affect the multimeter accuracy. A small current-draw will usually result in more precise measurements. With improper usage or too much current load, a multimeter may get damaged therefore making its measurements unreliable and substandard.

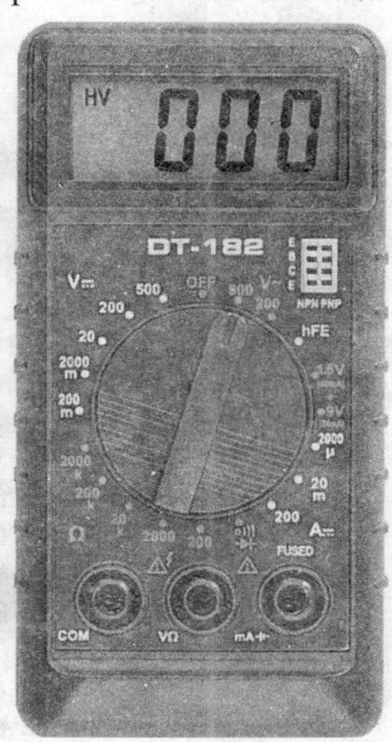

Fig. 5.29 *$3\frac{1}{2}$ Digit Compact, Pocket Sized DMM*

Another $3\frac{1}{2}$ digit multimeter is shown in Fig. 5.29. It is ideal for use in the field, office and home. This meter is extremely convenient for the on the go professionals that only requires to do basic measurements. The special features and technical specifications, this meter has, are as follows:

Features: Audible continuity test; Transistor test; Auto zero adjustment; Overrange indication; Low battery indication.

Technical Specifications

Maximum display : 1999

Power supply : 12 V Battery (included)

DC voltage : 200 mV to 500 V with accuracy ± (0.5% + 2 digit)

AC voltage : 1 V to 500 V with accuracy ± (1.2% + 10 digit)

DC current : 2 mA to 200 mA with accuracy ± (1% + 2 digit)

Resistance : 200 Ω to 2000 kΩ with accuracy ± (0.8% + 2 digit)

Diode test : Yes

Transistor test : Yes

Continuity test : Audible

Battery capacity check : 1.5 V (AA) and 9 V

$4\frac{1}{2}$ Digit True RMS Digital multimeter, which is a compact; high performance, easy to operate, desktop DMM, is shown in Fig. 5.30. It has 5 ranges for voltage measurement in dc/ac mode, 3 ranges for current measurement in dc/ac mode, 6 ranges for resistance measurement, 2 ranges up to 200 kHz for frequency measurement. It carries true rms ac measurements for ac voltages and currents. It makes use of dual slope integration ADC system for measurement with sampling rate of approximately 3 readings per second.

Fig. 5.30 $4\frac{1}{2}$ *Digit True RMS Digital Multimeter*

All voltage ranges have a high input impedance so as avoid circuit loading during measurements. Polarity of dc voltage and current is automatically displayed and the decimal point is placed at the appropriate position when a range is selected. Overrange is indicated by MSD "1" displayed. Protection against improper input is provided on all functions and all ranges and the ac power line. It has 19999 counts $4\frac{1}{2}$ digit 40 mm high characters LCD display. The technical specifications are as follows:

DC Voltage

Range	*Accuracy*	*Resolution*
200 mV	± (0.05% + 3*d*)	10 μV
2 V		100 μV
20 V		1 mV
200 V		10 mV
1000 V	± (0.1 % + 5*d*)	100 mV

Input Impedance : 10 MΩ, all ranges.
Overload Protection : 250 V DC/AC PEAK on 200 mV range; 1000 V DC or AC PEAK on other ranges.

AC Voltage (True RMS)

Range	*Input Frequency*	*Accuracy (Input > 10% of full scale)*	*Resolution*
200 mV	50 Hz – 50 kHz		10 μV
2 V	50 Hz – 20 kHz	± (0.8% + 80*d*)	100 μV
20 V			1 mV
200 V	50 Hz – 5 kHz		10 mV
750 V	50 Hz – 400 Hz	± (1% + 20*d*)	100 mV

Input value more than 10% of full scale range.
Input Impedance : 2 MΩ on all ranges.
Overload Protection : 1000 V DC/AC PEAK, except AC 200 mV range (250 V DC/AC PEAK).

DC Current

Range	*Accuracy*	*Resolution*
20 mA		1 μA
200 mA	± (0.3% + 10*d*)	10 μA
2 A		100 μA
20 A	± (0.8% + 10*d*)	1 mA

Overload Protection : 0.2 A/250 V, 20 A/250 V fast-fuse.
Max. Input Current : 20 A (up to 15 seconds).
Max. Measuring Voltage : 200 mV.

AC Current (True RMS)

Range	*Input Frequency*	*Accuracy (Input > 10% of full scale)*	*Resolution*
200 mA	50 Hz – 5 kHz	± (0.8% + 80*d*)	10 μA
2 A			100 μA
20 A	50 Hz – 400 Hz	± (1.0% + 20*d*)	1 mA

Overload Protection : 0.2 A/250 V, 20 A/250 V fast-fuse.
Max. Input Current : 20 A (up to 15 seconds).
Max. Measuring Voltage : 200 mV

Resistance

Range	*Accuracy*	*Resolution*
200 Ω		0.01 Ω
2 kΩ		0.1 Ω
20 kΩ	± (0.1% + 2*d*)	1 Ω
200 kΩ		10 Ω
2 MΩ		100 Ω
20 MΩ	± (0.2% + 5*d*)	1 kΩ

Overload Protection : 250 V DC/AC PEAK on all ranges.
Open Circuit Voltage : Less than 3 V.

Capacitance

Range	*Accuracy*	*Resolution*
20 nF		0.001 nF
2 μF	± (1.0% + 20*d*)	0.1 nF
200 μF		0.01 μF

Diode Test and Continuity Test

Range	*Description*	*Test Condition*
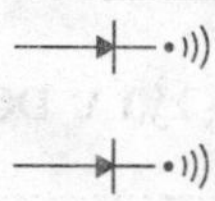	Display read approx. forward voltage of diode Buzzer sounds if resistance between terminals V/Ω and COM is less than about 70 Ω.	Forward DC Current approx. 1 mA. Reversed DC voltage approx. 3 V. Open circuit voltage 3 V.

Overload Protection : 250 V DC/AC PEAK.

Frequency

Range	*Accuracy*	*Resolution*
20 kHz	± (1.5% + 20*d*)	1 Hz
200 kHz		10 Hz

Input sensitivity : 100 mV RMS.
Overload Protection : 250 V DC/AC PEAK (max. 15 seconds).

General

Power Supply : 220-240 V/110-120 V AC (selectable) ± 10%, 50 Hz or 60 Hz.
Temperature for Guaranteed Accuracy : 23ºC ± 5ºC
Temperature Range : Operating 0ºC to 40ºC (32ºF to 104ºF), humidity < 75% RH.
Storage – 10ºC to 50ºC (14ºF to 122ºF), humidity < 75% RH.

5.11 FREQUENCY METER (COUNTER)

A frequency counter is a digital instrument that can measure and display the frequency of any periodic waveform. It operates on the principle of gating the unknown input signal into the counter for a predetermined time. For example if the unknown input signal were gated into the counter for exact 1 second, the number of counts allowed into the counter would be precisely the frequency of the input signal. The term gated stems from the fact that an AND or an OR gate is employed for allowing the unknown input signal into the counter to be accumulated.

One of the most straight forward methods of constructing a frequency counter is shown in Fig. 5.31 in simplified form. It consists of a counter with its associated decoder/display circuitry, clock oscillator, a divider and an AND gate. The counter is usually made up of cascaded BCD counters and the decoder/display unit converts the BCD outputs into a decimal display for easy monitoring. A GATE ENABLE signal of known time period t is generated with a clock oscillator and a divider circuit and is applied to one leg of an AND gate. The unknown signal is applied to the other leg of the AND gate and acts as the clock for the counter. The counter advances one count for each transition of the unknown signal and at the end of the known time interval, the contents of the counter will be equal to the number of periods of the unknown input signal that have occurred during time interval t. In other words, the counter contents will be proportional to the frequency of the unknown input signal. For instance if the gate signal is of a time of exactly 1 s and the unknown input signal is a 600 Hz square wave, at the end of 1 second the counter will count up to 600, which is exactly the frequency of the unknown input signal.

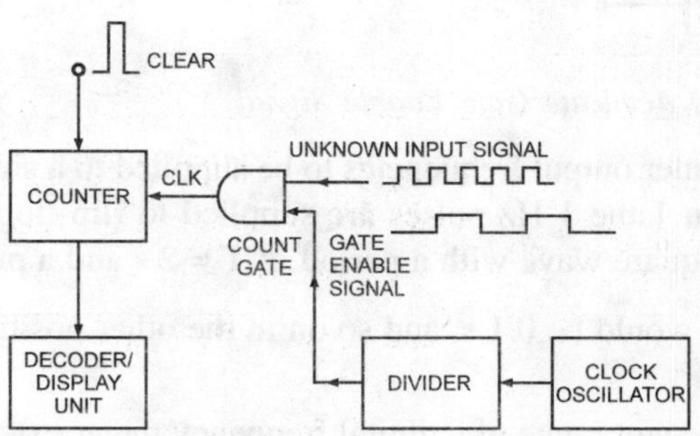

Fig. 5.31 *Basic Frequency Counter*

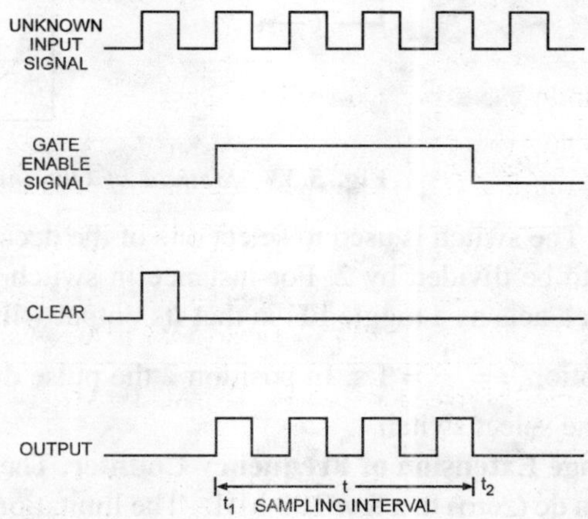

Fig. 5.32

The waveforms in Fig. 5.32 shows that a CLEAR pulse is applied to the counter at t_0 to start the counter at zero. Prior to t_1 the GATE ENABLE signal is LOW, and so the output of the AND gate will be LOW and the counter will not be counting. The GATE ENABLE signal goes HIGH from t_1 to t_2 and during this time interval t ($t = t_2 - t_1$) the unknown input signal pulses will pass through the AND gate and will be counted by the counter. After t_2 the AND gate output will be again LOW and the counter will stop counting. Thus, the counter will have counted the number of pulses that occurred during the time interval, t of the GATE ENABLE signal, and the resulting contents of the counter are a direct measure of the frequency of the input signal.

The accuracy of the measurement depends almost entirely on the time interval of the GATE ENABLE signal, which needs to be controlled very accurately. A commonly used method for obtaining very accurate GATE ENABLE signal is shown in Fig. 5.33. A crystal controlled oscillator is employed for generating a very accurate 100 kHz waveform, which is shaped into the square pulses and fed to a series of decade counters that are being used to successively divide this 100 kHz frequency by 10. The frequencies at the outputs of each decade counter are as accurate (percentage wise) as the crystal frequency.

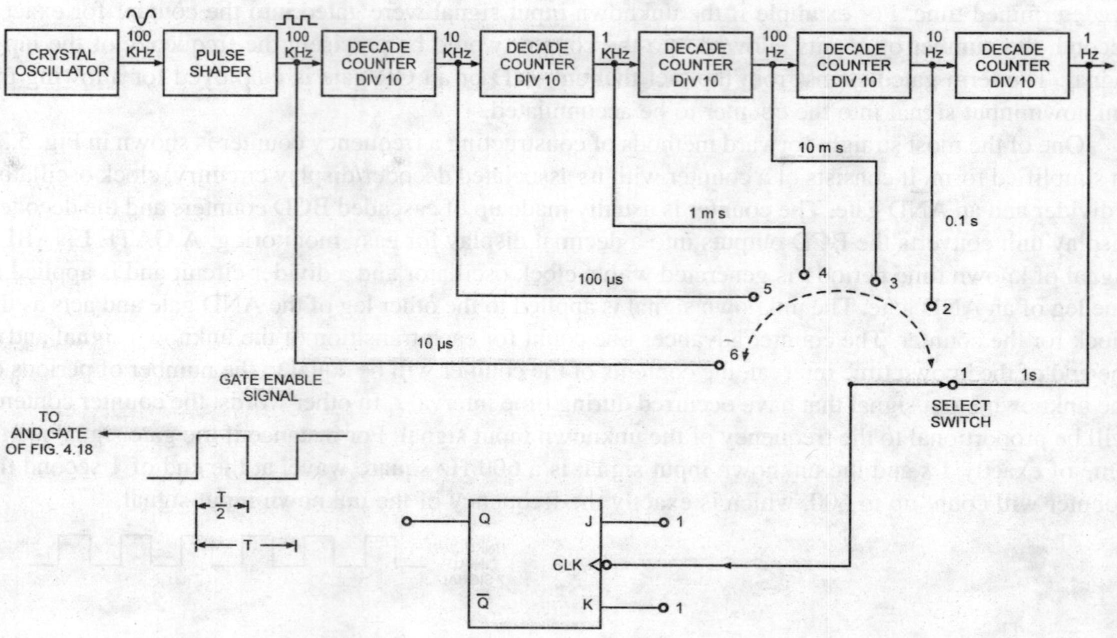

Fig. 5.33 *Method of Obtaining Very Accurate Gate Enable Signal*

The switch is used to select one of the decade-counter output frequencies to be supplied to a single FF to be divided by 2. For instance in switch position 1 the 1 Hz pulses are supplied to flip-flop Q, which acts as a toggle FF so that its output will be a square wave with a period of T = 2 s and a pulse duration, $t = \frac{T}{2} = 1$ s. In position 2 the pulse duration would be 0.1 s, and so on in the other positions of the select switch.

Range Extension of Frequency Counter. The direct count range of a digital frequency meter extends from dc (zero) to a few 100 MHz. The limitations arise because of the counters employed along with the digital frequency meter. The counter is unable to count at the speed required in high frequency measurement.

The range of a few MHz covers only a small portion of the frequency spectrum. Techniques other than direct counting are, therefore, used for extension of range of frequency counter to that exceeding 40 GHz. The input frequency is reduced before being applied to digital frequency counter. For accomplishing it some special techniques are used. Some of these techniques are enumerated below.

1. Prescaling. One technique is to use a prescaler, as illustrated in Fig. 5.34. A prescaler is a fast digital counter that, typically, divides the input frequency by 10. The prescaler does not drive a display, is not gated, nor are the output data strobed into the storage latch. So, the propagation delay of the prescaler is not significant as long as the prescaler can operate at the desired frequency. By using divide-by-10 prescaler ahead of a 100 MHz counter, the counter frequency range

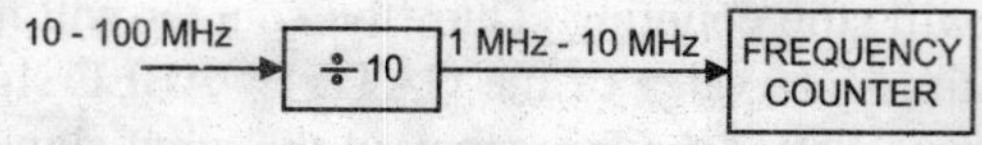

Fig. 5.34 *Use of Prescaler For Range Extension of Frequency Counter*

would be increased by a factor 10 and the system would be capable of counting to 1,000 MHz (or 1 GHz). The main drawback of using prescaler is that the resolution of the frequency counter is reduced by the same factor as the prescaler. For example if a counter had a resolution of 1 Hz, which is the value of the last digit, when multiplied by 10 the resolution would be reduced to 10 Hz. This drawback, however, can be overcome by using a longer time base and restoring the resolution. This can become a practical problem if the prescaler has a large division and very accurate frequency measurements are to be carried out.

2. Heterodyne Converter. For making frequency counter measurements at higher frequencies, heterodyning techniques are used. By using heterodyne converter shown in Fig. 5.35, the frequency of input high frequency signal is reduced so that it lies within the frequency range of the frequency counter. This converter is employed with a 50 MHz counter, which requires that the converter reduce the input frequency to 50 MHz or less, which it does with mixing frequencies every 100 MHz. Because both the sum and the differences are used, the converter frequency never exceeds 50 MHz. A 100 MHz source, which is derived from the frequency counter's time base, supplies a harmonics generator using a step recovery diode. The step recovery diode has a unique reverse recovery characteristic in that it stops conducting very abruptly and thus generating harmonics of the driving waveform to several GHz. The harmonic content of the diode generator extends well into the 5 GHz region. Harmonics from the fundamental at 100 MHz to 5 GHz are selected by a tuned cavity that tunes one of the harmonics. It is necessary to know which of the 50 harmonics is being tuned, and a calibrated dial is provided as a tuning meter to peak the desired signal. The setting of the harmonic tuner dial does not affect the accuracy of the measurement unless the incorrect harmonic is tuned. The 50 harmonics represent a 2 percent resolution, which can be easily obtained with a mechanical assembly. The selected harmonic is mixed with the input and the difference is filtered, amplified, and supplied to the counter. For selection of correct harmonic, the input frequency must be known to within 10 MHz or so. This can be accomplished with another measurement technique such as a wavemeter or spectrum analyzer. Since either the sum or difference between the selected harmonic and the input signal may be counted, the operator is required to carry out necessary calculations for determining the actual frequency. This involves adding or subtracting, depending on whether the sum or difference is counted, the harmonic frequency that is read from the harmonic tuner dial.

Modern frequency counters are capable of tuning the harmonic and making the necessary calculations automatically.

Effects on Accuracy of Both Prescaler and Heterodyning Techniques Used in Range Extension of Frequency Counter.

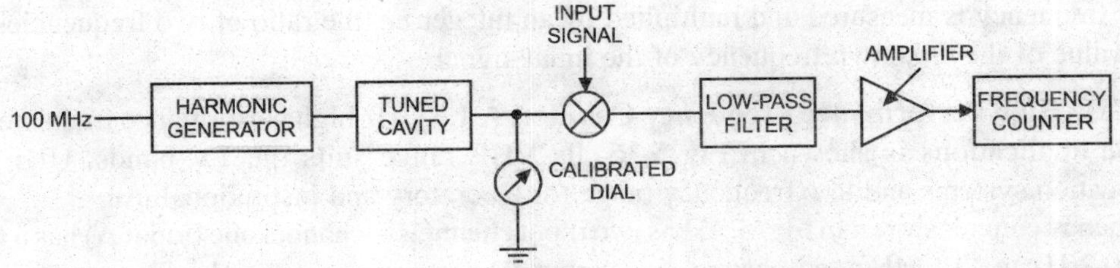

Fig. 5.35 *Manually Tuned Heterodyning Frequency Converters Used For Range Extension of Frequency Counter*

In the case of prescaler, assuming that the prescaler does not miss counts, and this is generally true, the output frequency is simply the input frequency divided by N, the prescaler ratio. The displayed frequency is the input frequency to the counter times the gate time, which is given as

$$\text{Displayed frequency} = \frac{\text{Input frequency, } f_{in}}{N} \times t \qquad \text{...(5.6)}$$

The accuracy of the display is thus simply a function of *t*, the gate time as N is a constant. Thus, the accuracy of the counter with a prescaler is exactly the same as the accuracy of the counter without a prescaler.

In case of a heterodyne converter in which the mixing signal is derived from the same clock as is used to drive the time base within the counter. The gate time is an integer number of cycles of the time-base clock, or

$$\text{Gate time} = \frac{Q}{f_c} \qquad \text{...(5.7)}$$

where Q is the division of time base and f_c is the time-base clock frequency. The mixing frequency in the converter is derived from the same source, and the output frequency of the converter is given as

$$f_{in} = f'_{in} \pm Nf_c \qquad \text{...(5.8)}$$

where f_{in} is the frequency input to the counter, f'_{in} is the frequency input to the converter and N is the multiplication between the internal time-base clock and the heterodyning signal.

The displayed frequency of the counter is the input frequency of the counter times the gate time, which is

$$\text{Displayed frequency} = f_{in}\frac{Q}{f_c} = \frac{f'_{in}Q}{f_c} + QN \qquad \text{...(5.9)}$$

The relationship between the actual input frequency and the displayed frequency is a function of f_c only, as the second term of Eq. (5.9) is a constant.

Thus neither the heterodyning nor the prescaling technique used for increasing the frequency range of a counter affects the accuracy.

Automatic divider and transfer oscillator can also be used for increasing the frequency range of a counter. In automatic divider the high frequency is reduced by some factor such as 1000 : 1, 100 : 1 or 10 : 1. This may be accomplished by making use of automatically tuned circuits which generate an output frequency equal to 1/1,000th, 1/100th or 1/10th of the input frequency.

In *transfer oscillator*, a harmonic or tunable low-frequency (LF) continuous wave oscillator is zero beat (mixed to generate zero frequency) with the unknown high frequency signal. Low-frequency oscillator frequency is measured and multiplied by an integer *i.e.* the ratio of two frequencies, so as to give the value of the unknown frequency of the input signal.

1.3 GHz High Performance Frequency Counter. A 1.3 GHz high performance frequency counter with wide applications is shown in Fig 5.36. Its VHE range suits the TV bands, UHF range to communication systems and low frequency range for laboratory and institutional use.

Frequency counter shown in Fig. 5.36 has two input channels—channel one (input A) has a frequency range of 0.2 Hz to 110 MHz and channel two (input B) has a pre-scaler and a frequency range of 50 MHz to 1.3 GHz. High input impedance of channel one provides minimum loading effect when 50 Ω input impedance of channel two matches with most of the source impedances in high frequency range. Selection of four gate times (*i.e.* 10 ms, 100 ms, 1 s and 10 s) provides the best compromise between time and resolution. It makes use of LSI technology and is therefore, very reliable. The accuracy of the instrument is high because of a high stability time base being employed as a reference. The gate times are derived from the source.

Fig. 5.36 *1.3 GHz Frequency Counter*

It has got features of wide frequency range (0.2 Hz to 1.3 GHz); High sensitivity; 8 digit green LED display; Microcontroller based advance technology; Reciprocal counting technique for high resolution at low frequency, Period measurement; AC/DC coupling; Attenuator 10 : 1; Low pass filter for noisy frequency; Slope selection facility on channel one; Sampling time control; High stability oven controlled crystal oscillator time base; Hold facility; Reset facility.

Technical specifications of this frequency counter are given below:

Frequency	:	
Input A	:	0.2 Hz to 110 MHz.
Input B	:	50 MHz to 1300 MHz.
Gate Time	:	10 ms, 100 ms, 1 s, 10 s selectable.
Resolution	:	100n Hz to 10 kHz
Accuracy	:	± 1 count ± time base accuracy.
Sensitivity		
Input A	:	0.2 Hz to 50 MHz – 30 mVrms
		50 MHz to 100 MHz – 50 mVrms
		100 MHz to 110 MHz – 100 mVrms
Input B	:	50 MHz to 1.1 GHz – 30 mVrms
		1.1 GHz to 1.3 GHz – 60 mVrms
Input Impedance	:	
Input A	:	1 M ohm // 25 pF nominal
Input B	:	50 ohms nominal
Max. Input Voltage	:	
Input A	:	DC to 5 kHz : 300 V (DC + AC pk).
		5 kHz to 110 MHz : $5.5\text{ V} + \dfrac{(1.25 \times 10^6 \text{ rms})}{(\text{Freq. in Hz})}$
Input B	:	DC – 50 V
		AC – 3 V rms
Attenuator	:	× 1 and × 10.

Period Measurement Range :

Input A	:	10 ns to 5 s
Annunciators	**:**	LED indication for overflow, Gate time (0.01, 0.1, 1, 10 in seconds), Display Hold, Freq. offset mode, Measurement mode (Freq. A conventional and normal, Freq. B, period A, ns, μs, ms., MHz, kHz and Hz).
Display	**:**	0.56″, 8 digit LED display.
Display Time	**:**	0.8 s to 5 s and Display hold.

Time Base Internal (For 1122X)

Time Base Freq.	:	10 MHz oven controlled.
Stability Over Temp. Range (0 to 50ºC)	:	$< \pm 1 \times 10^{-7}$
Short term stability	:	$< 5 \times 10^{-8}$/s.
Long term stability	:	$< 5 \times 10^{-7}$/day.
Warm-up Time	:	15 minutes.

Time Base Internal (For 1122)

Time Base Freq.	:	10 MHz
Ageing Rate	:	± 5 PPM/year.
Stability	:	± 1 PPM
Temp. Stability	:	± PPM from 0 to 50ºC
Internal Out	:	5 V p-p

Time Base External

Time Base Freq.	:	10 MHz
Sensitivity	:	2 V p-p (min.), 5 V p-p (max.)

General

Interface	:	RS232 (Optional)
Power	:	230 V AC ± 10%, 50 Hz and EXT BATT. operation

Example 5.12. The unknown input signal in Fig. 5.37 is a 2.5 kHz square wave. Determine the display indication if the GATE ENABLE time is (*i*) 0.1 s (*ii*) 1 s and (*iii*) 10 s.

Solution : (*i*) When $t = 0.1$ s, the counter will count up to 2,500 (transitions per second) × 0.1 (second) *i.e.* 250. **Ans.**

(*ii*) When $t = 1$ s, the counter will count or display 2,500 × 1 = 2,500 **Ans.**

(*iii*) When $t = 10$ s, the counter will count or display 2,500 × 10 = 25,000 **Ans.**

The contents of the counter, in above example, are always a number that is proportional to the frequency of the unknown input signal. In this case, the proportionality constant is either 10, 1 or $\frac{1}{10}$. So, it is a simple matter to insert a decimal point between the indications such that the unknown frequency is displayed directly. Fig. 5.37 illustrates how the decimal point moves in a five decimal-digit display as the gate width is changed. In the top display, the

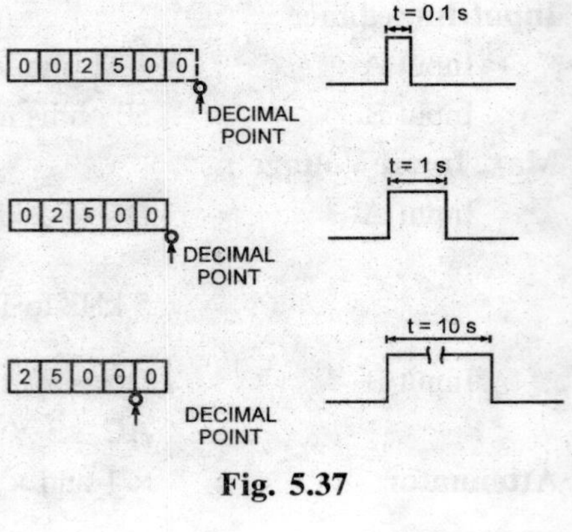

Fig. 5.37

unknown frequency is the display contents multiplied by 10, so the decimal point is moved one place to the right. The middle display gives the actual unknown frequency directly. In the bottom display, the contents must be divided by 10 to give the unknown frequency, so the decimal point is moved one place to the left.]

Example 5.13. A digital frequency counter which has a 3-digit display, a gated period of 10 milliseconds, is selected to measure an unknown frequency. The reading is 045. What is the frequency of the system ?

[U.P. Technical Univ. EMMI Third Semester 2001-02]

Solution : Count, N = 045

Gate enable time, t = 10 ms or 0.01 s

$$\text{Frequency, } f = \frac{N}{t} = \frac{045}{0.01} = 4{,}500 \text{ or } 4.5 \text{ kHz} \quad \textbf{Ans.}$$

Example 5.14. An electronic counter is connected to an unknown frequency. (*i*) If 174 pulses are counted in a gate interval of 100 μs, what is the frequency ? (*ii*) Determine the resolution of measurement.

Solution : $\text{Frequency} = \frac{1}{\text{Time period of one pulse}} = \frac{1}{100\times10^{-6}/174} = \frac{174\times10^{6}}{100} = 1.74 \text{ MHz}$ **Ans.**

Resolution of measurement is the pulses which can be counted in 1 μs. Thus resolution of measurement

$$= \frac{174}{100} = 1.74 \text{ per } \mu\text{s} \quad \textbf{Ans.}$$

Example 5.15. An electronic counter counts 500 cycles of a 2 MHz clock during the period of input signal when operated on its period mode. What is the frequency of the input signal ?

Solution : $\text{Time of one cycle of 2 MHz clock} = \frac{1}{2\times10^{6}} = 0.5 \ \mu\text{s}$

Time of 1 cycle counted by the electronic counter = 500 × 0.5 μs = 250 μs

$$\text{Frequency of input signal} = \frac{1}{\text{Counter time}}$$

$$= \frac{1}{250\times10^{-6}} = \frac{10^{6}}{250} = 4{,}000 \text{ Hz or } 4 \text{ kHz} \quad \textbf{Ans.}$$

EXERCISES

1. What are the advantages of digital instruments over analog instruments.
2. What is DAC ? What is its basic function. Give its transfer characteristics.
3. What are digital-to-analog converters ? Describe, with necessary diagrams, working of a 'binary weighted resistance D/A converter'. **[U.P.S.C. I.E.S. Electrical Engineering-I, 2001]**
4. Draw neat diagram for 4-bit R-2R ladder and weighted resistor types D/A converters and compare them. **[U.P.S.C. I.E.S. Electrical Engineering-I, 2002]**
5. What is DAC ? Give its performance characteristics.
6. What is A/D converter? Give its basic function. Explain its transfer characteristics also.
7. Describe the arrangement of A/D converters based on the following :
   (*i*) Successive approximation method (*ii*) Slope integration method (*iii*) Recirculating remainder system. **[U.P.S.C. I.E.S. Electrical Engineering-I, 1997]**
8. Explain the principle of operation of a dual slope ADC. How is it advantageous over ramp-type ADC ?

9. Differentiate clearly between D to A and A to D conversion. Name different methods of accomplishing each. With necessary diagram, explain working of simultaneous A/D converter. What are its advantages and disadvantages ? **[U.P.S.C. I.E.S. Electrical Engineering-I, 2004]**
10. Explain the principle of operation of a staircase ramp digital voltmeter with the help of block diagram. **[U.P.S.C. I.E.S. Electrical Engineering-I, 1997]**
11. Explain the operational principle of a successive approximation type DVM supported by necessary diagrams.
12. Explain with the help of a block diagram, the working of a potentiometric DVM. **[U.P.S.C. I.E.S. Electrical Engineering-I, 1994]**
13. Explain the following terms used in digital displays :

    (*i*) Resolution (*ii*) Difference between 3½ digit and 4 digit displays (*iii*) Sensitivity (*iv*) Accuracy of digital meters. **[U.P. Technical Univ. Elec. Measurements and Measuring Instruments 2005–06]**
14. What are the different methods of digital measurement of voltage and current ? Explain one method with neat diagram. **[U.P. Technical Univ. Elec. Measurements and Measuring Instruments 2004–05]**
15. Draw the block diagram and its working of an integrating type DVM. **[U.P.Technical Univ. Electronic Measurements and Instrumentation 2006-07]**
16. What is a digital volt meter (DVM) ? What are its advantages ? List different types of DVMs. How can a DVM be used for measurement of (*i*) current and (*ii*) resistance. **[U.P.S.C. I.E.S. Electrical Engineering-I, 1992]**
17. Explain with block diagrams how a DMM can measure dc and ac signals, and various electrical parameters.
18. Explain operating and performance characteristics of DVMs ?
19. Explain with the help of a functional block diagram, the principle of operation of a digital frequency meter. **[U.P.S.C. I.E.S. Electrical Engineering-I, 1993]**
20. Draw and explain the circuit of a digital frequency meter. **[U.P. Technical Univ. Elec. Measurements and Measuring Instruments 2005–06]**
21. Give the operating principle and basic circuit of a digital frequency meter. Set up this meter to measure frequency showing gate control electronics. **[U.P. Technical Univ. Elec. Measurements and Measuring Instruments 2005–06]**
22. Explain the principle of frequency measurements using digital techniques.
23. Draw the circuit of a digital frequency meter. Explain the difference between $3\frac{1}{2}$ digit and 4 digit displays. **[U.P. Technical Univ. Electronic Measurements and Instrumentation 2007-08]**
24. Write short notes on the following:
    (*i*) Advantages of digital instruments over analog instruments.
    (*ii*) DVM.
    (*iii*) DMM.
    (*iv*) Frequency meter.

SHORT ANSWER TYPE QUESTIONS WITH ANSWERS

Q. 1. Why are digital instruments becoming popular ?

Ans. Digital instruments are becoming more and more popular because of their inherent advantages over analog instruments such as greater speed, increased accuracy, better resolution, reduction in operator errors, and ability to provide automatic measurements in system applications.

Q. 2. Why conversion of an analog signal into an equivalent digital signal is essential ?

Ans. Because most sensors have analog output while much data processing is accomplished with digital computers, analog-to-digital conversion is essential.

Q. 3. What is digital-to-analog conversion ?

Ans. Basically, D/A conversion is the process of taking a value represented in digital code (such as simply binary or BCD) and converting it into a voltage or current which is proportional to the digital value.

Q. 4. What is A/D conversion ?

Ans. The A/D conversion is the process of converting an analog input voltage into an equivalent digital signal.

Q. 5. Mention the various types of analog to digital converters in the increasing order of speed of operation. **[U.P.S.C. I.E.S. 2003]**

Ans. A/D converters in the increasing order of speed of operation is given as dual slope type, successive approximation type, parallel-comparator type.

Q. 6. Give the advantages and disadvantages of an integrating type A to D converters under the following heads :

(*i*) speed of operation (*ii*) change in R-C values (*iii*) presence of periodic noise (*iv*) offset voltage of the op-amp.

Ans. (*i*) Speed of operation—Slow.

(*ii*) Change in R-C values—The accuracy of this converter is independent of R-C values or any change in them.

(*iii*) Presence of periodic noise—This converter is very less sensitive to periodic noise.

(*iv*) Offset voltage of op-amp—Accuracy depends upon the offset values, therefore, the reference voltage should be precise.

Q. 7. What are the advantages of integrating type A/D converters over the non-integrating type ? **[U.P.S.C. I.E.S. 1999]**

Ans. The integrating type ADC has the following advantages :

(*i*) It is relatively simple and inexpensive.

(*ii*) Conversion accuracies obtainable are very good.

(*iii*) Capacitors and resistors need not have high precision.

(*iv*) Errors due to propagation delay times are eliminated.

(*v*) Accuracy depends only on the stability of the reference voltage V_{REF}.

(*vi*) The comparator offset voltage and current are automatically compensated as the input signals cross zero voltage twice.

Q. 8. What is resolution for a DAC?

Ans. Resolution of a DAC defines the smallest increment in the voltage that can be discerned. Resolution is primarily a function of the number of bits in the digital input signal i.e. the smallest increment output voltage is determined by the LSB.

Q. 9. What is a digital voltmeter (DVM) ?

Ans. A digital voltmeter displays the value of ac or dc voltages being measured directly as discrete numerals in the decimal number system.

Q. 10. What are the advantages of a digital voltmeter (DVM) ?

Ans. A digital voltmeter displays the value of ac or dc voltage in decimal number system thus eliminating the errors due to parallax and observational errors. The use of DVMs increases the speed with which the readings can be taken. Also the output of DVMs can be fed to memory devices for future purpose which is not available in electronic voltmeter. On account of development of IC technology, it has been possible to reduce the size, cost of DVM. In fact DVM is less costly than analog type.

Q. 11. Why is dual slope ADC preferred in a digital voltmeter ?

Ans. Dual slope ADC is preferred in a DVM because it has low cost and high accuracy.

Q. 12. What is meant by resolution of a digital instrument ?

Ans. Resolution of a digital instrument is defined as the number of digit positions or simply the number of digits used in a meter. For example a three-digit meter display on the digital meter for 0-1 V range will be able to indicate from 000 to 999 mV, with the smallest increment (resolution) of 1 mV.

Q. 13. What is meant by 'sensitivity' of a digital instrument ?

Ans. Sensitivity of a digital meter is the smallest change in input which the meter is able to detect. Thus, it is the full-scale value of the lowest range multiplied by the resolution of the meter.

It is given as $S = (f_s)_{min} \times R$

where $(f_s)_{min}$ is the lowest full-scale value of the digital meter and R is the resolution in decimal.

Q. 14. What is difference between 3-digit and $3\frac{1}{2}$-digit digital meters ?

Ans. A three-digit digital meter has a resolution of $\frac{1}{1,000}$th of the range and has a maximum display of 999. In contrast, the $3\frac{1}{2}$-digit digital meter has resolution of $\frac{1}{2,000}$th of the range, and the maximum value displayed is 1999

Q. 15. What is DMM ?

Ans. DMM (digital multimeter) is basically a digital voltmeter and may be used for the measurement of voltage, current (dc or ac) and resistance.

Q. 16. What is frequency counter ?

Ans. A frequency counter is a digital instrument that can measure and display the frequency of any periodic waveform.

PROBLEMS

1. Determine the largest decimal value that can be represented using 10 bits. **[Ans.** 1,023]
2. A 6-bit DAC produces $V_{OUT} = 0.125$ V for a digital input of 000001. Determine the resolution and V_{OUT} for an input of 111111. **[Ans.** 0.125; 7.875 V]
3. An 8-bit DAC has a step size of 5 mV. Determine the full-scale output voltage and the percentage resolution. **[Ans.** 1.275 V; 0.392%]
4. What voltage resolution is possible using a 10-stage ladder network with a 5 V reference voltage ? **[Ans.** 4.88 mV]
5. In a digital counter time frequency, having a 3-digit display, a gated period of 10 ms is selected to measure an unknown frequency. The reading is 024. What is the likely value of frequency ?
[Pb. Univ. Elec. Measurements II December 1989]
[Ans. 2.4 kHz]

UNIT-III
Resistance Measurements and Inductance and Capacitance Measurements

CONTENTS IN THIS UNIT

Resistance Measurements: Voltmeter and ammeter methods, Wheatstone bridge, Low resistance measurements, Low resistance measuring instruments.

Inductance and Capacitance Measurements: AC bridge theory, Capacitance bridges, Inductance bridges, Q-meter.

CHAPTER 6 : Resistance Measurements.

CHAPTER 7 : Inductance and Capacitance Measurements.

6

Resistance Measurements

6.1 INTRODUCTION

The choice of a suitable method of measuring resistance depends on several different factors. The range of resistance to be measured, that is, whether low, medium, or high, is the most important factor. Other factors may be required accuracy of measurement (greater, say, than 1 percent, or not), working conditions etc. The second point can be rather important when we consider that the resistance of conducting materials increases with the rise in temperature as well as with the absorption of moisture, where these are hygroscopic.

In any particular case the choice of method will be limited to one or two, depending on the equipment available. The different methods of resistance measurement can be divided into two classes, laboratory and workshop methods. This division again depends on the accuracy with which the measurement is to be made. Laboratory measurements are expected to be more accurate with a given method. On the other hand the workshop measurements are less accurate, but are simple and, if possible, direct reading.

From the viewpoint of measurement, resistances are classified as (*i*) low (*ii*) medium and (*iii*) high resistances.

(*i*) *Low Resistances.* Resistances of about 1 Ω and under are included in this class. Measurement of low resistances is required for determination of resistance of armatures, and series field windings of large machines, ammeter shunts, cable lengths, contacts etc.

(*ii*) *Medium Resistances.* Resistances ranging from about 1 Ω to about 100 kilo ohms are included in this class. Most of the electrical apparatus employed are of medium resistance.

(*iii*) *High Resistances.* Resistances of 100 kilo ohms and above are usually termed as high resistances. Measurement of high resistances is required for determination of (*a*) Resistance of high resistance circuit elements; (*b*) Insulation resistance of components and built-up electrical equipment of all types; (*c*) Volume resistivity of a material and (*d*) Surface resistivity.

Although the ohmmeter method is very convenient for measurement of resistance, but it is not very accurate. Resistance can be measured by ammeter-voltmeter method which yields a moderately accurate value over a very wide range of resistances. Precise measurements of resistance in the medium range are best made by using a Wheatstone bridge. For measurement of very low resistance, Kelvin double bridge, a modified Wheatstone bridge, is used.

6.2 VOLTMETER AND AMMETER METHODS

This is most simple and quick method of measurement of resistance. It yields a moderately accurate value over a very wide range of resistances. In this method current through the resistor under test and the potential drop across it are simultaneously measured. The readings are obtained by ammeter and voltmeter respectively. The required range of instruments (voltmeter and ammeter) to be used and the voltage of the supply required, will depend on the size and rating of the resistance under test. A high-value resistor will require high-voltage source, a high-range voltmeter and a low-range ammeter whereas, a low-value resistor will require in most cases a low-voltage, high-current source, a low-range voltmeter and a high-range ammeter. The exact requirements will, of course, depend also on the rating of the resistor, as well as the instruments available.

It will be seen from Fig. 6.1 that there are two ways in which the ammeter and voltmeter can be connected for the measurement, and that, in either case, the resistance of one of the instruments will affect the measurement to some extent. If the voltmeter is connected directly across the resistor, the ammeter measures current flowing through the unknown resistance R_x and voltmeter. In this case,

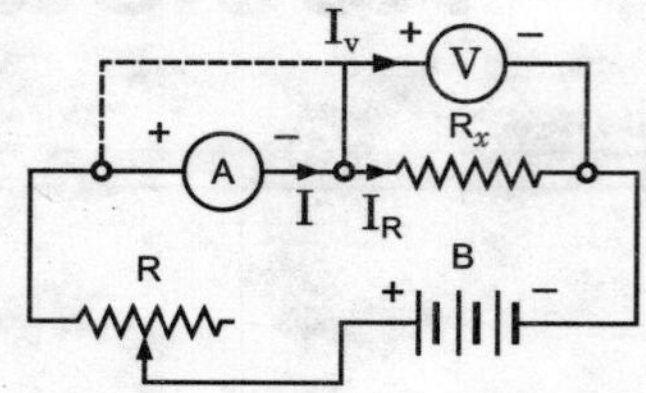

Fig. 6.1 *Ammeter-Voltmeter Method of Measuring Low Resistance*

Current through ammeter = Current through unknown resistance + current through voltmeter.

$$\text{or} \quad I = I_R + I_V$$

$$\text{or} \quad I_R = I - I_V$$

True value of unknown resistance, $$R_x = \frac{V}{I_R} = \frac{V}{I - I_V} = \frac{V}{I\left(1 - \dfrac{V}{IR_V}\right)} \quad ...(6.1)$$

where V is the voltmeter reading, R_V is the resistance of voltmeter and I is the current indicated by the ammeter.

Measured value of resistance, $R_m = \dfrac{V}{I}$

Substituting $\dfrac{V}{I} = R_m$ in Eq. (6.1) we have

$$R_x = R_m\left(\frac{1}{1 - R_m/R_V}\right) \quad ...(6.2)$$

From above Eq. (6.2) it is obvious that the true value of unknown resistance is equal to measured value of unknown resistance provided that voltmeter is of infinite resistance. However, if the voltmeter is of very large resistance as compared to the resistance under measurement *i.e.* $R_V >> R_m$ or $\dfrac{R_m}{R_V}$ is very small.

Now we have $$R_x = R_m\left(1 + \frac{R_m}{R_V}\right) \quad ...(6.3)$$

Thus the measured value of unknown resistance, R_m is lesser than its true value.

$$\text{Relative error, } \varepsilon_r = \frac{R_m - R_x}{R_x} = \frac{-R_m^2}{R_V R_x} = \frac{-R_x}{R_V} \quad \because R_m \simeq R_x \quad ...(6.4)$$

From above Eq. (6.4) it is obvious that the error in measurement would be small if the value of resistance under measurement is very small in comparison to the resistance of voltmeter. Hence this type of connection is used for measuring resistances of low values.

If the ammeter is connected so that it indicates only the current flowing through the unknown resistance, the voltmeter measures voltage drop across the ammeter and unknown resistance R_x. In this case true value of unknown resistance is given by the equation

$$R_x = \frac{V}{I} - R_A = R'_m - R_A = R'_m\left(1 - \frac{R_A}{R'_m}\right) \quad ...(6.5)$$

where R_A is the resistance of ammeter.

Thus the measured value of unknown resistance is larger than its true value.

It is also clear from above Eq. (6.5) that true value of unknown resistance will be equal to its measured value only if the value of ammeter resistance R_A is zero.

$$\text{Relative error, } \varepsilon_r = \frac{R'_m - R_x}{R_x} = \frac{R_A}{R_x} = \frac{R_A}{R_x} \quad ...(6.6)$$

i.e., the measurement error would be smaller provided that the resistance under measurement is large in comparison to the resistance of ammeter. So this arrangement should be used for measuring resistances of high values.

Thus the suitability of a particular method for measuring resistance depends upon the relative values of unknown resistance and the resistances of the instruments.

The division line between two methods is at resistance for which the relative errors given by the two methods are equal.

Now from Eqs. (6.4) and (6.6) $\quad \dfrac{R_x}{R_v} = \dfrac{R_A}{R_x}$

or when true value of unknown resistance, $R_x = \sqrt{R_A R_v}$ ...(6.7)

For resistances having values smaller than the value given by above Eq. (6.7), the voltmeter should be connected directly across the unknown resistance while for resistances of larger values voltmeter should be connected across the series combination of ammeter and unknown resistance, as shown dotted in Fig. 6.1.

The ammeter-voltmeter method of measuring resistance is capable of fair accuracy, depending on care in taking the readings and on the accuracy and ranges of the instruments used for measurement of voltage and current. If the two instruments used are of "1/2 percent accuracy", and their ranges are such that fairly large deflections (almost full-scale) are obtained, the instrumental error in the result may be anything between zero and 1 percent. If the deflections are near half-scale, the percentage error may be twice as great and for smaller deflections may mount correspondingly higher. With less accurate instruments the possible error is of course increased. It is difficult to secure much better than 1 percent accuracy in a resistance value under usual conditions, and the error may be considerable higher. However, this method is useful in some laboratory work in which high accuracy is not required.

Example 6.1. A voltmeter having a sensitivity of 1,000 Ω/V, reads 100 V on its 150 V scale when connected across an unknown resistor in series with a milliammeter. When the milliammeter reads 5 mA, calculate (*i*) the apparent resistance of the unknown resistor, (*ii*) the actual resistance of the unknown resistor, (*iii*) the error due to loading effect of voltmeter.

Solution: Voltmeter reading, V = 100 V

Ammeter reading, I = 5 mA = 0.005 A

Voltmeter resistance, R_V = Sensitivity of voltmeter × voltmeter range

= 1,000 Ω per volt × 150 V = 150,000 Ω = 150 kΩ

(*i*) Apparent value of unknown resistor, $R_m = \frac{V}{I} = \frac{100}{0.005} = 20{,}000\ \Omega$ or $20\ k\Omega$ **Ans.**

(*ii*) Actual value of unknown resistor, $R_x = \frac{V}{I\left(1-\frac{V}{IR_V}\right)}$

$$= \frac{100}{0.005\left(1-\frac{100}{5\times10^{-3}\times150\times10^{3}}\right)} = 23.077\ k\Omega$$

(*iii*) Error due to loading effect of voltmeter, $\varepsilon_r = \frac{R_m - R_x}{R_x}\times100 = \frac{20-23.077}{23.077}\times100 = -13.33\%$ **Ans.**

Example 6.2. A resistance of approximate value of 100 Ω is to be measured by ammeter-voltmeter method. The ammeter and voltmeter available are of ranges 0-1 A and 0–50 V and resistances 2.5 Ω and 6,000 Ω respectively. Determine which method will give more accurate results. Assuming that in measurement of resistance voltmeter is connected directly across the unknown resistance and the readings are found to be 0.4 A and 38.4 V, determine the resulting error. The instruments are of accuracy ± 1/2% of full-scale and the errors are standard deviations.

Solution : Resistance of ammeter, $R_A = 2.5\ \Omega$

Resistance of voltmeter, $R_V = 6{,}000\ \Omega$

Value of unknown resistance for which the errors are equal for the two types of connections

$$R_x = \sqrt{R_A R_V} = \sqrt{2.5\times6{,}000} = 122.5\ \Omega$$

Since the unknown resistance is of value smaller than 122.5 Ω, the voltmeter should be connected directly across the unknown resistor (Fig. 6.2) as it will give more accurate result. **Ans.**

When, for measuring the given unknown resistance, the voltmeter is connected directly across the resistance, as shown in Fig. 6.2.

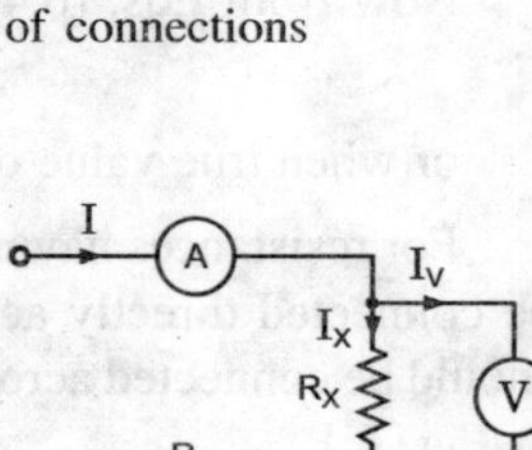

Fig. 6.2

Measured value of unknown resistance, $R_m = \frac{V}{I} = \frac{38.4}{0.4} = 96\ \Omega$

True value of unknown resistance $R_x = \frac{V}{I\left(1-\frac{V}{IR_V}\right)}$

$$= \frac{38.4}{0.4\left(1-\frac{38.4}{0.4\times6{,}000}\right)} = 97.56\ \Omega$$

Error in ammeter reading $= \frac{1}{2}\times\frac{1}{100}\times1 = 0.005$ A

$$\text{Error in voltmeter reading} = \frac{1}{2}\times\frac{50}{100} = 0.25 \text{ V}$$

$$\text{Percentage error at 0.4 A reading} = \frac{0.005}{0.4}\times 100 = 1.25\%$$

$$\text{Percentage error at 38.4 V reading} = \frac{0.25}{38.4}\times 100 = 0.651\%$$

Since the errors correspond to standard deviations,

$$\text{Error due to ammeter and voltmeter} = \sqrt{1.25^2 + 0.651^2} = \pm 1.409\%.$$

$$\text{Absolute error due to ammeter and voltmeter} = \pm \frac{1.409}{100}\times 97.56 = \pm 1.375\ \Omega$$

So the resistance is specified as (97.56 ± 1.375) Ω **Ans.**

Example 6.3. An unknown resistance is measured by ammeter-voltmeter method, the voltmeter being connected across the series combination of unknown resistance and ammeter, as shown in Fig. 6.3. The readings noted are 8 A on an ammeter of 0–10 A and 120 V on a voltmeter of range 0–150 V.

The instruments have uniform scales; ammeter can be read up to 0.01 A and voltmeter can be read up to 0.1 V. The constructional error of the ammeter is ± 0.25% and that of voltmeter ± 0.5%. The ammeter resistance is 0.3 Ω. Determine the value of unknown resistance and the limits of possible error in the results.

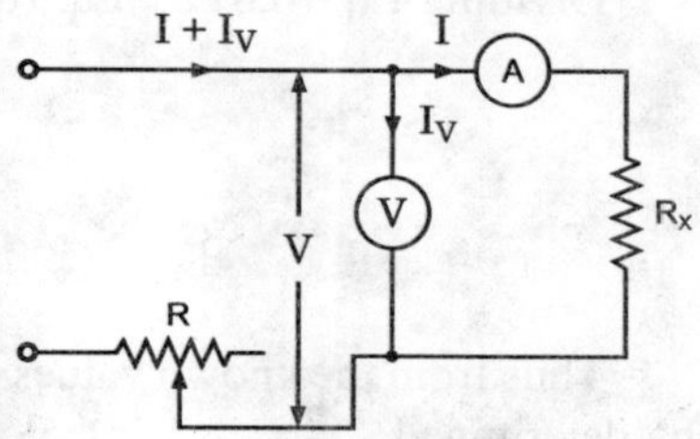

Fig. 6.3

Solution : Measured value of unknown resistance,

$$R_m = \frac{V}{I} = \frac{120}{8} = 15\ \Omega$$

$$\text{True value of unknown resistance, } R_x = R_m - R_A = 15 - 0.3 = 14.7\ \Omega$$

$$\text{Reading error of ammeter} = \pm \frac{0.01}{10}\times 100 = \pm 0.1\%$$

$$\text{Reading error of voltmeter} = \pm \frac{0.1}{150}\times 100 = \pm 0.067\%$$

$$\text{Total error in ammeter reading, } \delta I = \pm 0.1 \pm 0.25 = \pm 0.35\%$$

$$\text{Total error in voltmeter reading, } \delta V = \pm 0.067 \pm 0.5 = \pm 0.567\%$$

$$\text{Since resistance } R = \frac{V}{I},$$

$$\text{Total systematic error in measurement} = \pm \delta I \pm \delta V = \pm 0.35 \pm 0.567 = \pm 0.917\%$$

So value of R_x is specified as 14.7 ± 0.917% = 14.7 ± 0.135 Ω **Ans.**

6.3 WHEATSTONE BRIDGE METHOD

This is one of the most widely used methods for the measurement of resistances of medium values. As shown in Fig. 6.4 it is composed of known adjustable resistances P, Q and S. The unknown resistance R is connected between terminals A and D. The resistances are arranged in two parallel circuits through which the current from the battery can flow. A sensitive galvanometer G, shunted by a variable resistance N to avoid deflection of the galvanometer, when the bridge is out of balance, is bridged through switch S_2 across the two circuits between terminals B and D.

While balancing the bridge, the switch S_1 is closed, the values of resistances P, Q and S are so adjusted that the deflection of galvanometer is nil when switch S_2 is closed.

With the bridge balanced the current flowing through resistance P also flows through resistance Q (say I_1); the current flowing through resistance R also flows through resistance S (say I_2) and the points B and D are at the same potential *i.e.* the voltage drop from A to B must be equal to voltage drop from A to D and voltage drop from B to C must be also equal to voltage drop from D to C.

$$\text{Hence } I_1 P = I_2 R \qquad \ldots(6.8)$$

$$\text{and } I_1 Q = I_2 S \qquad \ldots(6.9)$$

Dividing Eq. (6.8) by Eq. (6.9) we have

$$\frac{P}{Q} = \frac{R}{S}$$

$$\text{or } R = \frac{P}{Q} S \qquad \ldots(6.10)$$

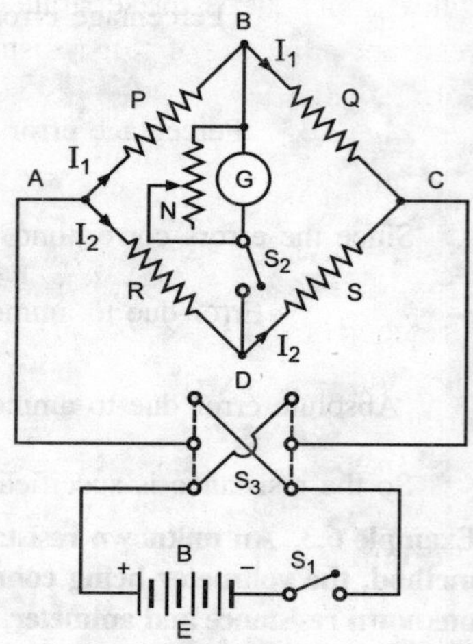

Fig. 6.4 *Wheatstone Bridge*

Thus from the known values of three resistances P, Q and S the value of the unknown resistance can be determined.

It should be noted that this relationship between resistances will also hold good if the battery and galvanometer are interchanged.

Advantages. 1. The method is a 'null one' therefore, errors due to inaccuracies in indicating instrument calibration are completely avoided as the galvanometer is merely used to indicate zero current.

2. The balance is quite independent of the source emf, so the inaccuracies in the measured values due to fluctuation of the emf of the source are completely avoided.

Range. From 1.0 up to 100,000 ohms, with a possible extension to this range at reduced sensitivity.

Accuracy. The accuracy obtainable depends almost solely on the quality of the resistors. The accuracy obtainable with a good commercial bridge is of the order of a few parts in 10,000.

Precautions. 1. The galvanometer should be well shunted during earlier stages of the balancing operation to avoid violent deflection and shunt resistance is increased as the bridge approaches balance position so that shunting is reduced and the galvanometer sensitivity is increased.

2. The battery key should always be closed before the galvanometer key and opened after it; this is to avoid violent deflection of the galvanometer, which may be caused by induced emfs due to rapidly changing currents if the resistor under test is inductive.
3. Measurements of the unknown resistance should also be made by reversing the battery connections after making one measurement of the resistance and mean of the two values thus measured be taken as correct value; this is to avoid the errors due to thermoelectric emfs.

Portable Form of Wheatstone Bridge Test Set-Post Office Box. The portable form of Wheatstone bridge set, known as the Post *Office Box,* contains three adjustable resistances and two switches in one box. In some a galvanometer is also included, with a compartment for housing and a 2-volt accumulator. The two resistances P and Q are termed the *'ratio-arms'* and the resistance S is called the *'adjustable'* arm. The battery, the galvanometer and the unknown resistance R are connected to the box externally. The box resistances are varied by removing the plug keys which short-circuit the separate resistance elements. Dial resistances instead of plug resistances are used in modern forms of the Wheatstone bridge.

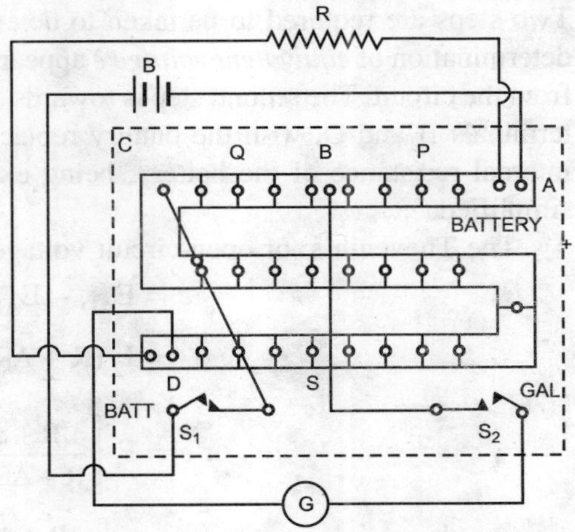

Fig. 6.5 *Post Office Box*

Slide-Wire Form of Wheatstone Bridge. A very simple and moderately accurate form of Wheatstone bridge can be made by using a slide-wire of the ratio-arms. The slide-wire may consist of a uniform wire stretched beside a linear scale (*e.g.* a metre scale divided into mm) or wound on a graduated drum. Such an arrangement is shown in Fig. 6.6.

In use, known standard resistance S, unknown resistance R, battery B, galvanometer G, switch S are connected, as shown in the figure. The slider is moved along the slide-wire until the bridge is balanced. The resistances of the slide-wire sections AB and BC are proportional of their lengths, and therefore, these lengths may be used for determination of value of unknown resistance R.

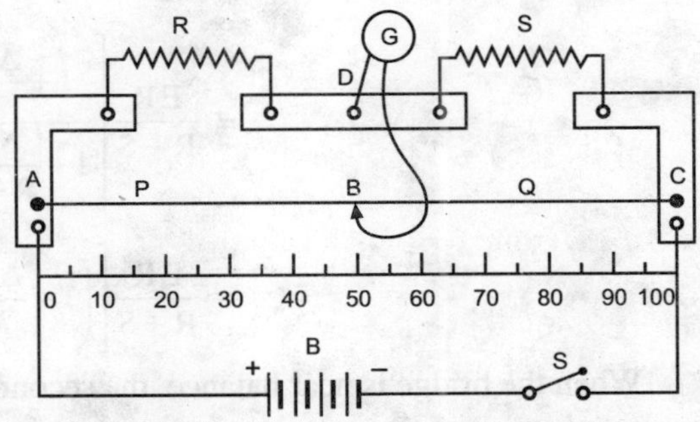

Fig. 6.6 *Slide-Wire Bridge*

Now for zero deflection of galvanometer unknown resistance,

$$R = S \times \frac{\text{Length AB}}{\text{Length BC}} \quad ...(6.11)$$

The accuracy of balance is greatest with the slide near the centre of the slide-wire. (*i.e.* with R nearly equal to S) and may be improved by taking the mean of two values of R, with R and S interchanged between the readings.

Sensitivity of Wheatstone Bridge. It is frequently desirable to know the galvanometer response to be expected in a bridge which is slightly unbalanced, so that a current flows in the galvanometer branch of the network. This may be used either for (*i*) selecting a galvanometer with which a given unbalance may be observed in a specified bridge arrangement (*ii*) determining the minimum unbalance which can be observed in the specified bridge with a given galvanometer or (*iii*) determining the deflection to be expected for a given unbalance.

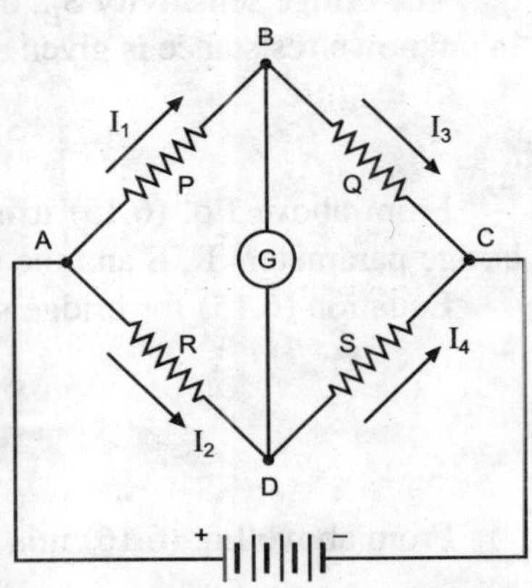

Fig. 6.7 (*a*) *Wheatstone Bridge Configuration*

The sensitivity to unbalance can be calculated by solving the bridge circuit for a small unbalance. The solution is approached by converting the Wheatstone bridge shown in Fig. 6.7 (*a*) to its Thevenin's equivalent. Assuming that the bridge given in Fig. 6.7 (*a*) is balanced so that $\frac{P}{Q} = \frac{R}{S}$ and that the resistance R is changed to R + ΔR so as to create small unbalance.

Since we are interested in knowing the current through galvanometer G, the Thevenin's equivalent circuit is determined by looking back into galvanometer terminals B band D in Fig. 6.7 (*a*).

Two steps are required to be taken to determine the Thevenin's equivalent. The first step is towards determination of *equivalent voltage e* appearing at terminals B and D when the galvanometer is removed from the circuit. The second step is towards determination of *equivalent resistance* looking back into the terminals B and D, with the battery replaced by its internal resistance. In most cases, however, the internal resistance of the battery, being extremely low, can be neglected so as to make calculation simplified.

The Thevenin's, or open-circuit voltage,

$$e = E_{AD} - E_{AB}$$

$$= I_2 (R + \Delta R) - I_1 P = \frac{E}{R + \Delta R + S}(R + \Delta R) - \frac{E}{(P + Q)} P$$

$$\text{or } e = E\left[\frac{R + \Delta R}{R + \Delta R + S} - \frac{P}{P + Q}\right] \quad \text{...(6.12)}$$

$$= E\left[\frac{R + \Delta R}{R + \Delta R + S} - \frac{R}{R + S}\right] \quad \because \frac{P}{Q} = \frac{R}{S} \text{ or } \frac{P}{P + Q} = \frac{R}{R + S}$$

$$= \frac{ER}{R + S}\left[\frac{1 + \frac{\Delta R}{R}}{1 + \frac{\Delta R}{R + S}} - 1\right]$$

$$= \frac{ER}{R + S}\left[\left(1 + \frac{\Delta R}{R}\right)\left(1 - \frac{\Delta R}{R + S} + \frac{(\Delta R)^2}{(R + S)^2} + \ldots\right) - 1\right] \simeq \frac{ES\Delta R}{(R + S)^2} \quad \text{...(6.13)}$$

When the bridge is near balance, the second and higher order terms in the power expansion may be neglected.

If S_v is the voltage sensitivity of the galvanometer, which is defined as the change in scale units per unit change in voltage in the galvanometer circuit when the total resistance, seen from the galvanometer circuit, is that required for the specified damping,

$$\text{Galvanometer deflection } \theta = S_v e = S_v \frac{ES\Delta R}{(R + S)^2} \quad \text{...(6.14)}$$

The bridge sensitivity S_B, defined as the deflection of the galvanometer per unit fractional change in unknown resistance is given as

$$S_B = \frac{\theta}{\Delta R / R} = \frac{S_v ESR}{(R + S)^2} \quad \text{...(6.15)}$$

From above Eq. (6.15) it is obvious that the bridge sensitivity depends upon bridge voltage E, bridge parameters R, S and the voltage sensitivity of the galvanometer S_v.

Equation (6.15) for bridge sensitivity can be rewritten as

$$S_B = \frac{S_v E}{(R + S)^2 / SR} = \frac{S_v E}{\frac{R}{S} + 2 + \frac{S}{R}} = \frac{S_v E}{\frac{P}{Q} + 2 + \frac{Q}{P}} \quad \text{...(6.16)}$$

From above Eq. (6.16) it is obvious that bridge sensitivity S_B will be maximum when $\frac{R}{S} = 1$. As the ratio becomes either smaller or larger, the sensitivity decreases. Since the measurement accuracy depends upon sensitivity a limit can be seen to be usefulness for a given bridge, battery and galvanometer combination. For a bridge with unity arm ratio *i.e.*, R = S = P = Q, Bridge sensitivity,

$$S_B = \frac{S_v E}{4} \qquad \text{...(6.17)}$$

As explained above, the bridge sensitivity is maximum for unity arm ratio. The sensitivity with ratio $\frac{P}{Q} = 100$ would be about $\frac{1}{25}$ of that for unity ratio. Similarly the sensitivity with ratio $\frac{P}{Q} = \frac{1}{100}$ would be about $\frac{1}{25}$ th of that for unit ratio.

Thus the bridge sensitivity decreases considerably when the ratio $\frac{P}{Q} = \frac{R}{S}$ is smaller or greater than unity. This reduction in sensitivity is accompanied by a reduction in accuracy with which a bridge can be balanced.

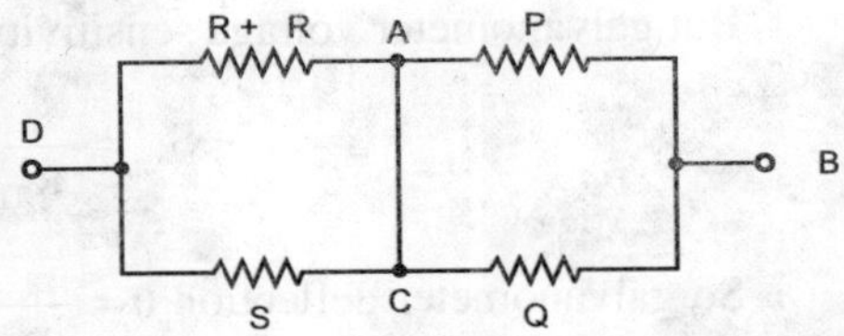

Fig. 6.7 ***(b) Thevenin's Resistance Looking into Terminals D and B***

The resistance of the Thevenin's equivalent circuit is determined by looking back into terminals B and D. Referring to Fig. 6.7 (*b*).

Thevenin's equivalent resistance of bridge,

$$R_{Th} = \frac{RS}{R+S} + \frac{PQ}{P+Q} \qquad \text{...(6.18)}$$

For an equal-ratio bridge *i.e.*

$$P = Q = R = S$$

$$R_{Th} = \frac{R}{2} + \frac{R}{2} = R \qquad \text{...(6.19)}$$

Thevenin's or open-circuit voltage, from Eq. (6.12),

$$E_{Th} = E\left[\frac{R+\Delta R}{R+\Delta R+S} - \frac{P}{P+Q}\right] \qquad \text{...(6.20)}$$

For an equal-ratio bridge, *i.e.*, P = Q = R = S

$$E_{Th} = E\left[\frac{R+\Delta R}{R+\Delta R+R} - \frac{R}{R+R}\right]$$

$$= E\left[\frac{R+\Delta R}{2R+\Delta R} - \frac{1}{2}\right] \simeq \frac{E\Delta R}{4R} \quad \text{as } \Delta R \ll R \qquad \text{...(6.21)}$$

The Thevenin's equivalent of the Wheatstone bridge circuit, therefore, reduces to a Thevenin's generator with an emf E_{Th} given by Eq. (6.20) and internal resistance R_{Th} given by Eq. (6.18). This is shown in Fig. 6.7 (*c*).

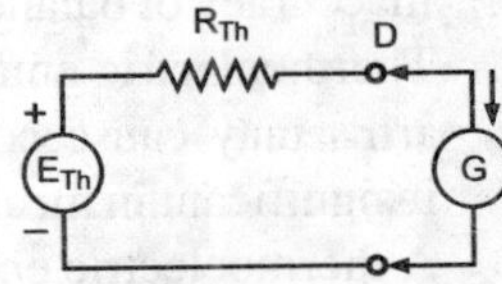

Fig. 6.7 ***(c) Complete Thevenin's Circuit of Wheatstone Bridge***

When the galvanometer is connected to the output terminals of the Thevenin's equivalent circuit, the galvanometer current is found to be

$$I_g = \frac{E_{Th}}{R_{Th} + R_g} \qquad \text{...(6.22)}$$

where I_g is the galvanometer current and R_g is galvanometer resistance.

For an equal-ratio bridge

$$\text{Galvanometer current, } I_g = \frac{E\Delta R/4R}{R + R_g} \qquad ...(6.23)$$

From Eq. (6.14) galvanometer deflection θ is given as

$$\theta = \frac{S_V E S \Delta R}{(R+S)^2} \qquad ...(6.24)$$

But galvanometer voltage sensitivity is given as

$$S_v = \frac{S_i}{R_{Th} + R_g}$$ where S_i is the current sensitivity of the galvanometer

$$\text{So galvanometer deflection } \theta = \frac{S_i E S \Delta R}{(R+S)^2 (R_{Th} + R_g)} \qquad ...(6.25)$$

$$\text{For an equal-ratio bridge, } \theta = \frac{S_i E \Delta R}{4R(R_{Th} + R_g)} \qquad ...(6.26)$$

$$\text{Also bridge sensitivity, } S_B = \frac{\theta}{\Delta R/R} = \frac{S_i E S R}{(R+S)^2 (R_{Th} + R_g)} \qquad ...(6.27)$$

$$\text{For an equal-ratio bridge } S_B = \frac{S_i E}{4(R_{Th} + R_g)} \qquad ...(6.28)$$

Errors in a Wheatstone Bridge. The Wheatstone bridge is a convenient and accurate method for measuring resistance. The main, or limiting source of error is inaccuracy in the three known resistors constituting the bridge. This, and other possible sources of errors are given below.

1. Discrepancies between the true and marked values of resistances of the three known arms. In case P, Q and S are decade resistance boxes guaranteed to be accurate within 0.1 percent of the marked values and if the marked values are all higher (or all lower), then the resistance determined will be 0.1 per cent higher (or lower) than the true value provided that the balance is perfectly made. However, if P and S should each be 0.1 percent high and Q 0.1 percent low, then the result for R will be in error by 0.3 per cent. This is the limiting value of error for this case, as determined by the equipment excluding personal errors, and other errors. The probable error is much smaller.
2. Personal errors in finding the balance point, in taking readings, and in making computations.
3. Inaccuracy of balance point due to insufficient sensitivity of the balance detector (galvanometer).
4. Thermoelectric emfs in the bridge or galvanometer circuit. Thermoelectric emfs in the bridge arms may cause considerable trouble, particularly in low resistance measurements and are insignificant in measurement of high resistances specially when the impressed voltage is higher. A thermoelectric emf in the galvanometer circuit may be significant, so care should be taken to minimize the effect in precision measurements. Some sensitive galvanometers have all-copper systems, *i.e.*, copper suspensions and copper coils, so that there is no junction of dissimilar metals to cause an emf. A galvanometer may be placed in a box, and surrounded by cotton, to avoid temperature inequalities caused by drafts. The effect of thermoelectric emfs can be balanced out by interposing a reversing switch S_3, as shown in Fig. 6.4, between the battery and the bridge, making a bridge balance for each polarity and averaging the results.

5. Changes of resistance due to self heating of one or more arms of the bridge. Great care should be taken to keep I^2R loss in the bridge arms to a safe amount. Moderate heating may simply introduce a small error in a particular measurement, due to a temporary change in resistance. Excessive heating, however, may cause a permanent change in resistance value. The voltage impressed across a bridge should be kept low to avoid the possibility of such an occurrence, otherwise a valuable piece of precision equipment may no longer be suitable for precise work. The current, or wattage, should be computed in advance, particularly when low resistance values are used, to determine whether the proposed set up is safe. Instrument manufacturers specify in their catalogues the safe current or power values for their precision resistors and decade boxes.
6. Errors due to resistance of leads and contacts. This effect is significant in measurement of low resistances.

Limitations of Wheatstone Bridge. There are limitations at both the high-resistance and low-resistance ends of the scales to the values that can be measured accurately by the Wheatstone bridge circuit. The high-resistance limit is set partly by the reduced sensitivity to unbalance. In terms of the Thevenin's circuit, this is due to the high value of R_{Th} in series with E_{Th} and the galvanometer. This limit can be extended to a certain extent by increasing the voltage applied to the bridge, provided that care is taken to avoid overheating of any arm of the bridge. Inaccuracy may also be introduced by leakage over insulation of the bridge arms when measuring a high resistance. The upper limit of the resistance measurement ranges from a few hundred kilo-ohms to several megohms or so. This limit, however, can be extended by special design features.

The lower limit for measurement is set by the resistances of the connecting leads and by contact resistance at the binding posts. The error due to the resistance of the leads may be corrected fairly well, but the contact resistance (at binding post etc.) is a source of uncertainty and is difficult to overcome. That is why the lower limit of resistances that can be measured by Wheatstone bridge has been put in the neighbourhood of 1 to 5 Ω.

Precision Measurement of Medium Resistances With Wheatstone Bridge. The possible sources of errors in the measurement of resistances by the Wheatstone bridge and limitations of Wheatstone bridge have already been discussed. So in precise measurement of medium resistances by Wheatstone bridge special care is to be taken to avoid the errors.

In precision measurement of medium resistances, the most accurate comparisons are made on an equal-ratio bridge with a fixed standard resistance nominally equal to resistance under measurement. The possible errors are reduced to the minimum possible by using equal leads, equal currents and equal heating of the ratio-arms. Now the problem arises of determination of the exact ratio of the unknown resistance R to the standard resistance S or the difference between them. This problem is solved by either making small known changes in the ratio-arms and then obtaining exact balance, balancing the bridge by connecting an adjustable high resistance in parallel with one of the bridge arms.

Built-up Wheatstone bridges have special arrangements of the ratio-arms for convenience and, in precision bridges, for aid in checking. The precision bridges usually have two arms each with values of 1,10,100, 1,000 and 10,000 ohms with plug arrangements so that any resistor may be used in either ratio arm. One of the best method is the Schone scheme illustrated in Fig. 6.8. (one resistor of each value is marked with a prime to distinguish it from the other). Any resistor can be connected to either the arm P or the Q side of the bridge by placing the plug in the proper position.

The above arrangement of the resistors makes it possible, when using equal-ratio-arms, to interchange the resistors. For instance, if we are using the ratio as 10,000/10,000, we can make one balance using the ratio P/Q as 10,000/10,000′ and then a second balance, 10,000′/10,000. If there is a small difference between the 10,000′ and 10,000 resistors, the two balance values are not exactly the same. One value is a little lower than it should be, but the balance with the inverted ratio is corresponding larger. The average of the two values eliminates the effect of errors in the ratio-arms for any reasonable discrepancy in their resistances. This is significant in precision measurement as it eliminates one of the potential sources of error. The only remaining calibration error in the bridge is in the variable arm S.

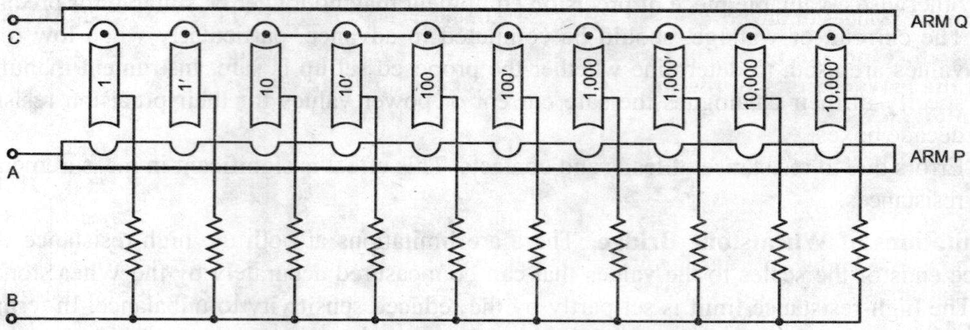

Fig. 6.8 *Schone Arrangement of Ratio-arms*

The direct interchange of arms cannot be made if other than equal arms are used. However, the arrangement given in Fig. 6.8 is still useful, as it makes available two arms of each value to check on each other. We might for example, take balances with the ratios set at 10,000/1,000, 10,000/1,000′, 10,000′/1,000 and 10,000′/1,000′. If the four results are closer, we at least gain confidence in the result. Anyhow, if one resistor is materially in error, we come to know the fact and can take steps to have it corrected.

Example 6.4. The Wheatstone bridge, shown in Fig. 6.4, when used for determining the value of unknown resistance R, is balanced when P = 100 Ω, Q = 10 Ω and S = 46 Ω. Determine the value of unknown resistance.

Solution : Unknown resistance, $R = \frac{P}{Q} \times S = \frac{100}{10} \times 46 = 460\ \Omega$ **Ans.**

Example 6.5. The slide-wire bridge, shown in Fig. 6.6 when used for determining the value of unknown resistance R, is balanced when S = 6 Ω, AB = 25 cm and BC = 75 cm.

Solution : Unknown resistance, $R = S \times \frac{\text{length AB}}{\text{length BC}} = 6 \times \frac{25}{75} = 2\ \Omega$ **Ans.**

Example 6.6. A resistance of approximately 5,000 Ω is required for balancing a bridge. It is obtained on a 5 dial resistance box having steps of 1,000, 100, 10, 1 and 0.1 ohms. The measurement is to be guaranteed to 0.1 per cent. Determine the dials required to be adjusted for obtaining the required accuracy.

Solution : Limiting value of 5,000 Ω resistor = 5,000 ± 0.1%

$$= 5{,}000 \pm 5{,}000 \times \frac{0.1}{100} = 4{,}995 \text{ to } 5{,}005\ \Omega$$

Thus dials of 1,000, 100, 10 and 1 Ω would have to be adjusted. **Ans.**

Example 6.7. In a Wheatstone bridge, for the measurement of resistance, derive the relation for sensitivity of bridge, with detecting galvanometer in operation.

For a bridge, as above, the ratio-arms are each 100 Ω and bridge is balanced with standard arm adjusted to 230 Ω. If each of the ratio-arms have an accuracy of ± 0.02% and standard has ± 0.01% accuracy guaranteed, what are the limiting values of unknown resistance ?

[U.P. Technical Univ. Electrical Measurements and Measuring Instruments 2002-03]

Solution : Unknown resistance, $R = \frac{P}{Q} \times S = \frac{100}{100} \times 230 = 230\ \Omega$

Relative limiting error of unknown resistance,

$$\frac{\delta R}{R} = \frac{\delta P}{P} + \frac{\delta Q}{Q} + \frac{\delta S}{S} = (\pm 0.02 \pm 0.02 \pm 0.01)\% = \pm 0.05\%$$

So limiting values of unknown resistance = 230 ± 0.05% = (230 ± 0.115) Ω *i.e.* 229.885 to 230.115 Ω **Ans.**

Example 6.8. Find the magnitude and direction of the current flowing through the galvanometer in the unbalanced Wheatstone's bridge circuit shown in Fig. 6.9.

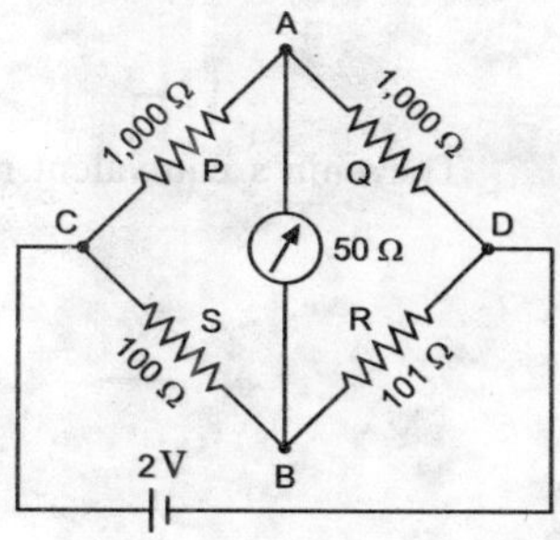

Fig. 6.9

Solution: Resistance required in arm BD for balance of bridge,

$$R = \frac{QS}{P} = \frac{1{,}000 \times 100}{1{,}000} = 100\ \Omega$$

In the actual bridge, the unknown resistor R is of 101 Ω, representing a small unbalance.

The deviation from the balance condition,

$$\Delta R = 101 - 100 = 1\ \Omega$$

Thevenin's open-circuit voltage, $E_{Th} = E\left[\frac{R+\Delta R}{R+\Delta R+S} - \frac{P}{P+Q}\right]$

$$= 2\left[\frac{100+1}{100+1+100} - \frac{1{,}000}{1{,}000+1{,}000}\right]$$

$$= 2\left(\frac{101}{201} - 0.5\right) = 0.004975\ \text{V}$$

Thevenin's equivalent resistance of the bridge looking back into the terminals A and B

$$R_{Th} = \frac{(R+\Delta R)S}{R+\Delta R+S} + \frac{PQ}{P+Q}$$

$$= \frac{(100+1)\times 100}{100+1+100} + \frac{1{,}000 \times 1{,}000}{1{,}000+1{,}000}$$

$$= 50.2487 + 500 = 550.2487\ \Omega$$

The galvanometer current, $I_g = \frac{E_{Th}}{R_{Th}+R_g} = \frac{0.004975}{550.2487+50} = 8.288\ \mu\text{A}$ **Ans.**

Since the point B is at higher potential with respect to point A, current will flow from terminal B to terminal A.

Example 6.9. In a Wheatstone bridge shown in Fig. 6.7 (a), the values of the bridge elements are P = 100 Ω, Q = 1,000 Ω, S = 2,000 Ω and R = 202 Ω. The battery voltage is 5 V and its internal resistance is negligible. The galvanometer used is of sensitivity 5 mm/μA and an internal resistance of 200 Ω. Determine the deflection of galvanometer caused by the 2 Ω unbalance in arm AD. Also determine the sensitivity of the bridge in terms of deflection per unit change in resistance.

Solution: Resistance in arm AB, P = 100 Ω

Resistance in arm BC, Q = 1,000 Ω

Resistance in arm CD, S = 2,000 Ω

Resistance required in arm AD for balance of bridge,

$$R = \frac{P}{Q} \times S = \frac{100}{1{,}000} \times 2{,}000 = 200\ \Omega$$

In the actual bridge, the unknown resistor R is of 202 Ω, representing a small unbalance. The deviation from the balance condition,

$$\Delta R = 202 - 200 = 2\ \Omega$$

Thevenin's or open-circuit voltage,

$$E_{Th} = E\left[\frac{R+\Delta R}{R+\Delta R+S} - \frac{P}{P+Q}\right]$$

$$= 5\left[\frac{202}{202+2{,}000} - \frac{100}{100+1{,}000}\right] = 4.13\ \text{mV}$$

Thevenin's equivalent resistance of the bridge looking back into the terminals B and D

$$R_{Th} = \frac{(R+\Delta R)S}{R+\Delta R+S} + \frac{PQ}{P+Q}$$

$$= \frac{(200+2)\times 2{,}000}{200+2+2{,}000} + \frac{1{,}000\times 100}{1{,}000+100} = 274.4\ \Omega$$

Thus galvanometer current, $I_g = \dfrac{E_{Th}}{R_{Th}+R_g} = \dfrac{4.13\times 10^{-3}}{274.4+200} = 8.7\ \mu\text{A}$

Deflection of the galvanometer, θ = Current sensitivity of the galvanometer, S_i × current through galvanometer, I_g

$$= \frac{5\,\text{mm}}{\mu\text{A}} \times 8.7\mu\text{A} = 43.5\ \text{mm}$$ **Ans.**

Sensitivity of the bridge, $S_B = \dfrac{\theta}{\Delta R} = \dfrac{43.5}{2} = 21.75\ \text{mm}/\Omega$ **Ans.**

Example 6.10. A Wheatstone bridge has ratio-arms of 100 Ω and 1,000 Ω and is being used to measure a resistance of 200 Ω. Two galvanometers are available - one of internal resistance 400 Ω and current sensitivity 10 mm/μA and the other of internal resistance 100 Ω and current sensitivity 5 mm/μ A. The galvanometer is connected from the junction of ratio-arms to the opposite corner, as illustrated in Fig. 6.7 (*a*). Which of the two galvanometers is more sensitive to a small unbalance on the above bridge ? Determine also the ratio of the galvanometer sensitivities.

Solution : Referring to Fig. 6.7 (a)

$$P = 1{,}000\ \Omega;\quad Q = 100\ \Omega;\quad R = 200\ \Omega$$

Resistance required in arm CD, $S = R \times \dfrac{Q}{P} = 200 \times \dfrac{100}{1{,}000} = 20\ \Omega$

Thevenin's equivalent resistance of the bridge looking back into the terminals B and D.

$$R_{Th} = \frac{RS}{R+S} + \frac{PQ}{P+Q} = \frac{200\times 20}{200+20} + \frac{1{,}000\times 100}{1{,}000+100} = 109.1\ \Omega$$

From Eq. (6.25), galvanometer deflection θ for small change in resistance is given as

$$\theta = \frac{S_i E S \Delta R}{(R+S)^2 (R_{Th}+R_{g1})}$$

Ratio of deflection on two galvanometers

$$\frac{\theta_2}{\theta_1} = \frac{S_{i_2} E S \Delta R}{(R+S)^2 (R_{Th}+R_{g_2})} \times \frac{(R+S)^2 (R_{Th}+R_{g_1})}{S_{i_1} E S \Delta R}$$

$$= \frac{S_{i_2}}{S_{i_1}} \times \frac{R_{Th}+R_{g_1}}{R_{Th}+R_{g_2}} = \frac{5}{10} \times \frac{109.1+400}{109.1+100} = 1.217$$ **Ans.**

The first galvanometer (internal resistance 400 Ω and sensitivity 10 mm/μA) is less sensitive to a small unbalance on the given bridge, though on its own it is more sensitive than the other galvanometer.

Example 6.11. An equal-ratio bridge has resistance of 500 Ω in each arm. The galvanometer used as a detector has a resistance of 100 Ω and voltage impressed across the bridge is of 10 V. Determine the smallest change in resistance that can be detected if the galvanometer is capable of detecting current upto 1 n A.

Solution : Resistance of each arm of bridge, $P = Q = R = S = 500\ \Omega$

Galvanometer resistance, $R_g = 100\ \Omega$

Battery voltage, $E = 10$ V

Thevenin equivalent resistance of bridge, $R_{Th} = R = 500\ \Omega$ See Eq. (6.19)

$$\text{Thevenin or open-circuit voltage, } E_{Th} = \frac{E\Delta R}{4R} = \frac{10\times\Delta R}{4\times 500} = 5\times 10^{-3}\ \Delta R$$

$$\text{Current through galvanometer, } I_g = \frac{E_{Th}}{R_{Th}+R_g} = \frac{5\times 10^{-3}\Delta R}{500+100} = 1\times 10^{-9}$$

$$\text{or the smallest change in resistance that can be detected, } \Delta R = \frac{1\times 10^{-9}\times 600}{5\times 10^{-3}} = 0.12\ \text{m}\Omega \quad \textbf{Ans.}$$

Example 6.12. In an equal-ratio bridge, the resistance in each arm is 200 Ω. The maximum allowable power dissipation in any arm of the bridge is 0.5 W. Determine the maximum permissible current through any arm of the bridge and maximum allowable voltage that can be applied across the bridge. If the available battery is of emf of 24 V and internal resistance 2 Ω, determine the value of series resistance to be connected between the battery and the bridge, so as to limit the current in the bridge arms to the permissible value.

Solution : Resistance of each arm of bridge, $R = S = P = Q = 200\ \Omega$

If I is the maximum current through each arm under balanced condition,

The maximum power dissipation $= I^2 R = P$

$$\text{or} \quad I = \sqrt{\frac{P}{R}} = \sqrt{\frac{0.5}{200}} = 0.05\ \text{A} \quad \textbf{Ans.}$$

The maximum voltage, that can be applied to the bridge,

$$V = I\times 2R = 0.05\times 2\times 200 = 20\ \text{V}$$

Current through series resistor = Current drawn from battery $= 2I = 2\times 0.05 = 0.1$ A

Since battery emf, $E = V + 2I(r + R)$,

where r is internal resistance of battery and R is series resistance.

$$\text{Series resistance, } R = \frac{E-V}{2I} - r = \frac{24-20}{0.1} - 2 = 38\ \Omega \quad \textbf{Ans.}$$

Example 6.13. A Wheatstone bridge having ratio-arms of 10 Ω and 10,000 Ω is used for measurement of high resistance. The adjustable arm has a maximum value of 5 k Ω. The battery is connected from the junction of ratio-arms to the opposite corner of 12 V. (*i*) Determine the maximum value of the resistance that can be measured with the given arrangement. How much unbalance is required to give a galvanometer deflection of 2.5 mm for the maximum resistance of part (*i*). Internal resistance of battery is negligible and the galvanometer available is of sensitivity 100 mm/μA and a resistance of 100 Ω.

Solution : (*i*) The maximum value of resistance that can be measured with the given arrangement,

$$R = \frac{P}{Q}\times S = \frac{10{,}000}{10}\times 5\times 10^3 = 5\ \text{M}\Omega \quad \textbf{Ans.}$$

Thevenin equivalent resistance of the bridge,

$$R_{Th} = \frac{RS}{R+S} + \frac{PQ}{P+Q} = \frac{5\times10^6 \times 5\times10^3}{5\times10^6 + 5\times10^3} + \frac{10,000\times10}{10,000+10} = 5,005\ \Omega$$

(*ii*) Change in resistance required to give deflection of 2.5 mm on a galvanometer of resistance 100 Ω and current sensitivity of 100 mm/μA,

$$\Delta R = \frac{\theta(R_{Th} + R_g)(R+S)^2}{S_i E S} \quad \text{Refer to Eq. (6.25)}$$

$$= \frac{2.5\times(5,005+100)(5\times10^6+5\times10^3)^2}{100\times10^6\times12\times5,000} = 53.28\ k\Omega \quad \textbf{Ans.}$$

6.4 LOW RESISTANCE MEASUREMENT

6.4.1. Special Features of Low Resistance Measurements. The effect of the resistance of connecting leads and contact resistances on the measurement results is liable to cause serious misjudgements when measuring low resistances such as those of the armature and series field windings of large machines, transformers, shunts or short lengths of wire.

The contact resistance is defined as the resistance offered to the flow of current from one conductor to another. This resistance varies with the contact area, the general condition of this area (smooth or rough, clean or dirty), and the contact density and pressure.

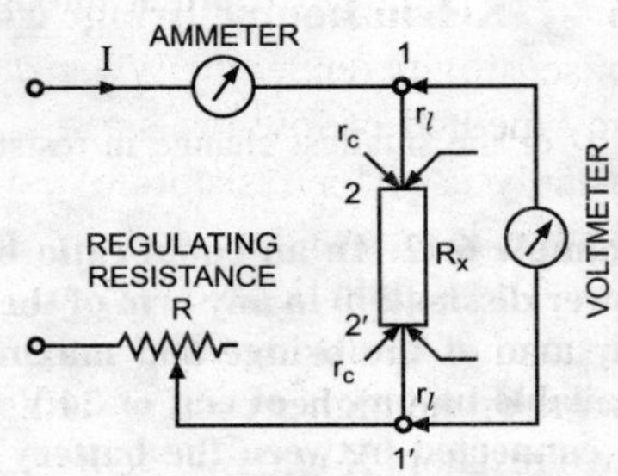

Fig. 6.10 *Incorrect Connection of Voltmeter on Measuring Low Resistance By Ammeter-Voltmeter Method*

The effect of the contact and lead resistances on the results of low resistance measurements can be evaluated and compared by making use of the circuit diagram illustrated in Figs. 6.10 and 6.11. On taking the readings of voltmeter and ammeter, the value of unknown resistance is given, in the first case, by equation

$$R_m = \frac{V_1}{I_1} = R_x + 2\,r_l + 2\,r_c \quad \text{...(6.29)}$$

where V_1 and I_1 are the voltmeter and ammeter readings in volts and amperes respectively.

From Eq. (6.29) it is obvious that the measured value of resistance R_m is equal to the sum of true value of unknown resistance R_x, the resistance of two leads $2\,r_l$ and contact resistance at two contacts $2r_c$.

Assuming that $R_x = 0.1\ \Omega$ and $r_e = r_c = 0.01\ \Omega$, we have $R_m = 0.1 + 2\times0.01 + 2\times0.01 = 0.14\ \Omega$

and error of measurement, $\varepsilon_r = \frac{R_m - R_x}{R_x}\times100 = \frac{0.14-0.1}{0.1}\times100 = 40\%.$

The relative error of measurement due to contact and lead resistances obviously increases with decreasing magnitude of the unknown resistance.

Connection of voltmeter to points 2 and 2′ (Fig. 6.10) *i.e.*, to the current terminals by means of which the resistance to be measured is inserted in the current circuit, gives a second value of the unknown resistance

$$R'_m = \frac{V_2}{I_2} = R_x + 2r_c \quad \text{...(6.30)}$$

i.e., measured resistance R'_m is less than by R_m by the resistance of connecting leads.

Connection of voltmeter to points 3 and 3′ (Fig. 6.11) *i.e.*, to the potential terminals located between the current terminals, gives a third value of the unknown resistance

$$R''_m = \frac{V_3}{I_3} = R_x \quad ...(6.31)$$

that is equal to the true value of resistance, since the voltmeter measures the actual value of the voltage across the potential terminals of the unknown resistance.

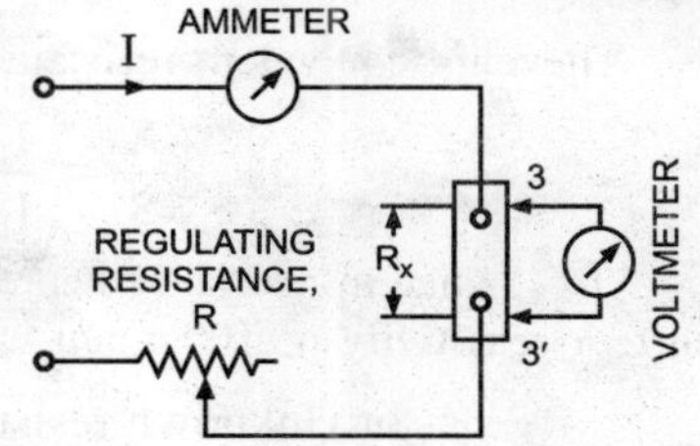

Fig. 6.11 *Correct Connection of Voltmeter on Measuring Low Resistance By Ammeter-Voltmeter Method*

The use of two pairs of terminals (current and potential terminals) is one of the means of eliminating the influence of the contact and lead resistances on the results of low-resistance measurements.

6.4.2. Kelvin Double Bridge Method. Errors owing to contact resistance, resistance of leads and lack of sensitivity render the Wheatstone bridge* unsuitable for the measurement of low resistances. Further, any method of low-resistance measurement must provide for the four terminal connections which is usually used for resistors of low-values. A network designed by Lord Kelvin, called Kelvin double bridge, is an example of such a method.

The Kelvin double bridge is shown schematically in Fig. 6.12. X is the resistor under test and S is a standard resistor of the same order of resistance and of the same or the higher current rating than the one under test. The two resistors X and S are connected in series with a short link of as low a resistance value as possible. P, Q, *p* and *q* are four known non-inductive resistances, one pair of which (P and *p* or Q and *q*) are variable. A current, preferably the rated current of resistor under test, is passed through the two resistors X and S from low-voltage, high current battery. A regulating rheostat and an ammeter are connected in the circuit for convenience only. A sensitive galvanometer G is connected across the dividing points of PQ and *pq*. The ratio P/Q is kept the same as *p*/*q*, these ratios being varied until the galvanometer reads zero.

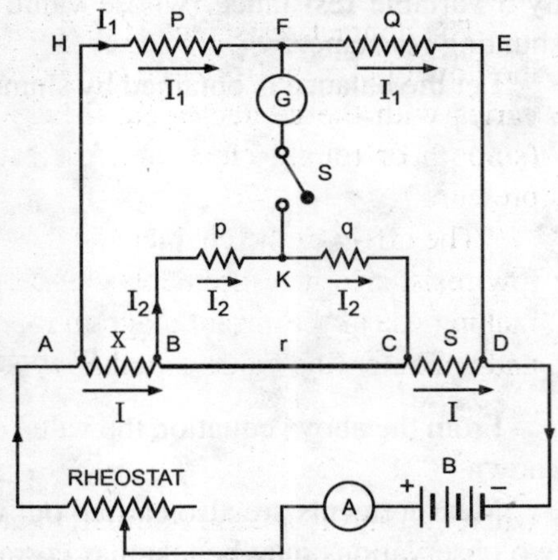

Fig. 6.12 *Kelvin Double Bridge*

In balance position of bridge, the currents in the bridge are, as shown in the figure. Applying Kirchhoff's second law to meshes AHFKBA and FEDCKF we get

$$I_1 P - I_2 p - I X = 0$$

$$\text{or } I X = I_1 P - I_2 p \quad ...(6.32)$$

$$\text{and } I_1 Q - I_2 q - I S = 0$$

$$\text{or } I S = I_1 Q - I_2 q \quad ...(6.33)$$

Dividing Eq. (6.32) by Eq. (6.33) we get

* Refer Art. 6.3.

$$\frac{X}{S} = \frac{I_1P - I_2p}{I_1Q - I_2q} = \frac{P\left(I_1 - \frac{p}{P}I_2\right)}{Q\left(I_1 - \frac{q}{Q}I_2\right)} = \frac{P}{Q} \qquad \text{Since } \frac{p}{P} = \frac{q}{Q}$$

$$\text{or Unknown resistance } X = S \times \frac{P}{Q} \qquad ...(6.34)$$

Thus by using the Kelvin's double bridge together with the procedure described below, the resistances of the leads between the potential terminals of the resistor and the bridge are eliminated from the measurement, and the four-terminal resistance of the unknown, X, is determined in terms of the four-terminal resistance of the standard S, and the ratio-arms P and Q.

Operation of Kelvin Double Bridge in Practice. The method of operating this bridge in practice slightly differs from that described above. In practice P, Q, p and q are known and fixed standard resistances and their temperature coefficients are also known. The resistances P, p, Q and q are of such values that $P/Q = p/q$. Usually P is taken equal to p, Q equal to q and the ratio $P/Q = p/q$ is made approximately equal to X/S.

If the approximate value of X is unknown, then it is determined by any other simple method (either by voltmeter-ammeter method or by potentiometer method). Now the unknown resistance X is shunted by a variable resistance (whose value is known) and balance is obtained by adjusting the value of shunting resistance.

Let the balance is obtained by shunting the unknown resistance X by a variable resistance of r then

$$X' = \frac{1}{\frac{1}{X} + \frac{1}{r}} = \frac{P}{Q}S$$

$$\text{or} \qquad \frac{Xr}{r + X} = \frac{P}{Q}S \qquad ...(6.35)$$

From the above equation the value of unknown resistance can be determined since P, Q, S and r are known.

Measurements are also carried out with the supply current reversed and the average value from the two observations may be taken as correct value. This is done to eliminate errors due to thermoelectric emfs.

The sensitivity of the bridge is determined by noting smallest variation of the shunting resistance r which causes an observable deflection of the galvanometer. The difference in X' for such a variation of r can be easily calculated which gives the sensitivity of the bridge.

The ratio of P/Q is made exactly equal to p/q by adjusting the lead resistances which are used for connections, since these are included in the arms as well as the resistances of the coils themselves.

The ratio of P/Q and p/q can also be made independent of temperature to a very close approximation by suitable proportioning the resistance of the copper leads.

Correction Factor. If the condition $P/Q = p/q$ cannot be very exactly fulfilled, the total resistance of the link BC and the end connections of the two resistors require consideration.

Let the total resistance of the link BC and the end connection of the two resistors be r.

$$\text{Now } I_2 = \frac{r}{p+q+r} I \approx \frac{r}{p+q} I \qquad ...(6.36) \text{ as } r << (p + q)$$

Substituting $I_2 = \frac{r}{p+q} I$ in Eqs. (6.32) and (6.33) we get

$$I_1 P - \frac{rp}{p+q} I - I X = 0 \quad ...(6.37)$$

$$\text{or } I_1 P = I X + \frac{pr}{p+q} I$$

$$\text{and } I_1 Q - \frac{qr}{p+q} I - I S = 0$$

$$\text{or } I_1 Q = I S + \frac{qr}{p+q} I \quad ...(6.38)$$

Dividing Eq. (6.37) by Eq. (6.38) we get

$$\frac{P}{Q} = \frac{X + \frac{pr}{p+q}}{S + \frac{qr}{p+q}}$$

$$\text{or } X = \frac{P}{Q} S + \frac{r}{p+q}\left(\frac{Pq}{Q} - p\right) = \frac{P}{Q} S + \frac{qr}{p+q}\left(\frac{P}{Q} - \frac{p}{q}\right) \quad ...(6.39)$$

If the ratio $\frac{P}{Q}$ is made very nearly equal to $\frac{p}{q}$ then $\left(\frac{P}{Q} - \frac{p}{q}\right)$ will be very small, and if r also is very small the whole of the second term becomes negligible and the balance condition simplifies to $X = \frac{P}{Q} . S$.

The Eq. (6.39) indicates the importance of making the two ratios P/Q and *p*/*q* equal and it may also be used to determine the error in Kelvin's double bridge measurement.

In a typical Kelvin's double bridge, the covered resistance range is 0.1 μΩ to 1.0 Ω. The accuracies are as given below:

From 1,000 μΩ to 1 Ω : 0.05%

From 100 μΩ to 1,000 μΩ : 0.2 to 0.05%

From 10 μΩ to 100 μΩ : 0.5 to 0.2%, limited by thermoelectric emfs.

This bridge consists of four internal resistance standards of 1 Ω, 0.1 Ω, 0.01 Ω and 0.001 Ω respectively.

Example 6.14. A four terminal resistance of approximately 50 μΩ was measured by Kelvin's double bridge. The bridge has the following component resistances.

Standard resistance = 100.03 μ Ω

Inner ratio-arms = 200.62 Ω and 400 Ω

Outer ratio-arms = 200.48 Ω and 400 Ω

The resistance of the link connecting the standard and unknown resistance = 700 μ Ω. Calculate the unknown resistance. *[U.P. Technical Univ. Elec. Measurements and Measuring Instruments 2005-06]*

Solution : p = 200.62 Ω; q = 400 Ω; P = 200.48 Ω; Q = 400 Ω; S = 100.03 μΩ and r = 700 μΩ.

Using Eq. (6.39) we have

$$\text{Unknown resistance, } X = \frac{P}{Q}S + \frac{qr}{p+q}\left[\frac{P}{Q} - \frac{p}{q}\right]$$

$$= \frac{200.48}{400} \times 100.03 \times 10^{-6} + \frac{400 \times 700 \times 10^{-6}}{200.62 + 400}\left[\frac{200.48}{400} - \frac{200.62}{400}\right]$$

$$= 50.135 \times 10^{-6} + \frac{280{,}000 \times 10^{-6}}{600.62} \times \frac{(-0.14)}{400}$$

$$= 50.135 \times 10^{-6} - 0.1631647 \times 10^{-6} = 49.971835\ \mu\Omega \quad \textbf{Ans.}$$

6.5 LOW-RESISTANCE MEASURING INSTRUMENTS

An ohmmeter makes use of a known constant current source and a voltmeter. The arrangement is shown in Fig. 6.13. Here a constant current of known magnitude, depending upon the range, supplied from a battery or a constant current source is passed through the resistance under measurement and the voltage developed across it is measured by a voltmeter. The scale is calibrated in ohms. For multirange ohmmeter, the value of the current is changed.

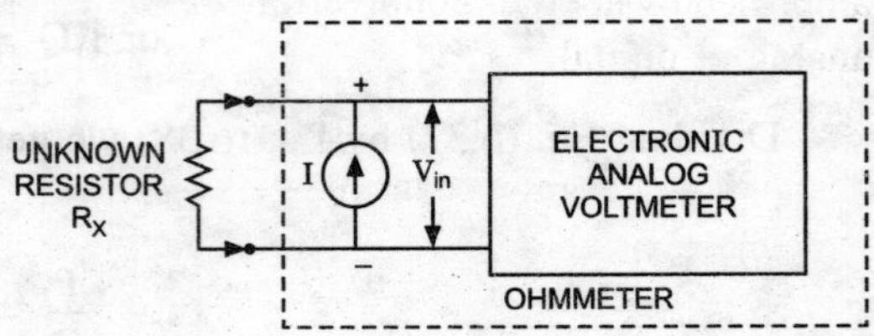

Fig. 6.13 *Block Diagram of an Ohmmeter*

Two-Wire Sensing: When the same meter terminals are used for measuring the voltage drop across the unknown resistor (Fig. 6.14), as are used to supply the current, a meter is said to use a 2-wire sensing technique. In this arrangement the wire or lead resistances, which are resistances due to connecting wires, are added to the unknown resistance value. This results in a large error in the measurement of the unknown resistor, which has the value of the order of lead resistors. This technique is often used due to simplicity.

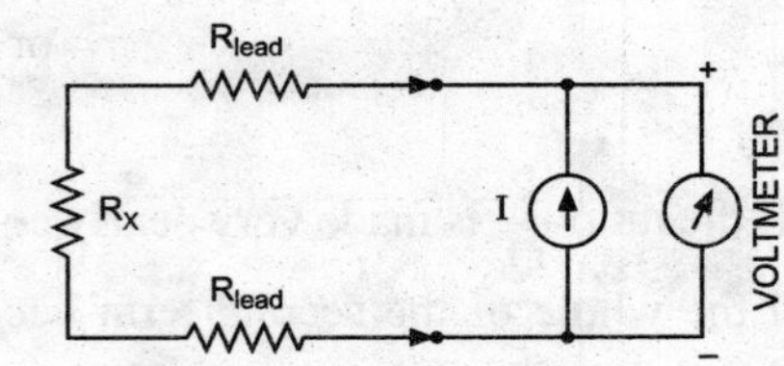

Fig. 6.14 *Simplified Schematic of Two-Wire Sensing*

Four-Wire Sensing: The lead resistance errors are removed in this technique, which provides the most accurate method of measuring small resistances. This technique is also called the Kelvin sensing technique. In this arrangement, shown in Fig. 6.15, lead and contact resistances are reduced.

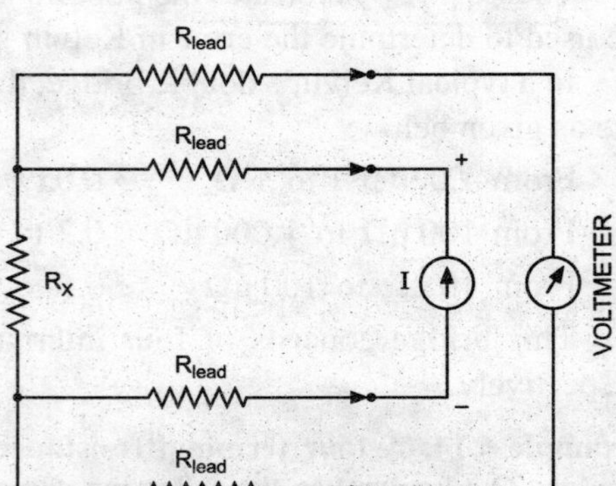

Fig. 6.15 *Simplified Schematic of 4-Wire Sensing*

This arrangement makes use of two separate pairs of connections to the unknown resistor R_X. One pair of connections, also known as the "source leads", connects the unknown resistor R_X to the constant current source while, the other pair of connections, also known as the "sense leads" connects the unknown resistor to the voltmeter, which has very high input impedance. The current supplied by current source develops voltage across unknown resistor R_X and R_{lead}. The voltmeter indicates the voltage developed across R_X, which is equal to IR_X because no current flows through R_{lead} of sense leads. The voltmeter indication is also independent of R_{lead} of source leads. Thus the error due to voltage drop in both the source leads and sense leads is completely removed in this arrangement.

This arrangement is widely used in applications where lead resistances become quite large and variable. The only drawback is in multichannel systems which need twice as many switches and twice as many wires as the two-wire technique.

Another possible circuit for measurement of low resistance is shown in Fig. 6.16. Noninverting op-amp A_1 and its associated components form a constant-current source. Because of circuit feedback, voltage V_2 applied to the op-amp noninverting input terminal also appears across resistor R_3. Thus, current through resistor R_3, being equal to $\frac{V_2}{R_3}$, is a constant quantity if V_2 is constant. This current is passed through resistance under measurement. The voltage drop across the unknown resistance R is amplified by another noninverting op-amp A_2 and passed to the voltmeter circuit, which might be either analog or digital.

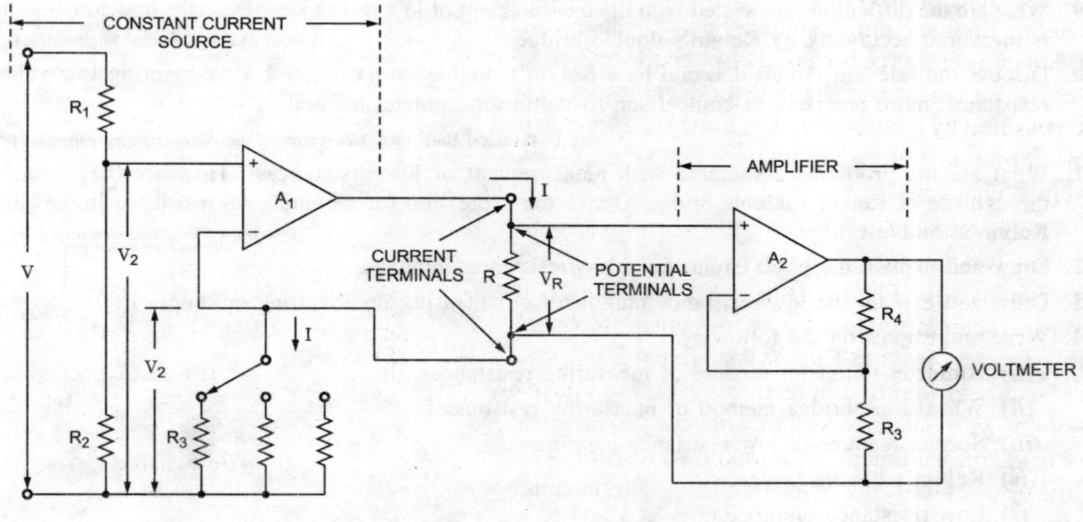

Fig. 6.16 *Low-Resistance Measurement Circuit Using Op-Amps*

Changing current I up or down by a factor of 10 alters the measured resistances for a given voltage drop by a factor of 10. The current magnitude is altered simply by selecting a new value of R_3. In case the voltage drop is measured digitally, the decimal point should be switched when the magnitude of current is changed. While making measurements with analog meter, the new scale range is to be noted.

EXERCISES

1. Explain what you mean by low, medium and high resistances.
2. Show how a voltmeter and ammeter can be used for measurement of resistance. Show the two possible connections and write the resistance equation for each. Discuss the errors involved.
3. Classify the resistances into low, medium and high resistances.

   Describe the method of measurement of medium resistances using Wheatstone bridge. Discuss the errors involved and describe how these errors can be minimized.

4. Explain the precision measurement of medium resistance with Wheatstone bridge. Also discuss the sensitivity of the bridge. Show when sensitivity becomes maximum and minimum.
5. In a Wheatstone bridge, for the measurement of resistance, derive the relation for sensitivity of the bridge, with detecting galvanometer in operation. **[U.P. Technical Univ. Elec. Measurements and Measuring Instruments 2002-03]**
6. Derive the expression for bridge sensitivity for a Wheatstone bridge with equal arms. Find also the expression for current through the galvanometer for a small unbalance. Also describe the working of a megohm bridge. **[U.P. Technical Univ. Elec. Measurements and Measuring Instruments 2006-07]**
7. State various methods of measurement of low resistance. Why is ammeter-voltmeter method not suitable for the precise measurement of low resistance ?
8. What modifications are made in the basic Wheatstone bridge to convert it into Kelvin's double bridge to make it suitable for low resistance measurement below the value of 0.1 Ω.
   Explain with the help of circuit diagram, how Kelvin bridge is used to measure low resistance. Give the necessary relationship used to determine unknown resistances in terms of known resistance of the bridge circuit.
9. What are the difficulties associated with the measurement of low resistance ? Describe how low resistance is measured accurately by Kelvin's double bridge. **[U.P.S.C. I.E.S. Electrical Engineering I, 1992]**
10. Discuss the relevant circuit diagram how Kelvin's double bridge is used for measuring low value of resistance, more precisely in comparison to voltmeter-ammeter method. **[U.P. Technical Univ. Elec. Measurements and Measuring Instruments 2005-06]**
11. What are the problems associated with measurement of low resistances ? How are they overcome through use of Kelvin's double bridge. Derive the expression for the unknown resistance in the case of Kelvin's double bridge. **[Pb. Univ. Elec. Measurement-I June 1992]**
12. Draw and explain the basic circuit for a low-resistance ohmmeter.
13. Draw and explain the low-resistance ohmmeter circuit using noninverting op-amps.
14. Write short notes on the following:
    (*i*) Ammeter-voltmeter method of measuring resistances.
    (*ii*) Wheatstone bridge method of measuring resistances.
    (*iii*) Special features of low resistance measurement.
    (*iv*) Kelvin's double bridge.
    (*v*) Low-resistance ohmmeter.

SHORT ANSWER TYPE QUESTIONS WITH ANSWERS

Q. 1. Name the factors that govern the choice of a suitable method of measuring resistance.

Ans. The choice of a suitable method of measuring resistance depends on several factors. Some of the more important are the range of resistance to be measured, required accuracy, the ease with which it can be made and whether the measurements are to be made with the item tested under its normal operating conditions or away from these.

Q. 2. What is meant by low, medium and high resistances ?

Ans. Resistances about 1 Ω or less are included in the category of low resistances while resistances of 100 kΩ and above are usually termed as high resistances. Resistances ranging from about 1 Ω to about 100 kΩ are considered as medium resistances.

Q. 3. Name the difficulty associated with the measurement of low resistances.

Ans. The difficulty associated with the measurement of low resistances is the effect of the resistance of connecting leads and contact resistances on the measurement results is liable to cause serious misjudgement when measuring low resistance.

Q. 4. Why is Kelvin's double bridge superior to the Wheatstone bridge for the purpose of low resistance measurement ?

Ans. Errors owing to contact resistance, resistance of leads and lack of sensitivity render the Wheatstone bridge unsuitable for the measurement of low resistances. Further, any method of low-resistance measurement must provide for the four terminal connections which is usually used for resistors of low-values. A network designed by Lord Kelvin's, called Kelvin double bridge, is an example of such a method.

PROBLEMS

1. In the measurement of a resistance by the voltmeter-ammeter method, the readings on the ammeter and voltmeter with the voltmeter connected across the resistance are 2 A and 180 V respectively. The readings on the ammeter and voltmeter with the voltmeter connected across the whole circuit (combination of ammeter and unknown resistance in series) are 2 A and 188.6 V respectively. Calculate (*a*) the true value of resistance (*b*) the percentage error as the quotient of the readings in both of the cases. The resistances of the ammeter and voltmeter are 0.01 Ω and 2,000 Ω respectively

[Ans. (*a*) 94.29 Ω (*b*) 4.56% low; 0.01% high]

2. A nonlinear resistance X is known to follow the law $X = K_1 + K_2 I$ where K_1 is zero current value of X, I is the current through X and K_2 is a constant. A Wheatstone bridge ABCD with arm AB = 200 Ω is used to determine K_1 and K_2. With arm BC = 100 Ω, a value of DA = 110 Ω is required for balance. With arm BC = 200 Ω the value of DA required for balance is 60 Ω. The supply voltage is 10 volts. Compute K_1 and K_2.

[Pb. Univ. April 1977]

[Ans. $K_1 = \frac{125}{3}$ and $K_2 = 220$]

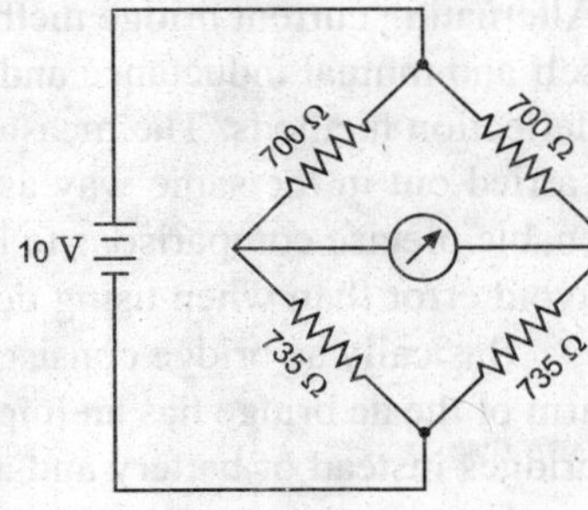

Fig. 1

3. Given a centre zero 200-0-200 μA movement having an internal resistance of 125 Ω. Calculate the current through the galvanometer given in the Fig. 1 by approximation method.

[Ans. 146.3 μA]

4. In a specific Kelvin bridge the ratio of the arms are 1:100. The standard resistance is 20 Ω. Find out the unknown resistance.

[Ans. (*a*) 0.2 Ω]

5. The resistance of a shunt is measured using a Kelvin's double bridge.
The following values give the resistance comprising the bridge at balance. Calculate the resistance of the shunt.

Standard resistance = 1 μΩ

Inner ratio arms = 2,402 Ω and 2,000 Ω

Outer ratio arms = 2,402 Ω and 2,000 Ω

Resistance of link between the shunt and the standard resistance = 0.1 Ω

2,000 Ω resistors of the inner and outer arms are on the standard resistor side of the bridge. Prove the formulae used.

[Bangalore Univ. August 1981]

[Ans. 1.201 μΩ]

❑❑❑

7

Inductance and Capacitance Measurements

7.1 INTRODUCTION

Alternating current bridge methods are the best and most usual methods for the precise measurement of self and mutual inductance and capacitance, since it is generally more difficult to obtain accuracy with deflection methods. The measurement of self and mutual inductances and capacitance by ac bridges is carried out in the same way as resistance is measured by dc bridge (Wheatstone bridge). AC bridges enable precise comparisons to be made, but adjustment is more complex, and great care is necessary to avoid error than when using dc bridge.

Basically ac bridge consists of four arms, a source of excitation and a balance detector. Each of the arm of the ac bridge has an impedance. AC source and a detector sensitive to ac voltages are used in ac bridges instead of battery and a galvanometer used in dc bridges.

For power frequency measurements the main supply is used as a source for ac bridge. For audio-frequency measurements electronic oscillators and amplifiers are used. Electronic oscillators can also be used for power frequency range when power requirement is low.

Electronic oscillators have many advantages :

(*i*) Frequency of oscillator remains constant throughout the test.

(*ii*) Frequency can be easily adjusted and can be determined very accurately in electronic oscillators.

(*iii*) Waveform of the output voltage of electronic oscillators is almost sine wave.

Commonly used detectors for ac bridges are telephone receivers or head phones, vibration galvanometers and tunable amplifier detectors.

7.2 AC BRIDGE THEORY

A bridge in general form is shown in Fig. 7.1, in which four impedances are shown connected in a bridge circuit, to an ac source and to a detector D. Balance is obtained by adjustment of one or more of the bridge arms and is indicated by zero response of the detector, which means that points B and D are at the same potential at all instances. Now equations for the bridge will be derived in complex notation.

When points B and D are at the same potential then voltage drop between points A and B will be equal to voltage drop between points A and D

So $I_1 Z_1 = I_3 Z_3$ ...(7.1)

In this condition voltage drop between points B and C will be equal to that between points D and C also.

i.e. $I_2 Z_2 = I_4 Z_4$ ...(7.2)

Dividing Eq. (7.1) by Eq. (7.2) we have

$$\frac{I_1 Z_1}{I_2 Z_2} = \frac{I_3 Z_3}{I_4 Z_4} \quad ...(7.3)$$

But in balance condition current through detector is zero

So $I_1 = I_2$ and $I_3 = I_4$

and, therefore, Eq. (7.3) becomes

$$\frac{Z_1}{Z_2} = \frac{Z_3}{Z_4}$$

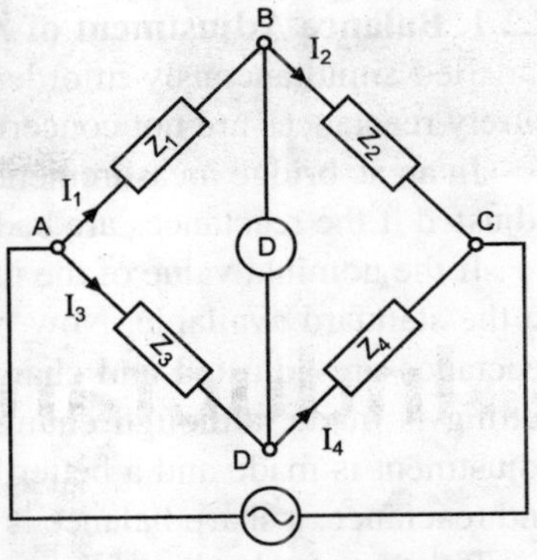

Fig. 7.1 *General Four-Arm AC Bridge*

or $Z_1 Z_4 = Z_2 Z_3$ ...(7.4)

This is the general equation for the four-arm bridge at balance. Just like the dc bridge, balance equation for the ac bridge are also not affected by interchange of supply and detector terminals, though the sensitivity might get affected.

Equation (7.4) indicates that the product of impedances of one pair of opposite arms is equal to the product of impedances of the other pair of opposite arms, when the impedances are expressed in complex form.

In the polar form the impedance Eq. (7.4) can be rewritten as

$$(Z_1 \angle\theta_1)(Z_4 \angle\theta_4) = (Z_2 \angle\theta_2)(Z_3 \angle\theta_3)$$

$$\text{or } Z_1 Z_4 \angle\theta_1 + \theta_4 = Z_2 Z_3 \angle\theta_2 + \theta_3 \quad ...(7.5)$$

Equation (7.5) indicates that two conditions must be satisfied simultaneously for balancing the ac bridge. The first condition is that the magnitude of impedances must satisfy the relationship

$$Z_1 Z_4 = Z_2 Z_3 \quad ...(7.6)$$

and second condition is that the phase angles of the impedances must satisfy the relationship

$$\angle\theta_1 + \theta_4 = \angle\theta_2 + \theta_3 \quad ...(7.7)$$

It is to be noted that it is not necessary for the four impedances to have identical phase angles or even for the impedances to be of the same kind, so long as the phase angle differences satisfy the above condition.

The individual branches may themselves be series or parallel combinations, and may include resistance, inductance and capacitance elements separately or in combination. This leads to a very large number of possible combinations, but not all of these form workable bridges. A particularly useful group of bridges is obtained if two of the four arms are purely resistive, there are other useful bridge methods which are more complicated than the four-arm ac bridge *e.g.* Anderson bridge, Carey-Foster's bridge etc.

The number of possible bridge circuits is obviously extremely large. Not all of these are of practical importance, but the number which have been developed for practical use is continuously increasing. The selection of bridge circuits described here is so made as to illustrate the main use of such networks, each is of theoretical interest and practical value.

7.2.1. Balance Adjustment of AC Bridges. It has been observed that there are two conditions to be satisfied simultaneously in order to obtain balance for an ac bridge if either only pure resistances or purely reactances are not concerned.

In an ac bridge measurement there will be little or no indication of a balance when the resistance is adjusted if the reactances are badly out of balance and *vice-versa.*

If the nominal value of the impedance under test is known then a suitable ratio is chosen according to the standard available. Now bridge arms are set to the estimated values for balance. Resistance or reactance are adjusted and change in indication is observed in the detector. This way a preliminary setting is made, although change in indication in the detector may be very small. After this other adjustment is made and a better balance is obtained. So by making alternate adjustments in resistance and reactance, a sharp balance is obtained.

If the nominal value of the test impedance is not known then an estimate is taken. Here one important point has to be noted that increasing the impedances to very high values reduces the detector current and this effect not be confused with those of the balance adjustment.

7.2.2. AC Bridge Sensitivity. The same considerations that determined the sensitivity of a Wheatstone bridge apply to ac bridge circuits. The bridge sensitivity may be defined as the smallest change in the measured quantity that causes the null detector to deflect from zero. Bridge sensitivity can be improved by using a more sensitive null detector and/or by increasing the supply voltage level. The sensitivity of the ac bridge is analyzed exactly in the same way as in case of the Wheatstone bridge, except that impedances are involved in place of resistances. Measurement accuracy is also determined in the same way as the Wheatstone bridge accuracy.

Example 7.1. The arms of an ac bridge are arranged as follows:

Arm AB: $Z_1 = 100 \angle 30^\circ\ \Omega$; Arm BC : $Z_2 = 50 \angle 0^\circ\ \Omega$ purely resistive, Arm AD: $Z_3 = 200 \angle -90^\circ\ \Omega$ purely capacitive, and Arm CD : $100 \angle 30^\circ\ \Omega$.

Determine whether it is possible to balance the bridge under above conditions.

Solution: Applying the first condition of balance required for magnitudes *i.e.*

$$Z_1 Z_4 = Z_2 Z_3$$

$$\text{Now } Z_1 Z_4 = 100 \times 100 = 10{,}000$$

$$\text{and } Z_2 Z_3 = 50 \times 200 = 10{,}000$$

Thus the first condition is satisfied.

Applying the second condition of balance required for phase angle *i.e*

$$\angle\theta_1 + \angle\theta_4 = \angle\theta_2 + \angle\theta_3$$

$$\text{Now } \angle\theta_1 + \angle\theta_4 = 30^\circ + 30^\circ = 60^\circ$$

$$\text{and } \angle\theta_2 + \angle\theta_3 = 0^\circ - 90^\circ = -90^\circ$$

i.e. second condition is not satisfied.

It means that bridge is unbalanced though the first condition for equality of magnitude products is satisfied. Obviously balance is not possible under above conditions.

Example 7.2. Consider the circuit of Fig. 7.2 (a) and determine whether or not the bridge is complete balance. If not, show two ways in which it can be made to balance and specify numerical values for any additional components. Assume that arm 4 is the unknown that cannot be modified.

Solution: $Z_1 = 1{,}000 \angle -90^\circ\ \Omega$ as it is pure capacitive

$Z_2 = 500 \angle 0^\circ\ \Omega$ as it is pure resistive.

$Z_3 = 1{,}000 \angle 0^\circ\ \Omega$ as it is pure resistive.

$$Z_4 = 100 + j\,500 = 509.9 \angle 78.69^\circ\ \Omega \quad \text{as it is inductive.}$$

From the inspection of the circuit it can be seen that the first condition for equality of magnitude products can easily be met by slightly increasing the resistance of R_3.

The second balance condition needs that

$$\theta_1 + \theta_4 = \theta_2 + \theta_3$$

$$\text{where } \theta_1 = -90^\circ,\ \theta_2 = 0^\circ,\ \theta_3 = 0^\circ \text{ and } \theta_4 = 78.69^\circ.$$

Obviously balance is not possible with configuration of Fig. 7.2 (*a*) because

$$\theta_1 + \theta_4 = -90^\circ + 78.69^\circ = -11.31^\circ \text{ and } \theta_2 + \theta_3 = 0^\circ$$

Balance can be restored by modifying the circuit so as to satisfy the phase angle condition. There are basically two methods to accomplish it. The first option is by modification of Z_1 in such a way that its phase angle is reduced to 78.69° (equal to θ_4) by placing a resistor R_1 in parallel with capacitor C_1, as shown in Fig. 7.19 (*b*). This modification results in Maxwell-Wien bridge configuration [Fig. 7.2 (*a*)]. The resistance R_1 can be determined as follows. Now impedance Z_1 becomes

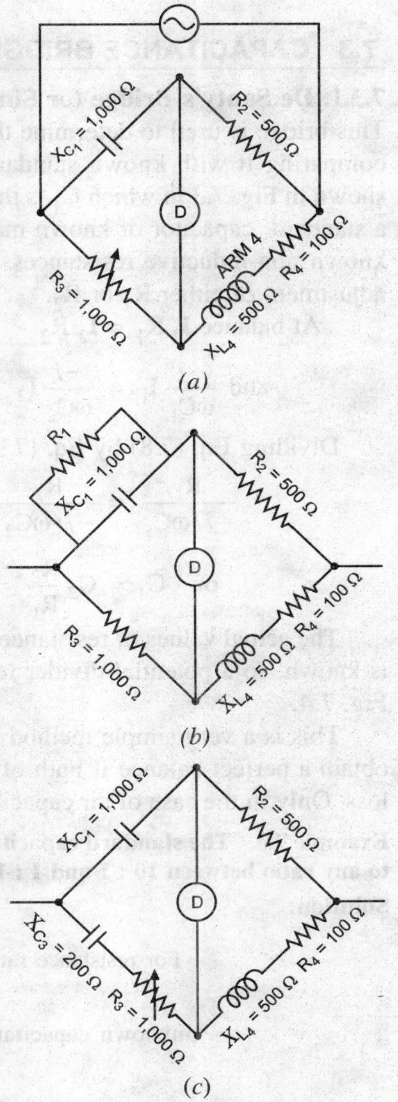

Fig. 7.2

$$\frac{1}{\frac{1}{R_1} + \frac{1}{jX_{C_1}}} \text{ or equal to } \frac{1}{\frac{1}{R_1} + j\omega C_1}$$

$$\text{or} \quad \frac{1}{Z_1} = \frac{1}{R_1} + j\,\omega\,C_1 \qquad \ldots(i)$$

and according to Eq. (7.4) for balanced condition of bridge

$$\text{or} \quad \frac{1}{Z_1} = \frac{Z_4}{Z_2 Z_3} = \frac{100 + j500}{500 \times 1{,}000}$$

$$= \frac{1}{5{,}000} + j\frac{1}{1{,}000} \qquad \ldots(ii)$$

Comparing real and *j*-components of Eqs. (*i*) and (*ii*) we have

$$R_1 = 5{,}000\ \Omega$$

$$\text{and } \omega\,C_1 = \frac{1}{1{,}000} \text{ or } \frac{1}{\omega C_1} = 1{,}000\ \Omega$$

Since $X_{C_1} = \dfrac{1}{\omega C_1}$ is already equal to 1,000 Ω, the bridge can be balanced simply by placing a resistance of 5,000 Ω across capacitor in arm 1.

The second option is by modification of phase angle of arm 2 or arm 3 by adding a series capacitor, as illustrated in Fig. 7.2 (*c*).

$$\text{Now impedance } Z_3 \text{ becomes } R_3 - \frac{j}{\omega C_3} \text{ or } R_3 - j\,X_{C_3} \qquad \ldots(iii)$$

and according to Eq. (7.4), for balanced condition of bridge

$$Z_3 = \frac{Z_1 Z_4}{Z_2} = \frac{1{,}000 \angle -90^\circ \times 509.9 \angle 78.69^\circ}{500 \angle 0^\circ}$$

$$= 1{,}019.8 \angle -11.31^\circ\ \Omega = (1{,}000 - j200)\ \Omega \qquad \ldots(iv)$$

Comparing real and *j* components of Eqs. (*iii*) and (*iv*) we have

$R_3 = 1{,}000\ \Omega$ and $X_{C_3} = 200\ \Omega$

Since R_3 is already of 1,000 Ω, the bridge can be balanced simply by adding a capacitor of reactance $X_{C_3} = 200\ \Omega$ in series with resistor R_3 of 1,000 Ω in arm 3, as illustrated in Fig. 7.2 (*c*).

Note. There was a possibility that with the addition of resistance R_1 in arm 1 as first option or with the addition of capacitor C_3 in arm 3 as second option the condition of phase angle for balancing bridge would have been satisfied but not the condition of equality of magnitude products. In that case for satisfying the first condition of equality of magnitude products we would have slightly adjusted resistance R_3 in arm 3 or resistance R_2 in arm 2.

7.3 CAPACITANCE BRIDGES

7.3.1. De Sauty's Bridge (or Simple Capacitance Bridge). This bridge is used to determine the unknown capacitance by comparing it with known standard capacitor. The circuit is shown in Fig. 7.3 in which C_1 is the capacitor under test, C_2 is a standard capacitor of known magnitude and R_1 and R_2 are known non-inductive resistances. The bridge is balanced by adjustment of either R_1 or R_2.

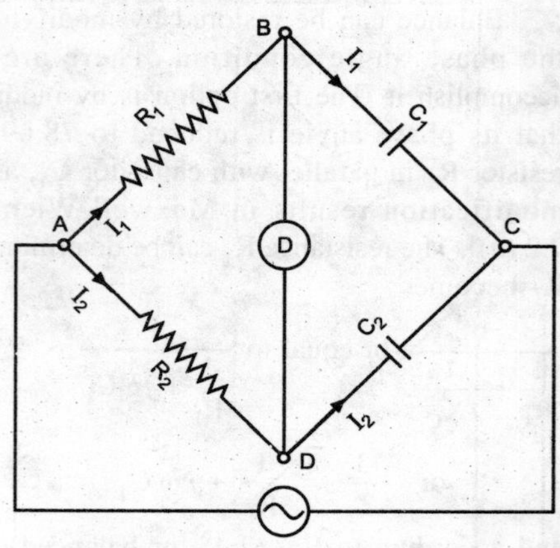

Fig. 7.3

At balance $\mathbf{I_1} R_1 = \mathbf{I_2} R_2$ ...(7.8)

and $\dfrac{-j}{\omega C_1}\mathbf{I_1} = \dfrac{-j}{\omega C_2}\mathbf{I_2}$ ...(7.9)

Dividing Eq. (7.8) by Eq. (7.9) we have

$$\frac{R_1}{-j/\omega C_1} = \frac{R_2}{-j/\omega C_2}$$

or $C_1 = C_2 \dfrac{R_2}{R_1}$ ...(7.10)

The actual values of resistances R_1 and R_2 are not significant if their ratio is known, so a potential divider resistane box could be used as depicted in Fig. 7.4.

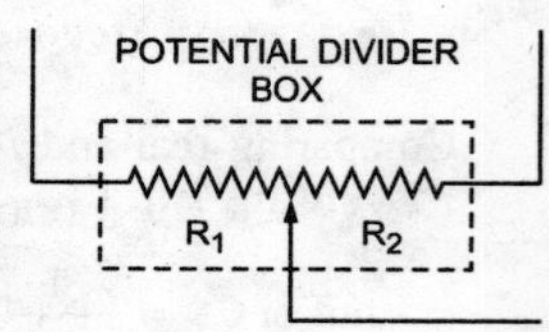

Fig. 7.4

This is a very simple method but it is very difficult rather impossible to obtain a perfect balance if both of the capacitors are not free from dielectric loss. Only in the case of air capacitors a perfect balance can be obtained.

Example 7.3. The standard capacitance C_2 used in circuit shown in Fig. 7.3 is 0.2 μF, and R_2/R_1 can be set to any ratio between 10 : 1 and 1 : 10. Determine the range of measurements of unknown capacitance C_1.

Solution: $C_2 = 0.2\ \mu F$

For resistance ratio, $\dfrac{R_2}{R_1} = 10 : 1$ or $\dfrac{10}{1}$

Unknown capacitance, $C_1 = C_2 \times \dfrac{R_2}{R_1}$ Refer to Eq. (7.10)

$= 0.2 \times 10 = 2\ \mu F$

For resistance ratio, $\dfrac{R_2}{R_1} = 1 : 10$ or $\dfrac{1}{10}$

Unknown capacitance, $C_1 = C_2 \times \dfrac{R_2}{R_1} = 0.2 \times \dfrac{1}{10} = 0.02\ \mu F$

So range of measurements of unknown capacitance is 0.02 μF to 2 μF. **Ans.**

7.3.2. Series Resistance-Capacitance or Modified De Sauty Bridge. For comparing two imperfect capacitors the bridge is modified by connecting resistances in series with them, as shown in Fig. 7.5 (*a*). R_3 and R_4 are the series resistances, where r_1 and r_2 are small resistances representing the loss components of the capacitors. Balance is obtained by variation of the resistances R_1, R_2, R_3, R_4.

$$\text{At balance } \mathbf{I}_1 R_1 = \mathbf{I}_2 R_2 \qquad \text{...(7.11)}$$

$$\text{and } \mathbf{I}_1\left(R_3 + r_1 - \frac{j}{\omega C_1}\right) = \mathbf{I}_2\left(R_4 + r_2 - \frac{j}{\omega C_2}\right) \qquad \text{...(7.12)}$$

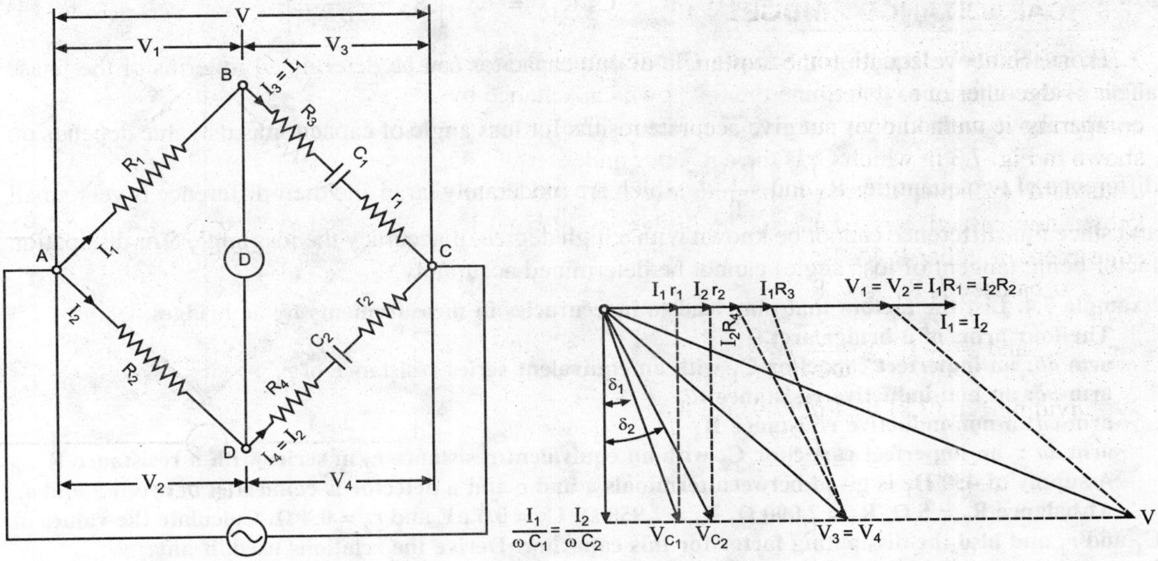

(*a*) Circuit Diagram (*b*) Phasor Diagram

Fig. 7.5 *Modified DC Sauty Bridge*

Dividing Eq. (7.12) by Eq. (7.11) we get

$$\frac{R_3 + r_1 - \dfrac{j}{\omega C_1}}{R_1} = \frac{R_4 + r_2 - \dfrac{j}{\omega C_2}}{R_2}$$

$$\text{or } R_1(R_4 + r_2) - \frac{jR_1}{\omega C_2} = R_2(R_3 + r_1) - \frac{jR_2}{\omega C_1}$$

Equating real and imaginary quantities separately we have

$$R_1(R_4 + r_2) = R_2(R_3 + r_1)$$

$$\text{and } \frac{R_1}{C_2} = \frac{R_2}{C_1}$$

$$\text{or } \frac{R_1}{R_2} = \frac{R_3 + r_1}{R_4 + r_2} = \frac{C_2}{C_1} \qquad \text{...(7.13)}$$

The phasor diagram is shown in Fig. 7.5 (*b*).

The angles δ_1 and δ_2 are the loss angles or phase angle errors of capacitors C_1 and C_2 respectively. Obviously

$$\delta_1 = \text{Tan } \delta_1 = \frac{r_1}{1/\omega C_1} = r_1 \omega C_1$$

$$\delta_2 = \text{Tan } \delta_2 = \frac{r_2}{1/\omega C_2} = r_2 \omega C_2$$

δ_1 and δ_2 being very small

$$\delta_2 - \delta_1 = \omega(C_2 r_2 - C_1 r_1)$$

$$= \omega (C_1 R_3 - C_2 R_4) = \omega C_1 \left(R_3 - \frac{R_1 R_4}{R_2} \right) \quad \text{...(7.14)}$$

From the above equation the loss angle of one capacitor can be determined in terms of the phase angle of the other one.

The above method does not give accurate results for loss angle of capacitor as its value depends on difference of two quantities R_3 and $\frac{R_1 R_4}{R_2}$ which are moderately large and their difference is very small and since this difference cannot be known with a high degree of accuracy the loss angle δ(or dissipation factor being tangent of loss angle) cannot be determined accurately.

Example 7.4. List the factors that may lead to inaccuracies in measurements by ac bridges.
The four arms of a bridge are:
arm *ab*: an imperfect capacitor C_1 with an equivalent series resistance of r_1
arm *bc*: an non-inductive resistance R_3
arm *cd*: a non-inductive resistance R_4
arm *da* : an imperfect capacitor C_2 with an equivalent resistance r_2 in series with a resistance R_2.
A supply of 450 Hz is given between terminals *a* and *c* and a detector is connected between *b* and *d*. At balance $R_2 = 5\ \Omega$, $R_3 = 2{,}000\ \Omega$, $R_4 = 2{,}950\ \Omega$, $C_2 = 0.5\ \mu F$ and $r_2 = 0.4\ \Omega$. Calculate the values of C_1 and r_1 and also the dissipating factor for this capacitor. Derive the relations used, if any.

[U.P.S.C. I.E.S. Electrical Engineering-I, 1995]

Solution: Under balanced conditions

$$Z_1 Z_4 = Z_2 Z_3$$

$$\left(r_1 + \frac{1}{j\omega C_1} \right) R_4 = \left(r_2 + R_2 + \frac{1}{j\omega C_2} \right) R_3$$

Equating the real parts we have

$$r_1 R_4 = (r_2 + R_2) R_3$$

$$\text{or } r_1 = \frac{R_3}{R_4} (r_2 + R_2)$$

$$= \frac{2{,}000}{2{,}950} (0.4 + 5) = 3.661\ \Omega \textbf{ Ans.}$$

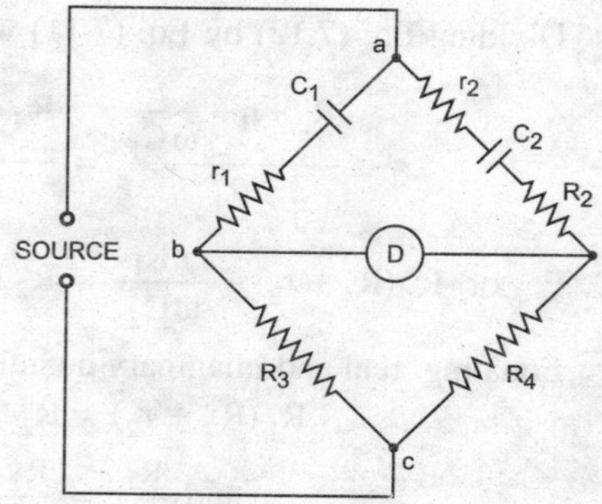

Fig. 7.6

Equating imaginary parts we have

$$\frac{R_4}{j\omega C_1} = \frac{R_3}{j\omega C_2}$$

$$\text{or } \quad C_1 = \frac{R_4}{R_3} \times C_2 = \frac{2{,}950}{2{,}000} \times 0.5 \times 10^{-6} = 0.7375\ \mu F \textbf{ Ans.}$$

Dissipation factor,

$$\text{Tan } \delta = \omega C_1 R_1 = 2\ \pi \times 450 \times 0.7375 \times 10^{-6} \times 3.661 = 7.634 \times 10^{-3} \textbf{ Ans.}$$

Example 7.5. In a modified De Sauty bridge measurement the following readings are obtained:
R_1 = 1,000 Ω; R_2 = 1,000 Ω; R_3 = 2,000 Ω; R_4 = 2,000 Ω;
C_1 = 1 μF; f = 1,000 Hz; r = 10 Ω (resistance of C_1)
Find out phase angle errors and unknown capacitance.

Solution: Refer circuit diagram shown in Fig. 7.5 (*a*)

$$R_1 = 1{,}000\ \Omega;\quad R_2 = 1{,}000\ \Omega;\ R_3 = 2{,}000\ \Omega;\ R_4 = 2{,}000\ \Omega;$$
$$C_1 = 1 \times 10^{-6}\ \text{F};\ \omega = 2\pi f = 2 \times 3.14 \times 1{,}000 = 6{,}280;\ r_1 = 10\ \Omega$$

Unknown capacitance, $C_2 = \dfrac{C_1 R_1}{R_2} = 1 \times 10^{-6} \times \dfrac{1{,}000}{1{,}000} = 1 \times 10^{-6}\ \text{F} = 1\ \mu\text{F}$ **Ans.**

$$r_2 = \frac{R_2(R_3 + r_1) - R_1 R_4}{R_1} = \frac{1{,}000\,(2{,}000 + 10) - 1{,}000 \times 2{,}000}{1{,}000} = 10\ \Omega$$

Phase angle error, $\delta_1 = \omega r_1 C_1 = 6{,}280 \times 10 \times 1 \times 10^{-6}$ radians

$$= 6{,}280 \times 10 \times 1 \times 10^{-6} \times \frac{180}{3.14} = 3.6^\circ \quad \textbf{Ans.}$$

Similarly, $\delta_2 = \omega r_2 C_2 = 6{,}280 \times 10 \times 1 \times 10^{-6} \times \dfrac{180}{\pi} = 3.6^\circ$ **Ans.**

Example 7.6. Figure 7.7 shows the values of different components in four arms of an ac bridge in its balanced state. The bridge is excited by a voltage source of 500 Hz. Calculate values of C_1 and r_1 and also the dissipation factor of the unknown capacitor.

Solution: From the circuit diagram shown in Fig. 7.7

$$C_2 = 0.5 \times 10^{-6}\ \text{F};\ r_2 = 0.4\ \Omega;\ R_2 = 4.8\ \Omega\ R_3 = 2\ \text{k}\Omega;$$
$$R_4 = 2.85\ \text{k}\Omega;\ \omega = 2\pi f = 2 \times 3.14 \times 500 = 3{,}140$$

Unknown capacitance,

$$C_1 = C_2 \times \frac{R_4}{R_3} = 0.5 \times 10^{-6} \times \frac{2.85 \times 10^3}{2 \times 10^3} = 0.7125\ \mu\text{F}\ \textbf{Ans.}$$

$$r_1 = \frac{R_3(R_2 + r_2)}{R_4} = \frac{2 \times 10^3}{2.85 \times 10^3} \times (4.8 + 0.4) = 3.649\ \Omega\ \textbf{Ans.}$$

Dissipation factor

$$= \tan \delta_1 = \omega C_1 r_1$$
$$= 3{,}140 \times 0.7125 \times 10^{-6} \times 3.649 = 0.00817\ \textbf{Ans.}$$

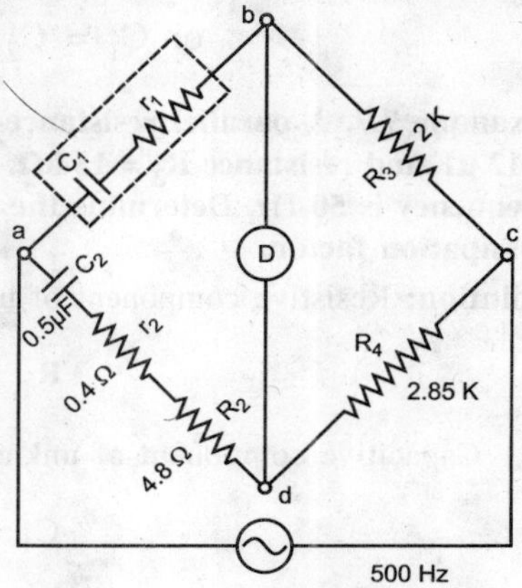

Fig. 7.7

7.3.3. Parallel Resistance-Capacitance Bridge. An alternative method of representing an inperfect capacitor diagrammatically is as a perfect capacitor C in parallel with a resistor R. The circuit of a *parallel resistance-capacitance bridge* is shown in Fig. 7.8 in which C_1 is the capacitor under test, R_1 is an imaginary resistor representing its dielectric loss component, R_3 and R_4 are resistors, either or both of which may be adjustable. Unknown impedance Z_1 is balanced by a standard capacitor C_2 in parallel with an adjustable resistor R_2. Bridge is balanced by adjustment of R_2 and either R_3 or R_4. The parallel resistance capacitance bridge is found to be most suitable for capacitors with a low resistance leakage (relatively high leakage current and high dissipation factor). Capacitors with a very high leakage resistance are measured as series R-C combinations.

The impedances of this bridge are:

$$Z_1 = \frac{1}{\dfrac{1}{R_1} + \dfrac{1}{-j/\omega C_1}} = \frac{1}{\dfrac{1}{R_1} + j\omega C_1} = \frac{R_1}{1 + j\omega C_1 R_1}$$

$$\mathbf{Z}_2 = \frac{R_2}{1 + j\omega C_2 R_2}$$

$$\mathbf{Z}_3 = R_3$$

$$\mathbf{Z}_4 = R_4$$

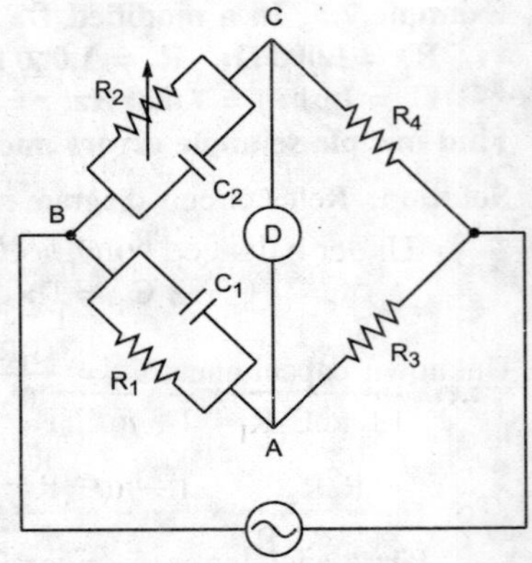

Fig. 7.8 *Parallel Resistance-Capacitance Bridge*

Under balanced conditions.

$$\mathbf{Z}_1\,\mathbf{Z}_4 = \mathbf{Z}_2\,\mathbf{Z}_3$$

$$\frac{R_1 R_4}{1 + j\omega C_1 R_1} = \frac{R_2 R_3}{1 + j\omega C_2 R_2}$$

$$\text{or } R_2 R_3 (1 + j\,\omega\, C_1 R_1) = R_1 R_4 (1 + j\,\omega\, C_2\, R_2)$$

$$\text{or } R_2 R_3 + j\,\omega\, C_1 R_1 R_2 R_3 = R_1 R_4 + j\,\omega\, C_2 R_1 R_2 R_4 \quad ...(7.15)$$

Equating the real terms of above Eq. (7.15) we have

$$R_2 R_3 = R_1 R_4$$

$$\text{or } R_1 = \frac{R_2 R_3}{R_4} \quad ...(7.16)$$

Equating the imaginary terms in Eq. (7.15) we have

$$\omega\, C_1\, R_1 R_2 R_3 = \omega\, C_2\, R_1\, R_2\, R_4$$

$$\text{or } C_1 = C_2 \times \frac{R_4}{R_3} \quad ...(7.17)$$

Example 7.7. A parallel resistance-capacitance bridge shown in Fig. 7.8 has a standard capacitor C_2 = 0.12 μF and resistance R_3 = 15 kΩ. Balance is achieved when R_2 = 330 kΩ and R_4 = 22 kΩ. Power supply frequency is 50 Hz. Determine the resistive and capacitive components of unknown capacitor C_1 and its dissipation factor.

Solution: Resistive component of unknown capacitor,

$$R_1 = \frac{R_2 R_3}{R_4} = \frac{330 \times 15}{22} = 225 \text{ k}\Omega \quad \textbf{Ans.} \qquad \text{Refer to Eq. (7.16)}$$

Capacitive component of unknown capacitor,

$$C_1 = C_2 \times \frac{R_4}{R_3} = \frac{0.12\,\mu\text{F} \times 22\text{ k}\Omega}{15\text{ k}\Omega} = 0.176\,\mu\text{F} \quad \textbf{Ans.} \qquad \text{Refer to Eq. (7.17)}$$

$$\text{Dissipation factor, } D = \frac{1}{\omega C_1 R_1} = \frac{1}{2\pi \times 50 \times 0.176 \times 10^{-6} \times 225 \times 10^{3}} = 0.08 \quad \textbf{Ans.}$$

7.3.4. Schering Bridges. The Schering bridge is one of the most important and useful circuits available for measurement of capacitance and dielectric loss. It is widely used, both for precision measurements of capacitors on low voltages and for study of insulation and insulating structures at high voltages. It is met in several arrangements, or modification to adapt to special applications.

The circuit is shown in Fig. 7.9 (*a*) in which C_1 is the capacitor under test, R_1 is an imaginary resistance representing its dielectric loss component; C_3 is a standard capacitor, C_4 is a variable capacitor and R_2 and R_4 are known non-inductive resistors. The resistance R_2 is variable.

The impedances of the Schering bridge arms are:

$$\mathbf{Z}_1 = \frac{1}{\dfrac{1}{R_1} + \dfrac{1}{-j/\omega C_1}} = \frac{1}{\dfrac{1}{R_1} + j\,\omega C_1} = \frac{R_1}{1 + j\,\omega\, C_1\, R_1}$$

$$\mathbf{Z}_2 = R_2$$

$$\mathbf{Z}_3 = -j/\omega C_3$$

$$\text{and } \mathbf{Z}_4 = \frac{R_4}{1 + j\,\omega C_4 R_4}$$

Under balanced conditions

$$\mathbf{Z}_1 \mathbf{Z}_4 = \mathbf{Z}_2 \mathbf{Z}_3$$

$$\text{i.e. } \frac{R_1}{1 + j\omega C_1 R_1} \times \frac{R_4}{1 + j\omega C_4 R_4} = \frac{-jR_2}{\omega C_3}$$

$$\text{or } \frac{R_1 R_4}{1 + j\omega C_1 R_1} \times \frac{1 - j\omega C_1 R_1}{1 - j\,\omega C_1 R_1} = \frac{-jR_2}{\omega C_3}(1 + j\omega C_4 R_4)$$

$$\text{or } \frac{R_1 R_4 (1 - j\omega C_1 R_1)}{1 + \omega^2 C_1^2 R_1^2} = \frac{-jR_2}{\omega C_3}(1 + j\omega C_4 R_4)$$

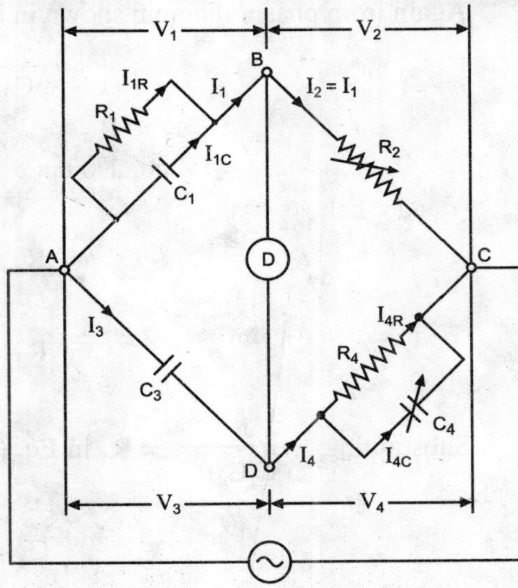

(a) Circuit Diagram

Equating real terms we have

$$\frac{R_1 R_4}{1 + \omega^2 C_1^2 R_1^2} = \frac{C_4 R_4 R_2}{C_3}$$

$$\text{or } \frac{R_1}{1 + \omega^2 C_1^2 R_1^2} = \frac{C_4 R_2}{C_3} \qquad \text{...(7.18)}$$

From phasor diagram shown in Fig. 7.9 (*b*) for the capacitor C_1 and R_1 in parallel

$$\cos\delta = \frac{I_{1C}}{I_1} = \frac{I_{1C}}{\sqrt{(I_{1C})^2 + (I_{1R})^2}}$$

$$= \frac{V_1 \omega C_1}{\sqrt{V_1^2 \omega^2 C_1^2 + \dfrac{V_1^2}{R_1^2}}}$$

$$\text{or } \cos\delta = \frac{\omega C_1 R_1}{\sqrt{1 + \omega^2 C_1^2 R_1^2}}$$

$$\text{or } \cos^2\delta = \frac{\omega^2 C_1^2 R_1^2}{1 + \omega^2 C_1^2 R_1^2}$$

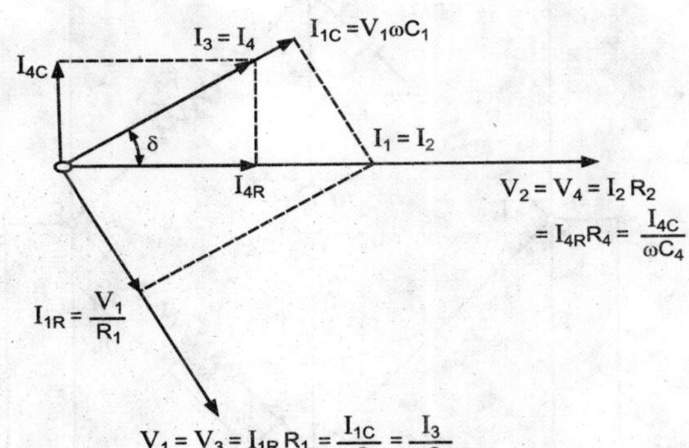

(b) Phasor Diagram

Fig. 7.9

$$\text{Substituting } \frac{\omega^2 C_1^2 R_1^2}{1 + \omega^2 C_1^2 R_1^2} = \cos^2\delta \text{ in Eq. (7.18)}$$

we have

$$\frac{\cos^2\delta}{\omega^2 C_1^2 R_1} = \frac{C_4 R_2}{C_3}$$

$$\text{or } C_1 = \frac{C_3 \cos^2\delta}{\omega^2 C_1 C_4 R_1 R_2} \qquad \text{...(7.19)}$$

Again from phasor diagram shown in Fig. 7.9 (*b*) for the bridge network under balance conditions,

$$\text{Tan } \delta = \frac{I_{4C}}{I_{4R}} = \frac{V_4 \omega C_4}{V_4 / R_4} = \omega C_4 R_4 \quad \text{...(7.20)}$$

$$\text{and also tan } \delta = \frac{I_{1R}}{I_{1C}} = \frac{V_1 / R_1}{V_1 \omega C_1} = \frac{1}{\omega C_1 R_1}$$

$$\therefore \omega C_4 R_4 = \frac{1}{\omega C_1 R_1}$$

$$\text{or } R_4 = \frac{1}{\omega^2 C_1 C_4 R_1} \quad \text{...(7.21)}$$

Substituting $\frac{1}{\omega^2 C_1 C_4 R_1} = R_4$ in Eq. (7.19) we have

$$C_1 = \frac{C_3 R_4 \cos^2 \delta}{R_2} \quad \text{...(7.22)}$$

where δ is known as a loss angle of capacitor and sin δ is known as a power factor of the capacitor. Tan δ is known as disipation factor of the capacitor.

(*i*) ***Imperfect Capacitor as a Series Circuit.*** An alternative method to that of representing an imperfect capacitor diagrammatically as a perfect capacitor C_1 in parallel with a resistance R_1, is to represent it as a perfect capacitor C_x in series with resistance R_x, as illustrated in Fig. 7.10 (*a*).

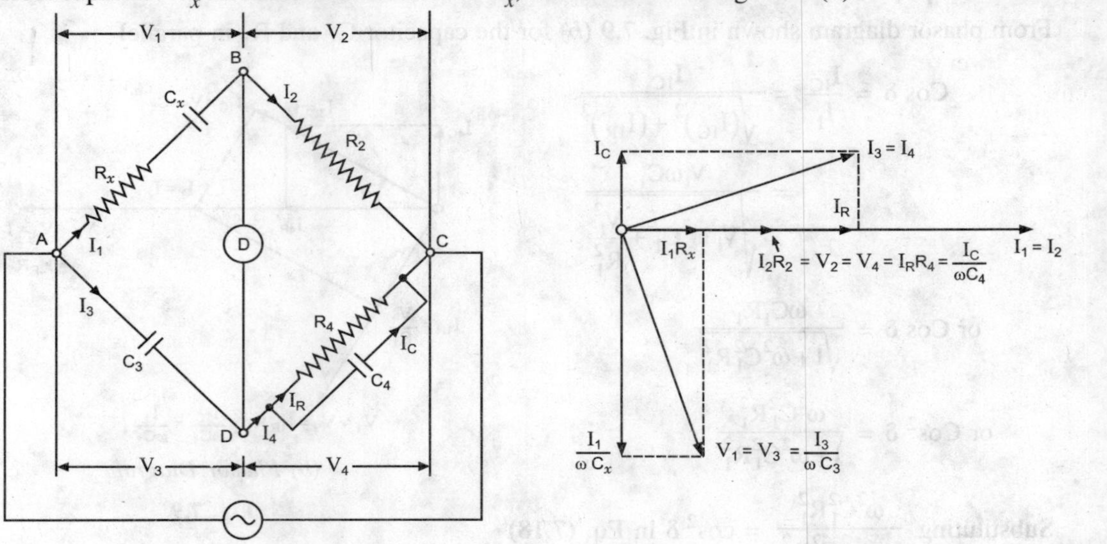

(*a*) *Circuit Diagram* (*b*) *Phasor Diagram*

Fig. 7.10

The standard capacitor C_3 is a high-quality mica capacitor (low-loss) for general measurements or an air capacitor (having a very stable value and a very small electric field) for insulation measurement.

The impedances of Schering bridge with series representation are:

$$\mathbf{Z}_1 = R_x - j/\omega C_x$$

$$\mathbf{Z}_2 = R_2$$

$$\mathbf{Z}_3 = -j/\omega C_3$$

$$\mathbf{Z}_4 = \frac{R_4}{1 + j\omega C_4 R_4}$$

Balance equations may be written as

$$\mathbf{Z}_1\mathbf{Z}_4 = \mathbf{Z}_2\mathbf{Z}_3$$

$$\left(R_x - \frac{j}{\omega C_x}\right)\left(\frac{R_4}{1 + j\omega C_4 R_4}\right) = \frac{-jR_2}{\omega C_3}$$

$$\text{or} \quad \left(R_x - \frac{j}{\omega C_x}\right)R_4 = \frac{-jR_2}{\omega C_3}(1 + j\omega C_4 R_4)$$

Equating real and imaginary components separately we have

$$R_x = \frac{C_4}{C_3} R_2 \qquad \text{...(7.23)}$$

$$\text{and } C_x = \frac{R_4}{R_2} C_3 \qquad \text{...(7.24)}$$

Generally C_4 and R_4 are chosen as the variable elements in order to get both of the balance equations independently.

This bridge is widely used for testing small capacitors at low voltages with very high precision.

The power factor of a series R-C combination is defined as the cosine of the phase angle θ of the circuit. So the power factor of the unknown capacitor

$$PF_x = \frac{R_x}{Z_x}$$

For phase angles very close to 90º, the reactance is almost equal to the impedance and power factor can be approximated by

$$PF_x = D_x = \frac{R_x}{X_x} = \omega C_x R_x \qquad \text{...(7.25)}$$

where D_x is the dissipation factor and is defined as the cotangent of the phase angle and, therefore, by definition, the dissipation factor

$$D_x = \frac{R_x}{X_x} = \omega C_x R_x$$

Incidentally, since the quality of a coil is defined by $Q = \frac{X_L}{R_L}$, we find that the dissipation factor, D, is the reciprocal of the quality factor, Q, and, therefore, $D = \frac{1}{Q}$.

The dissipation factor indicates the quality of a capacitor; *i.e.* how close the phase angle, θ, of the capacitor is to ideal value of 90º. By substituting the value of C_x from Eq. (7.24) and of R_x from Eq. (7.23) in Eq. (7.25) we have

$$\text{Dissipation factor, } D = \omega \times \frac{R_4}{R_2} \times C_3 \times \frac{C_4}{C_3} \times R_2 = \omega R_4 C_4 \qquad \text{...(7.26)}$$

Equation (7.26) not only is convenient for computation of dissipation factor but also suggests an arrangement that is used in several bridges. If R_4 is fixed, then for a particularly frequency PF_x depends directly on C_4. Accordingly, the dial of C_4 may be calibrated directly in terms of dissipation factor D. The calibration holds for the given frequency, but may be used at another frequency by multiplying the ratio of the frequencies.

The dissipation factor, D, of a series circuit, as already mentioned, is defined as the cotangent of the phase angle θ, or tangent $(90° - \theta)$ or tangent δ. The angle δ is called the phase-defect angle, loss angle or phase angle error as it is the angle by which an actual capacitor falls short of perfection. It is better to mark the loss dial of a Schering bridge as dissipation factor rather than power factor since this is true regardless of magnitude, whereas it equals power factor only for small loss-angles. The difference between the two terms is insignificant for small loss-angles but becomes appreciable for dissipation factors exceeding 0.1.

7.3.5. High Voltage Schering Bridge. This is one of the most important and useful bridge circuits available for measurement of capacitance, dielectric loss of a capacitor and power factor of cables. It is generally used, both for precision measurements of capacitors at low voltage and for study of insulation and insulating structures at high voltages.

The bridge is also used for measurement of capacitance and dissipation factor for insulation and in determining the general properties of condenser, bushings, insulating oil and other insulating materials which in any case have very small capacitances, so for such types of measurements high voltage Schering bridge shown in Fig. 7.11 is required. But certain modifications have to be made in Schering bridge to operate at high voltage.

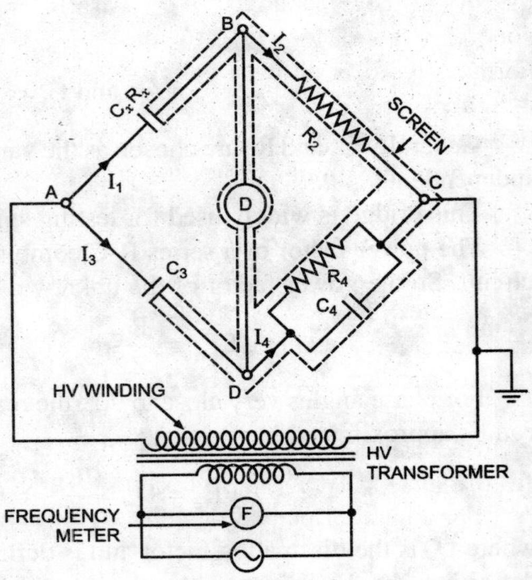

Fig. 7.11 *High Voltage Schering Bridge*

(*i*) In this case a high voltage supply is obtained from a transformer usually at the power frequency and the detector used is vibration galvanometer.

(*ii*) The Schering bridge is advantageous for high voltage tests due to the fact that the arms AB and AD, each contains only capacitors, which can be specially constructed for high voltage work. Also these two arms will have very high impedances, at power frequencies in comparison with that of the arms BC and CD. Thus the major portion of the source potential difference exists across the arms AB and AD and the potential difference across the arms BC and CD, which contains the control, is small. In order to afford safety to the operator from the high voltage hazards, while carrying out control, junction point C is earthed.

The arrangement is ideal from the point of view of safety but it is much less sensitive than its conjugate (with the source and detector interchanged). However, at high voltages sensitivity, with a good detector will usually be ample for all practical purposes. In precise measurements at low voltages the conjugate bridge is often used since in this case sensitivity will be much greater.

(*iii*) In order to prevent dangerous rises of voltage across these arms in case of breakdown of either of the high voltage capacitors; a spark (set to breakdown at about 100 volts) is connected across each of the arms BC and CD.

(*iv*) The impedances of arms AB and AD are high so current from the source is less. The arrangement is ideal from the point of view of power consumption but it is much less sensitive so a sensitive detector has to be used.

(*v*) In operating the Schering bridge at high voltages, and in the precise measurements at low voltages it is extremely important that the effect of stray capacitances between the bridge elements be eliminated. This is done by enclosing the vulnerable portions of the network within electrostatic shields and by maintaining these shields at suitable potentials. Such a shielded bridge suitable for high voltage operation is shown in Fig. 7.11. Earth capacitance effect on the galvanometer and leads is eliminated by using Wagner earth device.

(*vi*) A capacitor C_3 of fixed value is used which has normally air or compressed gas as dielectric. The dissipation factor with dry and clean gas is practically zero but loss in the insulating supports cannot be avoided. However, such losses can be prevented from influencing the measurement by the use of a guard ring from which both the high and low voltage electrodes are supported. Then the current through the high voltage supports passes direct to earth and as the potential difference between the low voltage electrode and the guard ring is very small, the insulation of the low voltage electrode has only a very small effect. In case, solid dielectric standard capacitor is used then the losses must be accurately known.

7.3.6. Measurement of Relative Permittivity With Schering Bridge. The Schering bridge circuit is also used for the measurement of relative permittivity of specimens of dielectric materials. This involves measuring the capacitance of a small capacitor with the specimen as dielectric. The electrode configuration normally used is either a parallel-plate or a concentric-cylinder, guard circuits being used to make the plate area definite.

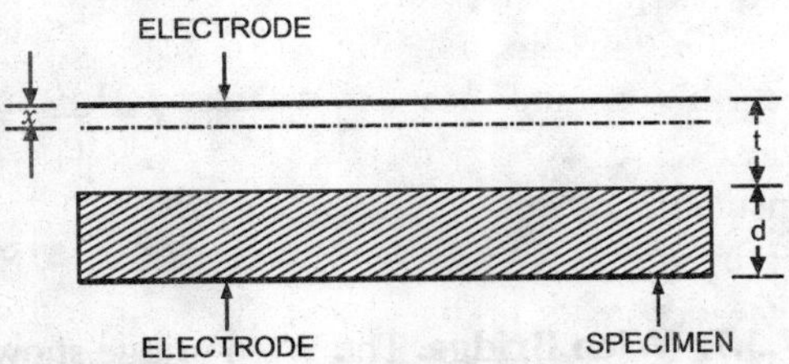

Fig. 7.12 *Relative Permittivity Measurement*

There are many different techniques which are employed, both in the type of specimen capacitor used and in the testing circuit. Usually a disc specimen with metal electrodes is used for testing of solid materials such as bakelite. The electrodes may consist of thin metal foil attached to the specimen by petroleum jelly or thin films of silver or aluminium applied by evaporation (both of such arrangements normally have solid metal backing electrodes) or mercury. Mercury electrodes can be had by floating (or supporting) the specimen on the surface of a mercury pool, the upper eletrodes consisting of a smaller merury pool held in place by a metal containing ring. Liquid specimens fill the space between the concentric cylindrical electrodes of a test cell.

For a parallel-plate arrangement, the capacitance is given as

$$C = \frac{\varepsilon_0 \varepsilon_r A}{d} \quad ...(7.27)$$

where ε_0 is the permittivity of the free space, ε_r the relative permittivity of the specimen, A is the effective area of electrodes and d is the spacing between electrodes.

Equation (7.27) may be rewritten as

$$\varepsilon_r = \frac{Cd}{\varepsilon_0 A} \quad ...(7.28)$$

Thus by measuring value of capacitance with specimen as dielectric and dimensions of electrodes (A, the area and d, the spacing), relative permittivity can be determined.

The necessity for close contact between the electrodes and specimen can be avoided if we use a pair of solid electrodes (one with a guard ring) between which the specimen of thickness lesser than that spacing between the electrodes is slipped. The capacitance of the arrangement is measured, the specimen is removed, and the spacing between the electrodes is adjusted (by means of micrometer adjustment), so as to restore the capacitance value (*i.e.* value of capacitance before removal of specimen).

The relative permittivity of the specimen then can be determined from the specimen thickness and the alteration in electrode spacing, as derived below.

The capacitance of an air capacitor with electrodes $(d + t)$ metres apart and specimen of d metre thickness between them is given as

$$C_1 = \frac{\varepsilon_0 A}{t + \frac{d}{\varepsilon_r}} \quad \text{...(7.29)}$$

and capacitance of air capacitor with specimen removed and having electrodes $(d + t - x)$ metres apart is given as

$$C_2 = \frac{\varepsilon_0 A}{d+t-x} \quad \text{...(7.30)}$$

Since $C_1 = C_2$,

$$\frac{\varepsilon_0 A}{t + \frac{d}{\varepsilon_r}} = \frac{\varepsilon_0 A}{d+t-x}$$

$$\text{or} \quad t + \frac{d}{\varepsilon_r} = d + t - x$$

$$\text{or} \quad \varepsilon_r = \frac{d}{d-x} \quad \text{...(7.31)}$$

7.3.7. Wien Bridge. The Wien bridge shown in Fig. 7.13 has a series R-C combination in one arm and a parallel combination in the adjoining arm.

This bridge circuit was widely used for measuring capacitance of capacitors and their losses, even at high voltages, until the advantages of the Schering bridge were generally realized. The bridge circuit is frequency sensitive, and nowadays it is employed for determination and control of frequency.

The impedances of Wien bridge are:

$$Z_1 = R_1 - \frac{j}{\omega C_1}$$

$$Z_2 = R_2$$

$$Z_3 = \frac{R_3}{1 + j\omega C_3 R_3}$$

$$\text{and } Z_4 = R_4$$

Under balanced conditions of bridge,

$$Z_1 Z_4 = Z_2 Z_3$$

$$\left(R_1 - \frac{j}{\omega C_1}\right) R_4 = R_2 \frac{R_3}{1 + j\omega C_3 R_3}$$

$$\text{or } R_4 \left(R_1 - \frac{j}{\omega C_1}\right)(1 + j\omega C_3 R_3) = R_2 R_3$$

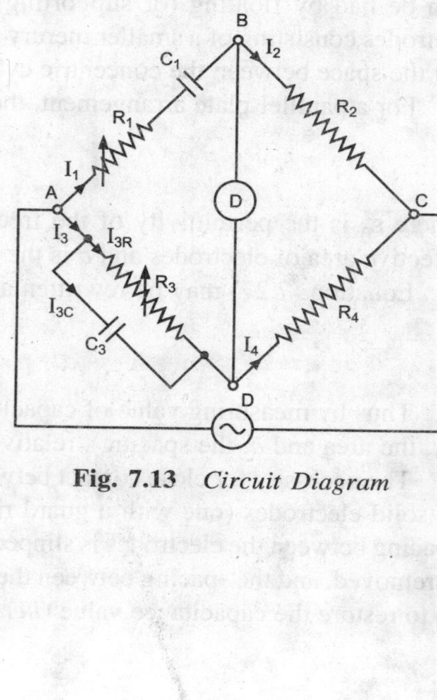

Fig. 7.13 *Circuit Diagram*

Separating real and imaginary terms we have

$$R_1 R_4 + R_3 R_4 \frac{C_3}{C_1} = R_2 R_3$$

$$\text{or}\quad \frac{C_3}{C_1} = \frac{R_2}{R_4} - \frac{R_1}{R_3} \qquad ...(7.32)$$

$$\text{and}\quad \frac{-R_4}{\omega C_1} + \omega C_3 R_1 R_3 R_4 = 0$$

$$\text{or}\quad \omega^2 = \frac{1}{R_1 R_3 C_1 C_3} \qquad ...(7.33)$$

$$\text{or}\quad \omega = \frac{1}{\sqrt{R_1 R_3 C_1 C_3}}$$

$$\text{or}\quad f = \frac{\omega}{2\pi} = \frac{1}{2\pi\sqrt{R_1 R_3 C_1 C_3}} \qquad ...(7.34)$$

The two conditions for bridge balance, Eqs. (7.32) and (7.34) result in an expression determining the required resistance ratio $\frac{R_2}{R_4}$ and another expression determining the frequency of the applied voltage. If satisfying Eq. (7.32) the bridge is excited with a frequency given by Eq. (7.34), the bridge will be balanced.

In most Wien bridge circuits, the components are chosen such that $R_1 = R_3 = R$ and $C_1 = C_3 = C$. Equation (7.32), therefore, reduces to $\frac{R_2}{R_4} = 2$ and Eq. (7.34) to $f = \frac{1}{2\pi RC}$, which is general equation for the frequency of the bridge circuit.

The bridge is used for measuring frequency in audio-range. Resistances R_1 and R_3 can be ganged together to have identical values. Capacitors C_1 and C_3 are normally of fixed values.

The audio-range is normally divided into 20–200–2 K–20 kHz ranges. In this case, the resistances can be used for range changing and capacitors C_1 and C_3 for fine frequency control within the range. The bridge can also be used for measuring capacitances. In that case, the supply frequency must be known.

At higher frequencies loss effects in the capacitors and stray capacitance effects make the method less satisfactory.

The bridge is also used in harmonic distortion analyzer, as a notch filter, and in audio-frequency and radio-frequency oscillators as a frequency determining element.

An accuracy of 0.5% – 1% can be readily achieved using this bridge. Because it is frequency sensitive, it is difficult to balance unless the waveform of the applied voltage is purely sinusoidal.

Example 7.8. A lossy capacitor is tested with a Schering bridge circuit. Balance is obtained with the capacitor under test in one arm, the succeeding arms being, a non-inductive resistor of 100 Ω, a non-inductive resistor of 309 Ω in parallel with a pure capacitor of 0.5 μF and a standard capacitor of 109 μμF. The supply frequency is 50 Hz. Calculate from the equation at balance the equivalent series capacitance and power factor (at 50 Hz) of the capacitor under test.[*Nagpur Univ. Elec. Measurements & Instruments, 1992*]

Solution: Capacitance of variable capacitor, $C_4 = 0.5 \times 10^{-6}$ F

Non-reactive resistance, $R_4 = 309\ \Omega$

Non-inductive resistance, $R_2 = 100\ \Omega$

Standard capacitor, $C_3 = 109 \times 10^{-12}$ F

Equivalent series capacitance, $C_x = \frac{R_4}{R_2} \times C_3 = \frac{309}{100} \times 109 \times 10^{-12} = 336.81$ μμF **Ans.**

$$\text{Series resistance, } R_x = \frac{C_4}{C_3} \times R_2 = \frac{0.5 \times 10^{-6}}{109 \times 10^{-12}} \times 100 = 0.4587 \text{ M}\Omega$$

Power factor of the specimen = Sin δ

$$= \text{Tan } \delta = \omega C_x R_x = 2\pi \times 50 \times 336.81 \times 10^{-12} \times 0.4587 \times 10^6 = 0.0485 \textbf{ Ans.}$$

Example 7.9. A Schering bridge is used for measuring the power loss in dielectries. The specimens are in the form of discs 0.3 cm thick and have a dielectric constant of 2.3. The area of each electrode is 314 cm² and the loss angle is known to be 9° for the frequency of 50 Hz. The fixed resistor of the network has a value of 1,000 Ω and the fixed capacitance is 50 pF. Determine the variable resistor and capacitor required. Derive the equations needed for balance. *[U.P. Technical Univ. Electrical Measurements and Measuring Instruments 2004-05; U.P.S.C. I.E.S Electrical Engineering-I, 2000]*

Solution: Refer Figs. 7.9 (*a*) and 7.9 (*b*)

Fixed resistance, $R_4 = 1{,}000\ \Omega$

Fixed capacitance, $C_3 = 50 \text{ pF} = 50 \times 10^{-12}$ F

$$\text{Unknown capacitance, } C_1 = \frac{\varepsilon_r \varepsilon_0 A}{d} = \frac{2.3 \times 8.854 \times 10^{-12} \times 314 \times 10^{-4}}{0.3 \times 10^{-2}} = 213.15 \text{ pF}$$

$$\text{From Fig. 7.9 (b), Tan } \delta = \frac{1}{\omega C_1 R_1}$$

$$\text{So unknown resistance, } R_1 = \frac{1}{\omega C_1 \tan \delta} = \frac{1}{2\pi \times 50 \times 213.15 \times 10^{-12} \times \tan 9^\circ} = 94.3 \text{ M}\Omega$$

From Eq. (7.21)

$$\text{Variable capacitor, } C_4 = \frac{1}{\omega^2 C_1 R_1 R_4} = \frac{1}{(2\pi \times 50)^2 \times 213.15 \times 10^{-12} \times 94.3 \times 10^6 \times 1{,}000} = 0.5\ \mu\text{F } \textbf{Ans.}$$

From Eq. (7.22)

$$\text{Variable resistance, } R_2 = \frac{C_3 R_4 \cos^2 \delta}{C_1} = \frac{50 \times 10^{-12} \times 1{,}000 \times (\cos 9^\circ)^2}{213.15 \times 10^{-12}} = 230\ \Omega \textbf{ Ans.}$$

Example 7.10. Find the equivalent parallel-resistance and capacitance that causes Wien bridge to null with the following component values:

$R_1 = 3.1$ kΩ, $C_1 = 5.2$ μF, $R_2 = 25$ kΩ, $R_4 = 100$ kΩ; $f = 2.5$ kHz

[U.P. Technical Univ. Electronics Measurements and Instrumentation, 2005-06]

Solution: The circuit diagram is shown in the Fig. 7.14. In this problem

$R_1 = 3.1 \text{ k}\Omega = 3{,}100\ \Omega$

$C_1 = 5.2\ \mu\text{F} = 5.2 \times 10^{-6}$ F

$R_2 = 25 \text{ k}\Omega = 25{,}000\ \Omega$

$R_4 = 100 \text{ k}\Omega = 100{,}000\ \Omega$

$f = 25 \text{ kHz} = 25{,}000$ Hz

From Eqs. (7.32) and (7.33) we have

$$\frac{C_3}{C_1} = \frac{R_2}{R_4} - \frac{R_1}{R_3}$$

$$\text{or } C_3 = C_1 \left[\frac{R_2}{R_4} - \frac{R_1}{R_3}\right] = 5.2 \times 10^{-6} \left[\frac{25{,}000}{100{,}000} - \frac{3{,}100}{R_3}\right]$$

$$= 5.2 \times 10^{-6} \left[0.25 - \frac{3{,}100}{R_3}\right] \quad ...(i)$$

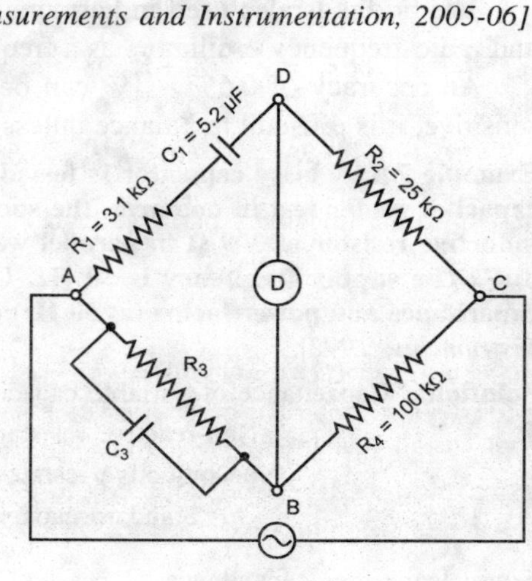

Fig. 7.14

$$\text{and } \omega = \frac{1}{\sqrt{R_1R_3C_1C_3}}$$

$$\text{or } R_3C_3 = \frac{1}{\omega^2R_1C_1} = \frac{1}{(2\pi\times 25{,}000)^2\times 3100\times 5.2\times 10^{-6}} = 1.5797\times 10^{-8}$$

$$\text{or } \quad R_3 = \frac{1.5797\times 10^{-8}}{C_3}$$

Substituting $R_3 = \dfrac{1.5797\times 10^{-8}}{C_3}$ in Eq. (*i*) we have

$$\text{or } C_3 = 5.2\times 10^{-6}\left[0.25 - \frac{3{,}100\times C_3}{1.5797\times 10^{-8}}\right]$$

$$\text{or } C_3 = 1.274\times 10^{-12}\text{ F} = 1.274\ \mu\mu\text{F} \quad \textbf{Ans.}$$

$$R_3 = \frac{1.5797\times 10^{-8}}{1.274\times 10^{-12}} = 12{,}400\ \Omega \text{ or } 12.4\text{ k}\Omega \quad \textbf{Ans.}$$

Example 7.11. List the factors that may lead to inaccuracies in measurements by ac bridges. The four arms of a bridge are:

arm *ab*: an imperfect capacitor C_1 with an equivalent series resistance of r_1

arm *bc*: an non-inductive resistance R_3

arm *cd*: a non-inductive resistance R_4

arm *da* : an imperfect capacitor C_2 with an equivalent resistance r_2 in series with a resistance R_2.

A supply of 450 Hz is given between terminals *a* and *c* and a detector is connected between *b* and *d*. At balance $R_2 = 5\ \Omega$, $R_3 = 2{,}000\ \Omega$, $R_4 = 2{,}950\ \Omega$, $C_2 = 0.5\ \mu F$ and $r_2 = 0.4\ \Omega$. Calculate the values of C_1 and r_1 and also the dissipating factor for this capacitor. Derive the relations used, if any.

[*U.P.S.C. I.E.S. Electrical Engineering-I, 1995*]

Solution: Under balanced conditions

$$\mathbf{Z_1Z_4 = Z_2Z_3}$$

$$\left(r_1 + \frac{1}{j\omega C_1}\right)R_4 = \left(r_2 + R_2 + \frac{1}{j\omega C_2}\right)R_3$$

Equating the real parts we have

$$r_1R_4 = (r_2 + R_2)R_3$$

$$\text{or } r_1 = \frac{R_3}{R_4}(r_2 + R_2)$$

$$= \frac{2{,}000}{2{,}950}(0.4 + 5) = 3.661\ \Omega \quad \textbf{Ans.}$$

Equating imaginary parts we have

$$\frac{R_4}{j\omega C_1} = \frac{R_3}{j\omega C_2}$$

$$\text{or } \quad C_1 = \frac{R_4}{R_3}\times C_2 = \frac{2{,}950}{2{,}000}\times 0.5\times 10^{-6} = 0.7375\ \mu F \quad \textbf{Ans.}$$

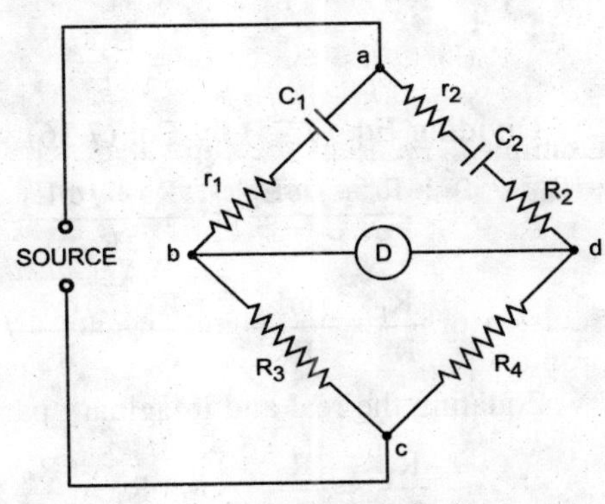

Fig. 7.15

Dissipation factor,

$$\tan\delta = \omega C_1R_1 = 2\pi\times 450\times 0.7375\times 10^{-6}\times 3.661 = 7.634\times 10^{-3} \quad \textbf{Ans.}$$

7.4 INDUCTANCE BRIDGES

7.4.1. Inductance Comparison or Maxwell's Bridge. This method is very suitable for accurate measurement of medium inductances. In this method unknown inductance is determined by comparing it with a standard self inductance. Such a bridge circuit is shown in Fig. 7.16 (*a*) in which L_1 is an unknown self-inductance of resistor R_1, L_3 is a known variable inductance of resistor R_3 whose resistance is constant, R_2 and R_4 are pure resistances and D is a detector. The magnitude of L_3 should be of the same order as that of L_1. The bridge is balanced by varying L_3 and one of the resistances R_2 or R_4. The bridge can also be balanced by keeping R_2 and R_4 constant and by varying the resistance of any one of the other two arms by connecting an additional resistance in that arm.

When the bridge is balanced, the current flowing through detector D is zero and

$$\mathbf{I_1} = \mathbf{I_2};\ \mathbf{I_3} = \mathbf{I_4}$$

PD across arm AB = PD across arm AD = $\mathbf{V_1}$

$$i.e.\ \mathbf{I_1}\,\mathbf{Z_1} = \mathbf{I_3}\,\mathbf{Z_3} = \mathbf{V_1}$$

$$\text{or } \mathbf{I_1}\,(R_1 + j\,\omega\,L_1) = \mathbf{I_3}\,(R_3 + j\,\omega\,L_3) = \mathbf{V_1} \quad ...(7.35)$$

and PD across arm BC = PD across arm CD = $\mathbf{V_2}$

$$\text{or } \mathbf{I_2}\,R_2 = \mathbf{I_4}\,R_4 = \mathbf{V_2}$$

$$\text{or } \mathbf{I_1}\,R_2 = \mathbf{I_3}\,R_4 \quad ...(7.36)$$

$$\because \quad \mathbf{I_1} = \mathbf{I_2} \text{ and } \mathbf{I_3} = \mathbf{I_4}$$

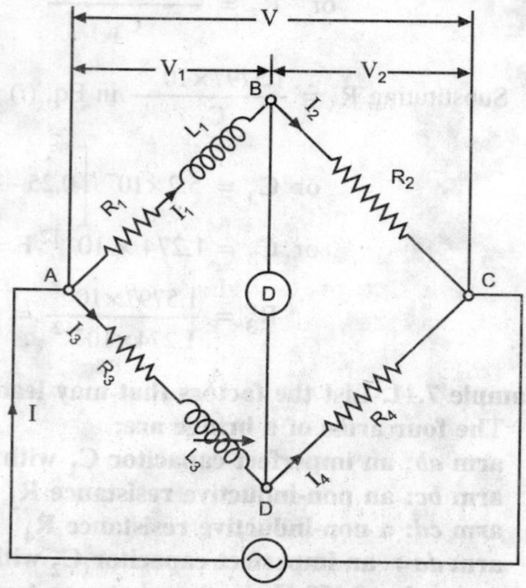

(*a*) *Circuit Diagram*

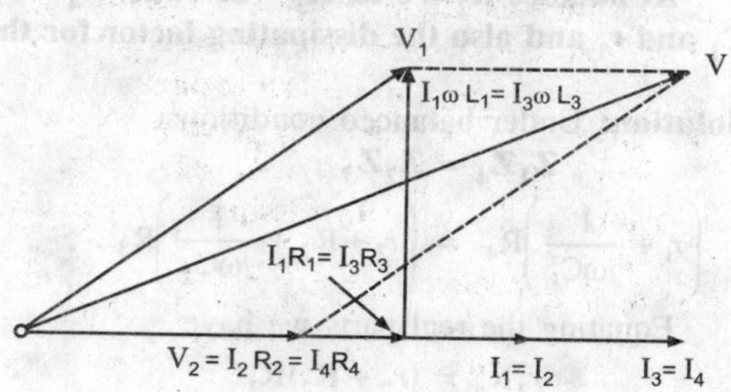

(*b*) *Phasor Diagram*

Fig. 7.16 *Maxwell's Bridge*

Dividing Eq. (7.35) by Eq. (7.36) we have

$$\frac{R_1 + j\omega L_1}{R_2} = \frac{R_3 + j\omega L_3}{R_4}$$

$$\text{or} \quad \frac{R_1}{R_2} + \frac{j\omega L_1}{R_2} = \frac{R_3}{R_4} + \frac{j\omega L_3}{R_4}$$

Equating the real and imaginary parts of both sides separately we have

$$\frac{R_1}{R_2} = \frac{R_3}{R_4} \quad \text{or} \quad R_1 = \frac{R_2}{R_4}R_3 \quad ...(7.37a)$$

$$\text{and} \quad \frac{\omega L_1}{R_2} = \frac{\omega L_3}{R_4} \quad \text{or} \quad L_1 = \frac{R_2}{R_4}L_3 \quad ...(7.37b)$$

Thus value of unknown self inductance L_1 can be determined.

Phasor diagram for the balanced condition of the bridge is shown in Fig. 7.16 (*b*), in which currents I_1 and I_2 are made in phase with currents I_3 and I_4 by adjusting the impedances of the various branches. Here one point is to be noted that inductances L_1 and L_3 should be placed at a distance from each other and leads used in the bridge arms should be twisted together properly to avoid loops, otherwise loops introduce inductance in the bridge circuit which may cause error during measurement.

Example 7.12. In a bridge shown in figure below the balance was obtained with L_1 = 52.6 mH and R_2 = 1.68 Ω. Find the resistance and inductance of coil.

[U.P. Technical Univ. Electronic Measurements and Instrumentation 2007-08]

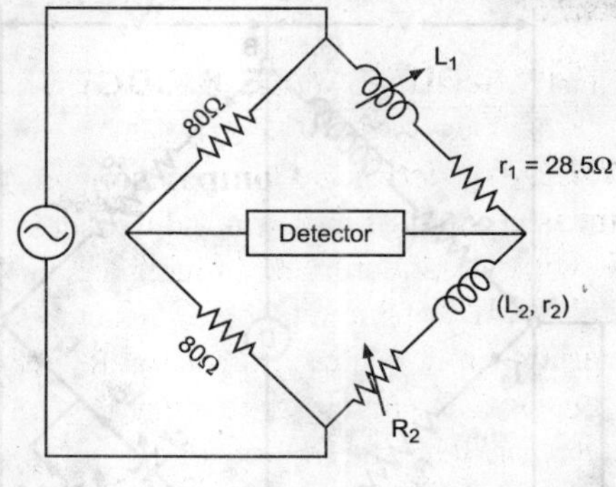

Fig. 7.17

Solution: Resistance of coil, $r_2 = r_1 \times \frac{80}{80} - R_2$

$$= 28.5 - 1.68 = 26.82\ \Omega \quad \textbf{Ans.}$$

Inductance of coil, $L_2 = L_1 \times \frac{80}{80}$

$$= L_1 = 52.6\ \text{mH} \quad \textbf{Ans.}$$

Example 7.13. The four arms of a Maxwell bridge network are as follows :

AB and BC are non-inductive resistors of 100 Ω each, DA is a standard variable inductor L of resistance 32.7 Ω and CD comprises a standard variable resistor R in series with a coil of unknown impedance. Balance is obtained when L = 47.8 mH and R = 1.36 Ω. Find resistance and inductance of the coil.

[Pb. Univ. Elec. Measurements-I, December 1992]

Solution: Circuit diagram is shown in Fig. 7.18.

When the bridge is balanced, the product of the resistances of opposite arms are equal

i.e. $(R_1 + R)R_4 = R_2 R_3$

or Resistance of coil,

$$R_1 = \frac{R_2 R_3}{R_4} - R = \frac{100 \times 32.7}{100} - 1.36$$

$$= \mathbf{31.34\ \Omega}\ \textbf{Ans.}$$

In case of balanced position of bridge also

$$L_1 = \frac{R_2}{R_4} \times L = \frac{100}{100} \times 47.8 = \mathbf{47.8\ mH\ Ans.}$$

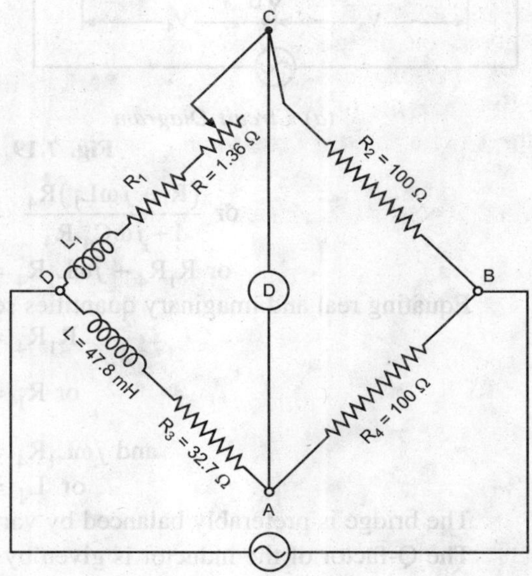

Fig. 7.18

7.4.2. Maxwell Wien Bridge. This is also known as Maxwell's inductance-capacitance bridge.

In this method of measurement of self-inductance, the unknown self inductance is compared with a standard variable capacitance, the circuit being shown in Fig. 7.19 (*a*). In the circuit L_1 is unknown self-inductance and R_1 is unknown resistance of the inductor, R_2, R_3 and R_4 are known non-inductive resistances and C_4 is a standard variable capacitor.

$$\text{Impedance } \mathbf{Z_1} = R_1 + j\,\omega\,L_1$$

$$\text{Impedance } \mathbf{Z_2} = R_2$$

$$\text{Impedance } \mathbf{Z_3} = R_3$$

$$\text{Impedance } \mathbf{Z_4} = \frac{1}{\frac{1}{R_4} + \frac{1}{-jX_c}} = \frac{1}{\frac{1}{R_4} + j\omega C_4} = \frac{R_4}{1 + j\omega C_4 R_4}$$

For balanced condition of bridge,

$$\mathbf{Z_1 Z_4 = Z_2 Z_3}$$

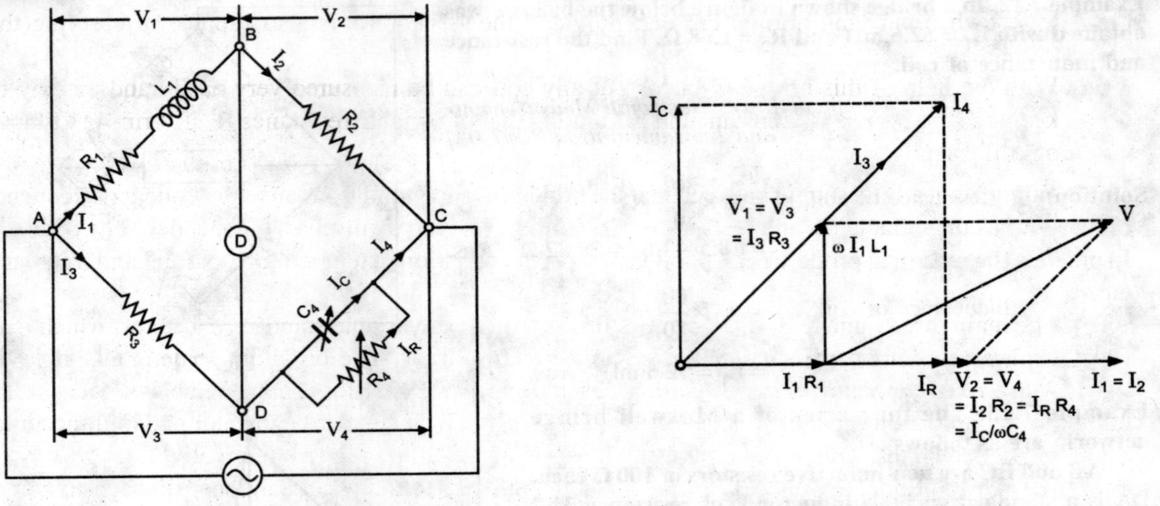

(a) *Circuit Diagram* (b) *Phasor Diagram*

Fig. 7.19 *Maxwell Wein Bridge*

$$\text{or} \quad \frac{(R_1 + j\omega L_1)R_4}{1 + j\omega C_4 R_4} = R_2R_3$$

$$\text{or} \quad R_1R_4 + j\omega L_1R_4 = R_2R_3 + j\omega C_4R_4R_2R_3$$

Equating real and imaginary quantities separately we have

$$R_1R_4 = R_2R_3$$

$$\text{or } R_1 = \frac{R_2R_3}{R_4} \qquad \text{...(7.38)}$$

$$\text{and } j\omega L_1R_4 = j\omega C_4R_4R_2R_3$$

$$\text{or } L_1 = C_4R_2R_3 \qquad \text{...(7.39)}$$

The bridge is preferably balanced by varying C_4 and R_4 which gives independent settings. The Q-factor of the inductor is given by $\omega L_1/R_1$ and at balance condition

$$Q = \frac{\omega L_1}{R_1} = \omega C_4R_2R_3 \times \frac{R_4}{R_2\,R_3} = \omega C_4R_4 \qquad \text{...(7.40)}$$

The phasor diagram for the balanced condition of the bridge is shown is Fig. 7.19 (*b*).

Advantages. (*i*) This bridge permits the measurement of inductance in terms of capacitance and a capacitor has always advantage in comparison to an inductor as it gives practically no external field, so it is more compact and is easier to shield.

(*ii*) This ac bridge possesses most of the desirable features *i.e.* independence of balance equations when R_4 and C_4 are chosen as variable elements.

(*iii*) The bridge has simple expressions for both L_1 and R_1 in terms of the values of bridge components. Physically R_2 and R_3 each will be say 10, 100, 1,000 or 10,000 ohms to give a suitable value to the product R_2 R_3 which appears in both Eqs. (7.38) and (7.39), C_4 will be a decade capacitor box and R_4 will be a decade resistance box.

(*iv*) In this bridge, balance conditions are independent of frequency as the term frequency does not appear in any of the balance equations.

(*v*) An additional advantage possessed by this bridge lies in the fact that, at balance, the only component in series with the coil under test is a resistor. Thus heavy current carrying capacity coils may be tested when carrying normal current, provided the resistor R_2 is also capable of carrying this current.

(*vi*) With the help of this bridge Q-factor* of any coil can be measured very easily and quickly by keeping capacitance C_4 constant and calibrating dial of variable resistance R_4 in terms of Q-factor of coil directly.

Disadvantages. (*i*) It should be noted that the bridge-balance conditions are independent of frequency only, so far as the inductance and resistance of coil under test are themselves independent of frequency. In practice the electrical properties of a coil vary with the variation in temperature, current and frequency etc.

(*ii*) The main disadvantage of this bridge is that it requires a variable standard capacitor, which is an expensive component when calibrated to a high accuracy. The bridge is sometimes employed with a fixed capacitor, either because a variable standard capacitor is not available or because it is desired to obtain the higher accuracy possible with a fixed standard capacitor and high-accuracy resistance boxes. This is done by using any one of the following two alternatives.

(*a*) An additional resistance may be connected in series with the coil, and the balance is obtained by adjusting this resistance and R_4.

(*b*) The bridge may be balanced by adjustment of R_3 and R_4. It experiences sometimes difficulty in obtaining balance condition since the balancing conditions are not independent.

(*iii*) This is very convenient and useful bridge for the determination of inductance having medium Q-factor *i.e.* it should be $1 < Q < 10$.

When value of Q becomes higher then resistance R_4 becomes very high to achieve the balance condition (as it can be seen from the derivation) such that sometimes higher than available decade resistance boxes, and also resistance balance becomes very difficult to be determined precisely.

This bridge also works well for low values of Q except for very low values of Q when $Q < 1$. This condition arises only when Q-factors of inductive resistors or RF coils operating on audio-frequency are measured. In this type of measurement of very low value of Q-factor, when balance is obtained by adjusting capacitor C_4 and resistor R_4 then there is no interaction but usually capacitor C_4 is taken of fixed value because of cost and accuracy involved and balance is obtained by varying R_4 with resistances R_2 and R_3. In this new condition when R_2 or R_3 is used with resistance R_4 to get balance, Maxwell bridge has very poor convergence of balance and gives the effect known as *sliding balance.* This term sliding balance describes a condition of interaction between the controls, so that, when resistance R_4 is adjusted and then resistance R_2 (or R_3) is adjusted and then again R_4 is adjusted, a new apparent balance point is found *i.e.* the balance point appears to move or slide and settles only gradually to its final point after many adjustments. So a general problem of convergence to balance is found while measuring the Q-factor of very low value.

---

* Q-factor of a coil, also known as quality factor or figure of merit of a coil, is the reciprocal of its power factor

i.e. Q-factor of a coil $= \dfrac{1}{\cos \phi} = \dfrac{Z}{R} = \dfrac{\omega L}{R}$ because R is very small in comparison to its inductive reactance.

Time constant of a coil, $\tau = \dfrac{L}{R}$ is defined as the time required for the current to rise to its final steady-state value if it is continued rising at its initial rate or it may also be defined as the time required for the current to reach 0.632 of its final steady value in an inductive (R-L) circuit.

But the ratio of $\dfrac{L}{R}$ is called the time constant of the coil, represented by τ or T.

Example 7.14. The four arms of a Maxwell's capacitance bridge at balance are: arm *ab*, an unknown inductance L_1, having an inherent resistance R_1; arm *bc*, a non-inductive resistance of 1000 Ω; arm *cd*, a capacitor of 0.5 μF in parallel with a resistance of 1000 Ω; arm *da*, a resistance of 1000 Ω. Derive the equations of balance for the bridge and determine the value of R_1 and L_1. Draw the phasor diagram of the bridge under balanced conditions.

[U.P. Technical Univ. Elec. Measurements and Measuring Instruments 2006-07]

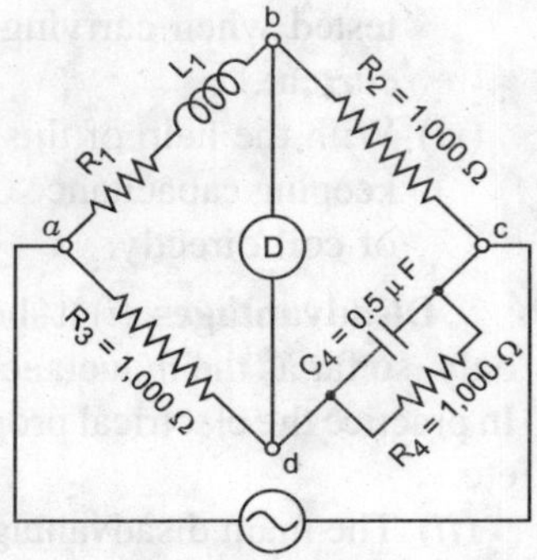

Fig. 7.20

Solution: Resistance of inductor,

$$R_1 = \frac{R_2 R_3}{R_4} = \frac{1{,}000 \times 1{,}000}{1{,}000} = 1{,}000\ \Omega \textbf{ Ans.}$$

Inductance of inductor,

$$L_1 = C_4 R_2 R_3 = 0.5 \times 10^{-6} \times 1{,}000 \times 1{,}000 = 0.5\ \text{H} \textbf{ Ans.}$$

Phasor diagram is shown in Fig. 7.16 (*b*).

7.4.3. Anderson Bridge.

This method is one of the commonest and best bridge method for precise measurement of inductance over a wide range. In this method the unknown self-inductance is measured in terms of a known capacitance and resistance by comparison. It is actually a method derived from Maxwell's method. The circuit diagram is shown is Fig. 7.21(*a*) in which L_1 is the self- inductance to be measured, R_1 is the resistance of arm 1 (including the resistance of the self-inductance), R_2, R_3, R_4 and *r* are known non-inductive resistances and C is a standard known capacitor.

The bridge is preliminary balanced for steady current by adjusting R_2, R_3 and R_4 and using an ordinary galvanometer as a detector, and then bridge is balanced in alternating current by varying *r* and using vibration galvanometer or telephone depending upon the supply frequency.

When the bridge is balanced

$$\mathbf{I_2 = I_1, I_r = I_c, I_3 = I_4 + I_r = I_4 + I_c}$$

$$\mathbf{V_1 = V_3 + I_c r \text{ and } V_2 = V_4 - I_c r}$$

From the theory of star-delta transformation

$$R_{AE} = R_3 r\left(\frac{1}{R_4} + \frac{1}{R_3} + \frac{1}{r}\right)$$

$$R_{CE} = R_4 r\left(\frac{1}{R_4} + \frac{1}{R_3} + \frac{1}{r}\right)$$

$$R_{AC} = R_3 R_4\left(\frac{1}{R_4} + \frac{1}{R_3} + \frac{1}{r}\right)$$

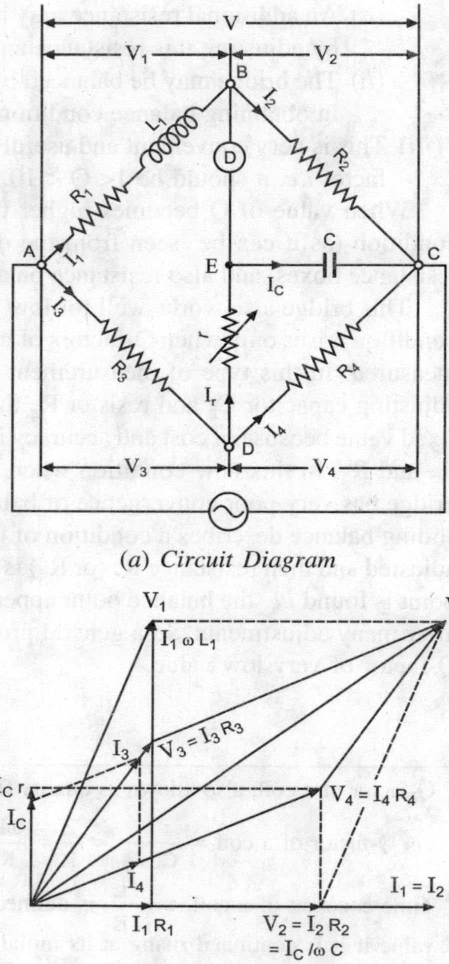

(a) Circuit Diagram

(b) Phasor Diagram

Fig. 7.21 *Anderson Bridge*

Hence the Anderson bridge shown in Fig. 7.21 (*a*) is reduced to an equivalent Maxwell Wien bridge shown in Fig. 7.23. Using the relation derived for the Maxwell Wien bridge we have

$$R_1 = \frac{R_2 R_{AE}}{R_{CE}} = \frac{R_2 R_3 r\left(\frac{1}{R_4}+\frac{1}{R_3}+\frac{1}{r}\right)}{R_4 r\left(\frac{1}{R_4}+\frac{1}{R_3}+\frac{1}{r}\right)}$$

$$= \frac{R_2 R_3}{R_4} \quad \ldots(7.41)$$

$$\text{and } L_1 = C\,R_2\,R_{AE} = C\,R_2\,R_3\,r\left(\frac{1}{R_4}+\frac{1}{R_3}+\frac{1}{r}\right)$$

$$= \frac{C\,R_2}{R_4}\left[r(R_3+R_4)+R_3\,R_4\right] \quad \ldots(7.42)$$

Note. The resistance R_{AC} in parallel with the ac source of supply contributes nothing to the balance bridge.

Advantages. 1. In an Anderson bridge a fixed capacitor is used while in case of Maxwell Wien bridge variable capacitor (which is an expensive component when calibrated to a high frequency) is used.

2. From the study of convergence conditions, it would be revealed that it is much easier to balance Anderson bridge than Maxwell bridge in case of low Q-coils.

Disadvantages. 1. The Anderson bridge is more complicated than the Maxwell bridge. It has more components and is more complicated to set up and manipulate.

2. The balance equations are much more tedious.

3. Difficulty in shielding of bridge is experienced due to additional junction.

This bridge is used for precise measurement of inductance over a wide range. In all cases where a variable capacitor use is permissible the more simple Maxwell bridge is employed.

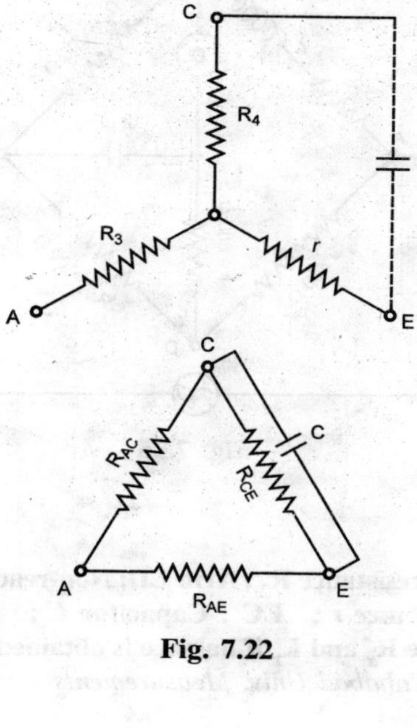

Fig. 7.22

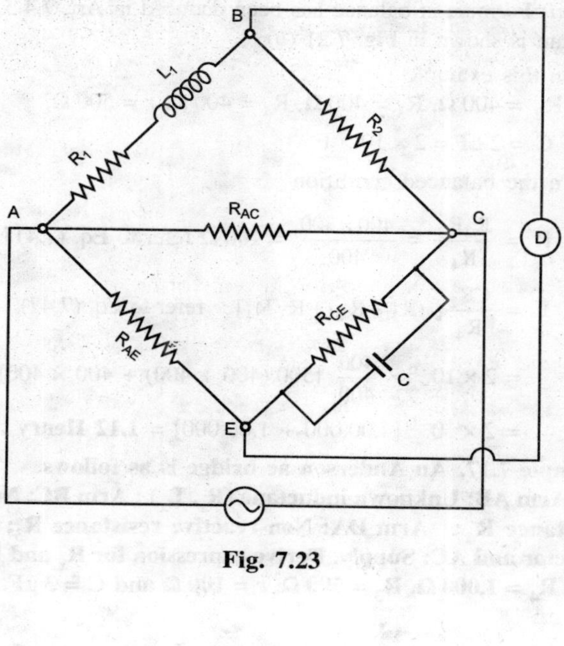

Fig. 7.23

Example 7.15. A balanced bridge has the following components connected between its five nodes A, B, C, D and E:

Between A and B: 1,000 Ω resistance; Between B and C: 1,000 Ω resistance.

Between C and D: an inductor; Between D and A: 218 Ω resistance.

Between A and E: 469 Ω resistance; Between E and B: 10 μF capacitance.

Between E and C: a detector; Between B and D: a power supply ac.

Obtain the values of resistance and inductance of the unknown inductor. Derive the formula used, if any.

[*U.P. Technical Univ. Elec. Measurements and Measuring Instruments 2001-02*]

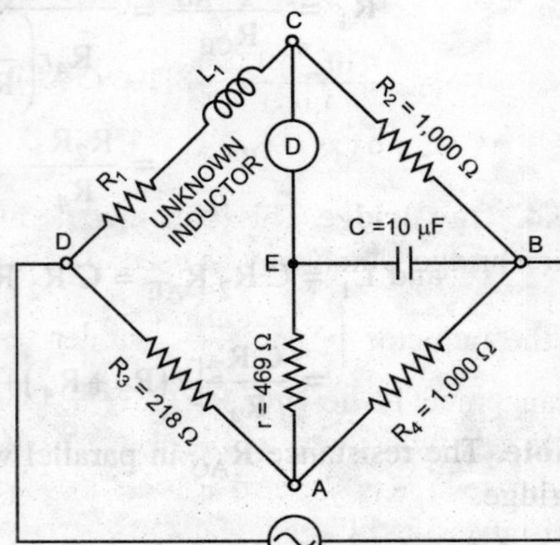

Fig. 7.24

Solution: From Eqs. (7.41) and (7.42) we have

$$R_1 = \frac{R_2R_3}{R_4} = \frac{1{,}000 \times 218}{1{,}000} = 218\ \Omega \quad \textbf{Ans.}$$

$$L_1 = \frac{CR_2}{R_4}\,[r\,(R_3 + R_4) + R_3R_4]$$

$$= \frac{10 \times 10^{-6} \times 1{,}000}{1{,}000}\,[469\,(218 + 1{,}000) + 218 \times 1{,}000]$$

$= 7.892$ H **Ans.**

Example 7.16. An ac bridge is connected as follows :

Branch AB is an inductive resistor, branches BC and ED are variable resistors, branches CD and DA are non-reactive resistors of 400 Ω each and branch CE is a condenser of 2 μF capacitance. The supply is connected to A and C and the detector to B and E. Balance is obtained when the resistance of BC is 400 Ω and that of ED is 500 Ω. Determine the resistance and inductance of AB.

(*i*) Identify the bridge and mention its advantages.

(*ii*) Deduce the formula at balance and draw the phasor diagram at balance.

(*iii*) Determine the resistance and inductance of AB.

Solution : (*i*) As per bridge configuration given in Fig. 7.25, it is Anderson bridge. Advantages of this bridge are given in Art. 7.4.3.

(*ii*) Formula at balance has been deduced in Art. 7.4.3. Phasor diagram is shown in Fig. 7.21 (*b*).

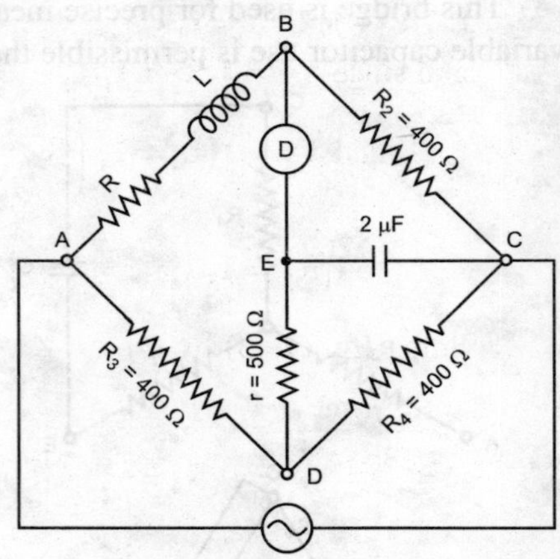

Fig. 7.25

In this example

$R_2 = 400\ \Omega$, $R_3 = 400\ \Omega$, $R_4 = 400\ \Omega$, $r = 500\ \Omega$,

$C = 2\ \mu F = 2 \times 10^{-6}$ F

In the balanced condition

$$R = \frac{R_2R_3}{R_4} = \frac{400 \times 400}{400} = 400\ \Omega \text{ refer to Eq. (7.41)}$$

$$L = \frac{CR_2}{R_4}[r(R_3 + R_4) + R_3\,R_4] \quad \text{refer to Eq. (7.42)}$$

$$= 2 \times 10^{-6} \times \frac{400}{400}\,[500\,(400 + 400) + 400 \times 400]$$

$= 2 \times 10^{-6}$ [4,00,000 + 1,60,000] = **1.12 Henry** **Ans.**

Example 7.17. An Anderson ac bridge is as follows.

Arm AB: Unknown inductane (R_x, L_x) ; Arm BC: Non-reactive resistance R_2 ; Arm CD: Non-reactive resistance R_4 ; Arm DA: Non-reactive resistance R_3; DE : Resistance r ; EC : Capacitor C ; EB: Detector and AC: Supply. Derive expression for R_x and L_x. Calculate R_x and L_x if balance is obtained for $R_2 = R_4 = 1{,}000\ \Omega$, $R_3 = 500\ \Omega$, $r = 100\ \Omega$ and C = 3 μF. [*Allahabad Univ. Measurements-I 1980*]

Solution : From Eqs. (7.41) and (7.42) we have

$$R_x = \frac{R_2 R_3}{R_4} = \frac{1{,}000 \times 500}{1{,}000} = \mathbf{500\ \Omega} \quad \textbf{Ans.}$$

$$L_x = \frac{CR_2}{R_4}\,[r\,(R_3 + R_4) + R_3 R_4]$$

$$= \frac{3 \times 10^{-6} \times 1{,}000}{1{,}000}\,[100\,(500 + 1{,}000) + 500 \times 1{,}000]$$

$$= 1.95 \text{ H} \quad \textbf{Ans.}$$

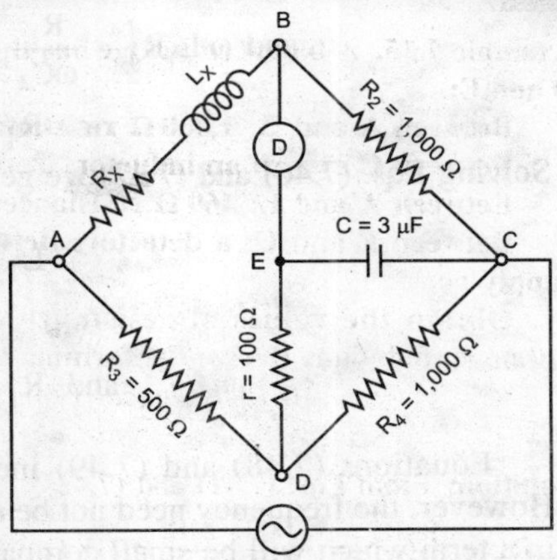

Fig. 7.26

7.4.4. Hay Bridge. This is another modification of the Maxwell-Wien bridge which may be used to advantage if the phase angle of the inductor $\left(\text{Tan}^{-1}\dfrac{\omega L}{R}\right)$ under test is large. The circuit arrangement is shown in Fig. 7.27(*a*). In this arrangement L_1 is self-inductance and R_1 is the resistance of the coil under test, R_2, R_3 and R_4 are known non-inductive resistances and C_4 is a standard variable capacitor. The bridge is balanced by varying R_4 and C_4. It is often more convenient to use a capacitor of fixed value and to make R_4 and either R_2 or R_3 adjustable.

When the bridge is balanced

$$\mathbf{I}_2 = \mathbf{I}_1;\ \mathbf{I}_4 = \mathbf{I}_3;\ \mathbf{V}_1 = \mathbf{V}_3$$

and $\mathbf{V}_2 = \mathbf{V}_4$

Since $\mathbf{V}_1 = \mathbf{I}_1\mathbf{Z}_1 = \mathbf{I}_1(R_1 + j\omega L_1)$

and $\mathbf{V}_3 = \mathbf{I}_3 R_3$,

$$\therefore\ \mathbf{I}_1\,(R_1 + j\omega L_1) = \mathbf{I}_3 R_3 \qquad \text{...(7.43)}$$

and since $\mathbf{V}_2 = \mathbf{I}_2 R_2 = \mathbf{I}_1 R_2$

and $\mathbf{V}_4 = \mathbf{I}_4\mathbf{Z}_4 = \mathbf{I}_3\left(R_4 - \dfrac{j}{\omega C_4}\right)$,

$$\mathbf{I}_1 R_2 = \mathbf{I}_3\left(R_4 - \frac{j}{\omega C_4}\right) \qquad \text{...(7.44)}$$

Dividing Eq. (7.43) by Eq. (7.44) we have

$$\frac{R_1 + j\omega L_1}{R_2} = \frac{R_3}{\left(R_4 - \dfrac{j}{\omega C_4}\right)}$$

or $$R_1 R_4 + \frac{L_1}{C_4} + j\left(\omega L_1 R_4 - \frac{R_1}{\omega C_4}\right) = R_2 R_3 \qquad \text{...(7.45)}$$

Separating reals and imaginaries, we have

$$R_1 R_4 + \frac{L_1}{C_4} = R_2 R_3 \qquad \text{...(7.46)}$$

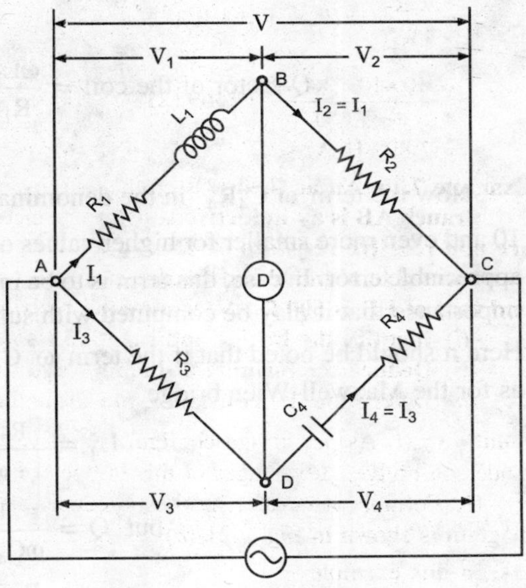

(*a*) *Circuit Diagram*

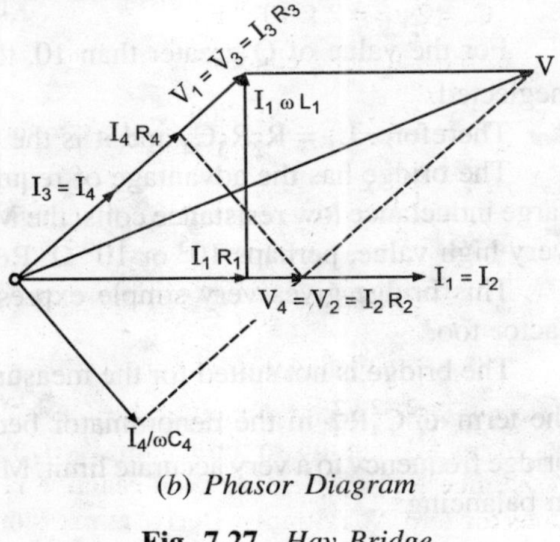

(*b*) *Phasor Diagram*

Fig. 7.27 *Hay Bridge*

$$\text{and } \omega L_1R_4 - \frac{R_1}{\omega C_4} = 0$$

$$\text{or } R_1 = \omega^2 C_4 L_1 R_4 \qquad \text{...(7.47)}$$

Solving Eqs. (7.46) and (7.47) we get

$$L_1 = \frac{R_2R_3C_4}{1+\omega^2C_4^2R_4^2} \qquad \text{...(7.48)}$$

$$\text{and } R_1 = \omega^2C_4R_4 \times \frac{R_2R_3C_4}{1+\omega^2C_4^2R_4^2} = \frac{R_2R_3R_4C_4^2\omega^2}{1+\omega^2C_4^2R_4^2} \qquad \text{...(7.49)}$$

Equations (7.48) and (7.49) indicate that the balance condition is a function of frequency. However, the frequency need not be accurately known to determine inductance since ω appears only in a term which will be small compared to unity in cases where use of Hay bridge is indicated (*i.e.* where Q-factor of coil is large).

$$\text{Q-factor of the coil} = \frac{\omega L_1}{R_1} = \frac{\dfrac{\omega R_2R_3C_4}{1+\omega^2C_4^2R_4^2}}{\dfrac{R_2R_3C_4^2\omega^2R_4}{1+\omega^2C_4^2R_4^2}} = \frac{1}{\omega R_4C_4} \qquad \text{...(7.50)}$$

Now the term $\omega^2C_4^2R_4^2$ in the denominators of Eqs. (7.48) and (7.49) has the value of 0.01 if Q = 10 and even more smaller for higher values of Q so the term $\omega^2C_4^2R_4^2$ can be droped without causing an appreciable error. In case, this term is to be included in calculations of L_1 and R_1 then it is of such minor importance that it may be computed with sufficient accuracy from an approximate value of frequency. Here it should be noted that if the term $\omega^2 C_4^2 R_4^2$ is excluded from the equation, then it is same for L_1 as for the Maxwell-Wien bridge.

$$L_1 = \frac{R_2R_3C_4}{1+\omega^2C_4^2R_4^2} \qquad \text{...(7.51)}$$

$$\text{but } Q = \frac{1}{\omega C_4R_4}$$

$$\text{so } L_1 = \frac{R_2R_3C_4}{1+(1/Q)^2} \qquad \text{...(7.52)}$$

For the value of Q greater than 10, the term $\omega^2C_4^2R_4^2$ will be smaller than 1/100 and so can be neglected.

Therefore, $L_1 = R_2R_3C_4$ and it is the same as for the Maxwell-Wien bridge.

The bridge has the advantage of requiring only a relatively low-value resistor for R_4 whereas, for large inductance low resistance coils, the Maxwell-Wien bridge would require a parallel resistance R_4 of very high value, perhaps 10^5 or 10^6 Ω. Resistance boxes of such high values are very costly.

This bridge gives very simple expressions for unknown inductance for high-Q coils and for Q-factor too.

The bridge is not suited for the measurement of low Q-factor of the inductors, because in that case, the term $\omega^2C_4^2R_4^2$ in the denominator becomes very important. And then it is required to know the bridge frequency to a very accurate limit. Moreover, with low value of Q-factor, it gives poor convergence in balancing.

Example 7.18. Four arms of Hay's ac bridge are arranged as follows:

AB-Coil of unknown impedance; BC-Non-reactive resistance of 1,000 Ω; CD-Non-reactive resistance of 833 Ω in series with a standard capacitor of 0.38 μF; DA is a non-reactive resistance of 16,800 Ω. If the supply frequency is 50 Hz, determine the inductance and resistance at the balance condition.

[Kanpur Univ. Elec. Measurement 1985]

Solution: In the balance condition of bridge, from Eqs. (7.48) and (7.49) we have

$$L_1 = \frac{R_2R_3C_4}{1+\omega^2C_4^2R_4^2}$$

$$= \frac{1,000\times 16,800\times 0.38\times 10^{-6}}{1+(2\pi\times 50)^2\times(0.38\times 10^{-6})^2(833)^2} = 6.32 \text{ H}$$

$$R_1 = \frac{R_2R_3R_4\,\omega^2C_4^2}{1+\omega^2C_4^2R_4^2}$$

$$= \frac{1,000\times 16,800\times 833\times(2\pi\times 50)^2\times(0.38\times 10^{-6})^2}{1+(2\pi\times 50)^2\times(0.38\times 10^{-6})^2\,(833)^2}$$

$= 197.49\ \Omega$ **Ans.**

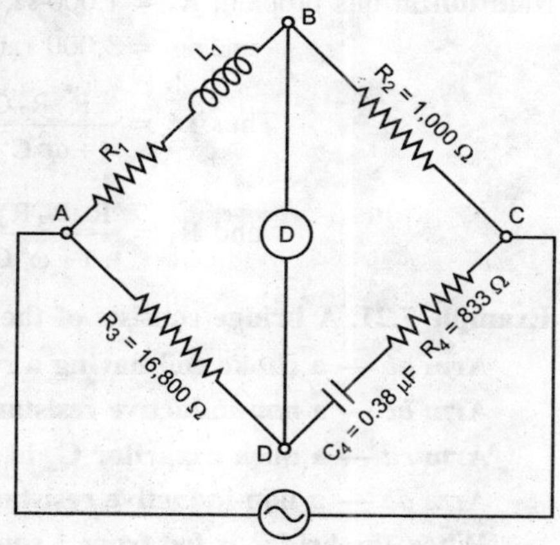

Fig. 7.28

Example 7.19. In a balanced network, AB is a resistance of 500 Ω in series with an inductor of 0.18 H, BC and DA are non-inductive resistances of 1,000 Ω each and CD consists of a resistance R in series with a capacitor C. A potential difference of 5 V at a frequency of 5,000/2π is applied between points A and C. Determine the values of R and C.

Solution : From Eqs. (7.48) and (7.49)

$$L_1 = \frac{R_2R_3C_4}{1+\omega^2C_4^2R_4^2} \qquad (i)$$

$$\text{and } R_1 = \frac{R_2R_3R_4\omega^2C_4^2}{1+\omega^2C_4^2R_4^2}$$

$$\text{or } R_1 = \omega^2R_4C_4L_1 \qquad \because \frac{R_2R_3C_4}{1+\omega^2C_4^2R_4^2} = L_1$$

$$\text{or } R_4C_4 = \frac{R_1}{\omega^2L_1} \qquad ...(ii)$$

In this problem

$$R_1 = 500\ \Omega,\ L_1 = 0.18\text{ H},\ R_2 = R_3 = 1,000\ \Omega,\ f = \frac{5,000}{2\pi}$$

$$\therefore\ R_4C_4 = \frac{R_1}{\omega^2L_1} = \frac{500}{\left(2\pi\times\frac{5,000}{2\pi}\right)^2\times 0.18} = 1.1111\times 10^{-4}$$

From Eq. (i)

$$C_4 = \frac{L_1(1+\omega^2C_4^2R_4^2)}{R_2R_3} = \frac{0.18\left[1+\left(2\pi\times\frac{5,000}{2\pi}\right)^2\times(1.1111)^2\times 10^{-8}\right]}{1,000\times 1,000} = 0.2355\ \mu\text{F} \textbf{ Ans.}$$

$$\text{and } R_4 = \frac{R_4C_4}{C_4} = \frac{1.1111\times 10^{-4}}{0.2355\times 10^{-6}} = 472\ \Omega \textbf{ Ans.}$$

Example 7.20. Find the series equivalent inductance and resistance of the network that causes such a bridge to null with the following bridge arms.

ω = 3,000 radians/second, 2 kΩ and 1 μF in series in one arm, two other arms have 10 kΩ in one and 1 kΩ in the other.

Solution: In this problem R_2 = 1,000 Ω; R_3 = 10,000 Ω; $C_4 = 1 \times 10^{-6}$ F; R_4 = 2,000 Ω

and ω = 3,000 radians per second

$$\text{Thus } L_1 = \frac{R_2 R_3 C_4}{1+\omega^2 C_4^2 R_4^2} = \frac{1{,}000\times 10{,}000\times 1\times 10^{-6}}{1+(3{,}000\times 1\times 10^{-6}\times 2{,}000)^2} = \frac{10}{37} = \textbf{0.27 H Ans.}$$

$$\text{and } R_1 = \frac{R_2 R_3 R_4 \omega^2 C_4^2}{1+\omega^2 C_4^2 R_4^2} = \frac{1{,}000\times 10{,}000\times 2{,}000\times (3{,}000)^2\times(1\times 10^{-6})^2}{1+(3{,}000\times 1\times 10^{-6}\times 2{,}000)^2} = \textbf{4.865 k}\Omega\textbf{ Ans.}$$

Example 7.21. A bridge consists of the following:

Arm *ab* — a choke coil having a resistance R_1 and inductance L_1.

Arm *bc* — a non-inductive resistance R_3

Arm *cd* — a mica capacitor C_4 in series with a non-inductive resistance R_4

Arm *da* — a non-inductive resistance R_2

When the bridge is fed from a source of 500 Hz balance is obtained under following conditions: R_2 = 2,410 Ω; R_3 = 750 Ω; C_4 = 0.35 μF, R_4 = 64.5 Ω. The series resistance of capacitor is 0.4 Ω. Calculate the resistance and inductance of the choke coil. The supply is connected between *a* and *c* and the detector is between *b* and *d*.

[U.P. Technical Univ. Electrical Measurements & Measuring Instruments 2006-07]

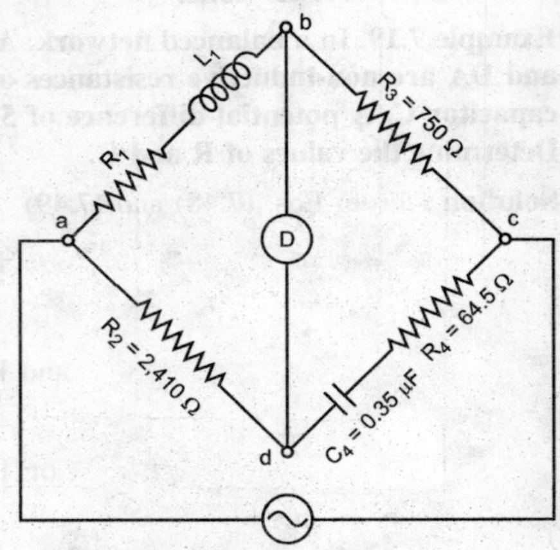

Fig. 7.29

Solution: $R'_4 = R_4$ + series resistance of capacitor

= 64.5 + 0.4 = 64.9 Ω

In balance condition of bridge

$$R_1 = \frac{R_2 R_3 R_4 \omega^2 C_4^2}{1+\omega^2 C_4^2 R_4^2}$$

$$= \frac{2{,}410\times 750\times 64.9\times(2\pi\times 500)^2\times(0.35\times 10^{-6})^2}{1+(2\pi\times 500)^2\times(0.35\times 10^{-6})^2\times(64.9)^2}$$

$$= \frac{141.827}{1+0.00007136} = 141.82\ \Omega\ \textbf{Ans.}$$

$$L_1 = \frac{R_2 R_3 C_4}{1+\omega C_4^2 R_4^2} = \frac{2{,}410\times 750\times 0.35\times 10^{-6}}{1+0.00007136} = 0.6326\text{ H }\textbf{Ans.}$$

7.4.5. Own Bridge. The bridge circuit also determines the unknown inductance in terms of resistance and capacitance. The advantage of this method is that the inductances over a very wide range can be determined by this method, employing capacitors of reasonable size.

The arrangement of the bridge is shown in Fig. 7.30 (*a*) in which the unknown inductance L_1, whose resistance is R_1, is connected in series with another variable non-inductive resistor R. The capacitance C_2 is also connected in series with a non-inductive variable resistor R_2, R_3 is a known non-inductive resistor and C_4 is known standard capacitor. The bridge is balanced by successively varying R and R_2 in the circuit. The phasor diagram is shown in Fig. 7.30 (*b*).

$$\text{At balance } \frac{R_1 + j\omega L_1 + R}{R_2 - \dfrac{j}{\omega C_2}} = \frac{R_3}{-\dfrac{j}{\omega C_4}}$$

$$\text{or} \quad -\frac{j}{\omega C_4}(R + R_1) + \frac{L_1}{C_4} = R_2 R_3 - \frac{jR_3}{\omega C_2}$$

Equating real and imaginary quantities separately, we have

$$L_1 = R_2 R_3 C_4 \qquad \text{...(7.53)}$$

$$\text{and } R_1 = \frac{R_3 C_4}{C_2} - R \qquad \text{...(7.54)}$$

Since the expressions obtained for unknown quantities R_1 and L_1 do not have ω, so the balancing of bridge is independent of frequency and waveform.

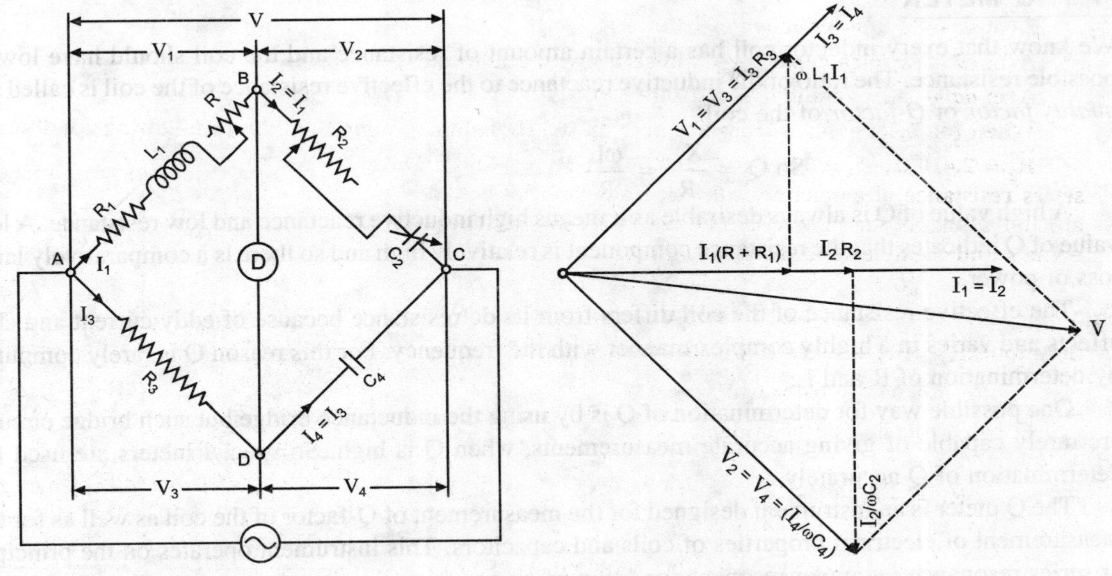

(a) *Circuit Diagram* (b) *Phasor Diagram*

Fig. 7.30 *Owen Bridge*

The main drawback of Owen's bridge is that it needs a variable standard capacitor which is an expensive component and also its accuracy is about 1 percent. The other drawback is that the value of variable capacitance C_2 tends to become rather large in case of measuring inductance of high-Q coil.

Example 7.22. An Owen's bridge is used to measure the properties of a sample of sheet steel at 2 kHz. At balance arm *ab* is test specimen, arm *bc* is R_3 = 100 Ω, arm cd is C_4 = 0.1 μF and arm da is R_2 = 834 Ω in series with C_2 = 0.124 μF. Derive the balance condition and calculate the effective impedance of the specimen under test conditions.

[U.P. Technical Univ. Electrical Measurements & Measuring Instruments 2007-08]

Solution: At balance, $$\frac{R_1 + j\omega L_1}{R_3} = \frac{R_2 - \dfrac{j}{\omega C_2}}{\dfrac{-j}{\omega C_4}}$$

or $\frac{-jR_1}{\omega C_4} + \frac{L_1}{C_4} = R_2 R_3 - \frac{jR_3}{\omega C_2}$

Equating real and imaginary quantities separately, we have

$$L_1 = R_2 R_3 C_4$$
$$= 834 \times 100 \times 0.1 \times 10^{-6} = 8.34 \text{ mH } \textbf{Ans.}$$

and $R_1 = R_3 \frac{C_4}{C_2} = 100 \times \frac{0.1 \times 10^{-6}}{0.124 \times 10^{-6}} = 80.65\ \Omega$ **Ans.**

Effective impedance of test specimen,

$$Z = R + j2\pi fL$$
$$= (80.65 + j2\pi \times 2{,}000 \times 8.34 \times 10^{-3})\ \Omega$$
$$= (80.65 + j104.8)\ \Omega = 132.24 \angle 52.42^\circ\ \Omega \textbf{ Ans.}$$

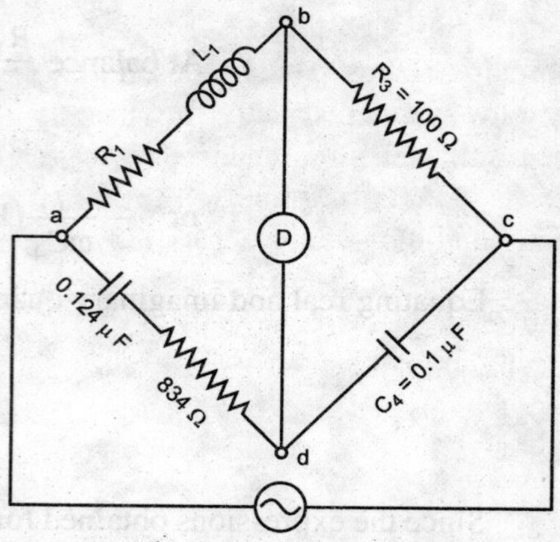

Fig. 7.31

7.5 Q METER

We know that every inductor coil has a certain amount of resistance and the coil should have lowest possible resistance. The ratio of the inductive reactance to the effective resistance of the coil is called the *quality factor* or *Q-factor* of the coil.

$$\text{So } Q = \frac{X_L}{R} = \frac{\omega L}{R}$$

A high value of Q is always desirable as it means high inductive reactance and low resistance. A low value of Q indicates that the resistance component is relatively high and so there is a comparatively large loss of power.

The effective resistance of the coil differs from its dc resistance because of eddy current and skin effects and varies in a highly complex manner with the frequency. For this reason Q is rarely computed by determination of R and L.

One possible way for determination of Q is by using the inductance bridge but such bridge circuits are rarely capable of giving accurate measurements, when Q is high. So special meters are used for determination of Q accurately.

The Q meter is an instrument designed for the measurement of Q-factor of the coil as well as for the measurement of electrical properties of coils and capacitors. This instrument operates on the principle of series resonance *i.e.* under resonant condition of an ac series circuit voltage across the capacitor is equal to the applied voltage times of Q of the circuit. If the voltage applied across the circuit is kept constant then voltmeter connected across the capacitor can be calibrated to indicate Q directly.

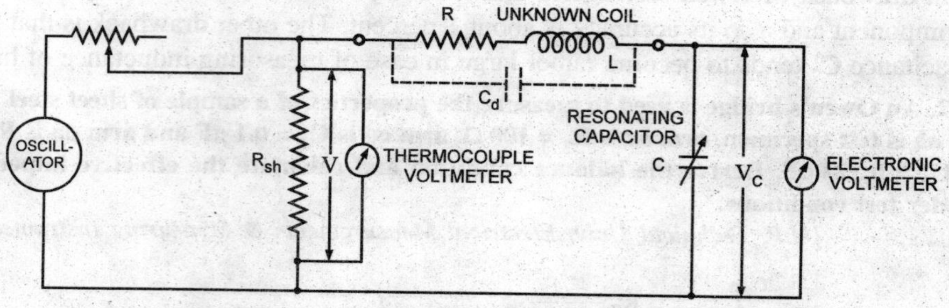

Fig. 7.32 *Circuit Diagram of a Q Meter*

Circuit diagram of a Q-meter is shown in Fig. 7.32. A wide-range *oscillator* with frequency range from 50 kHz to 50 MHz is used as a power supply to the circuit. The output of the oscillator is shorted by a low-value resistance, R_{sh} usually of the order of 0.02 Ω. So it introduces almost no resistance into the oscillatory circuit and represents a voltage source with a very small or of almost negligible internal resistance. The voltage across the low-value shunt resistance R_{sh}, V is measured by a thermocouple meter and the voltage across the capacitor, V_c is measured by an electronic voltmeter.

For carrying out the measurement, the unknown coil is connected to the test terminals of the instrument, and the circuit is tuned to resonance either by varying the frequency of the oscillator or by varying the resonating capacitor C. Readings of voltages across capacitor C and shunt resistance R_{sh} are obtained and Q-factor of the coil is determined as follows :

By definition Q-factor of the coil, $Q = \frac{X_L}{R}$

and when the circuit is under resonance condition

$$X_L = X_C$$

$$\text{or} \quad I X_L = I X_C = V_C$$

and the voltage applied to the circuit, $V = I R$

$$\text{So } Q = \frac{X_L}{R} = \frac{I X_L}{I R} = \frac{V_C}{V} \qquad ...(7.55)$$

This Q-factor is called the circuit Q because this measurement includes the losses of the resonating capacitor, voltmeter and the shunt resistance R_{sh}. So the actual Q-factor of the coil will be somewhat greater than the calculated Q-factor. This difference is usually very small and may be neglected, except when the resistance of the coil under test is relatively small in comparison to the shunt resistance, R_{sh}.

The inductance of the coil can also be computed from the known values of frequency *f* and resonating capacitor C as follows.

$$\text{At resonance } X_L = X_C$$

$$\text{or } 2\pi f L = \frac{1}{2\pi f C}$$

$$\text{or } L = \frac{1}{(2\pi f)^2 C} \text{ henry} \qquad ...(7.56)$$

7.5.1. Measurement of Low Impedance Components. Low impedance components, such as low-value resistors, small coils or high-value capacitors, are measured in series with the measuring circuit, as shown in the circuit diagram (Fig. 7.33). The component under measurement, Z is connected in series with the standard coil whose R and L are accurately known.

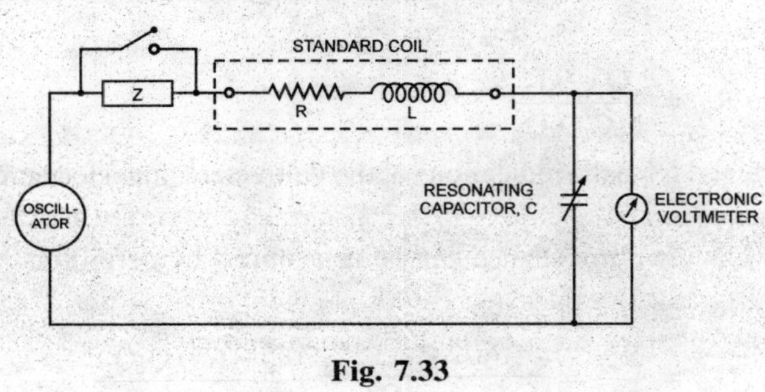

Fig. 7.33

Firstly the measurement is carried out by shorting the terminals of low impedance component under measurement and the resonating capacitor is set to C_1 for the resonance condition of the circuit.

In this condition, $X_L = X_{C1}$

$$\text{or} \quad \omega L = \frac{1}{\omega C_1}$$

$$\text{and} \quad Q_1 = \frac{\omega L}{R} = \frac{1}{\omega C_1 R} \quad \text{(neglecting the resistance of measuring circuit)}$$

$$\text{Also } R = \frac{X_L}{Q_1} = \frac{X_{C1}}{Q_1}$$

Second measurement is carried out by opening the switch across the impedance component under measurement, Z and the resonating capacitor is set to C_2 for resonance condition of the circuit.

In this condition

$$X_L + X_m = X_{C2} \quad \text{where } X_m \text{ is the reactance of the unknown impedance}$$

$$\text{or} \quad X_{C1} + X_m = X_{C2} \qquad \because X_{C1} = X_L$$

$$\text{So } X_m = X_{C2} - X_{C1} = \frac{1}{\omega C_2} - \frac{1}{\omega C_1} = \frac{C_1 - C_2}{\omega C_1 C_2} \quad \text{...(7.57)}$$

If the unknown impedance is a low inductance then

$$L_m = \frac{C_1 - C_2}{\omega^2 C_1 C_2} \quad \text{...(7.58)}$$

X_m will be inductive if $C_1 > C_2$ and capacitive if $C_1 < C_2$.

Value of resistance R_m of unknown impedance can also be determined as

$$R_2 = \frac{X_{C2}}{Q_2} \quad \text{where } R_2 \text{ is total resistance of the circuit}$$

$$\text{and} \quad R = \frac{X_{C1}}{Q_1}$$

$$\text{Now } R_2 = R + R_m$$

$$\text{So } R_m = R_2 - R = \frac{X_{C2}}{Q_2} - \frac{X_{C1}}{Q_1} = \frac{1}{\omega C_2 Q_2} - \frac{1}{\omega C_1 Q_1} = \frac{C_1 Q_1 - C_2 Q_2}{\omega C_1 C_2 Q_1 Q_2} \quad \text{...(7.59)}$$

If the unknown impedance is purely resistive then tuning of capacitor will remain the same for both measurements *i.e.* $C_1 = C_2$

$$\text{So } R_m = \frac{Q_1 - Q_2}{\omega C_1 Q_1 Q_2}$$

Q_1 and Q_2 are determined from the indications of the voltmeter connected across the tuning capacitor C and supply voltage.

The Q-factor of the unknown impedance can be determined by definition

$$i.e. \quad Q_m = \frac{X_m}{R_m} = \frac{\dfrac{(C_1 - C_2)}{\omega C_1 C_2}}{\dfrac{C_1 Q_1 - C_2 Q_2}{\omega C_1 C_2 Q_1 Q_2}} = \frac{(C_1 - C_2) Q_1 Q_2}{C_1 Q_1 - C_2 Q_2} \quad \text{...(7.60)}$$

If the unknown impedance is a large capacitor then

$$C_m = \frac{1}{\omega X_m} = \frac{\omega C_1 C_2}{\omega (C_1 - C_2)} = \frac{C_1 C_2}{(C_1 - C_2)}$$

The Q of the capacitor may be determined from the expression

$$Q = \frac{(C_1 - C_2)\, Q_1 Q_2}{C_1 Q_1 - C_2 Q_2} \quad \text{... (7.61)}$$

7.5.2. Measurement of High Impedance Components. High impedance components, such as high value resistors, inductors and low-value capacitors are measured by connecting them in parallel with the measuring circuit, as shown in Fig. 7.34.

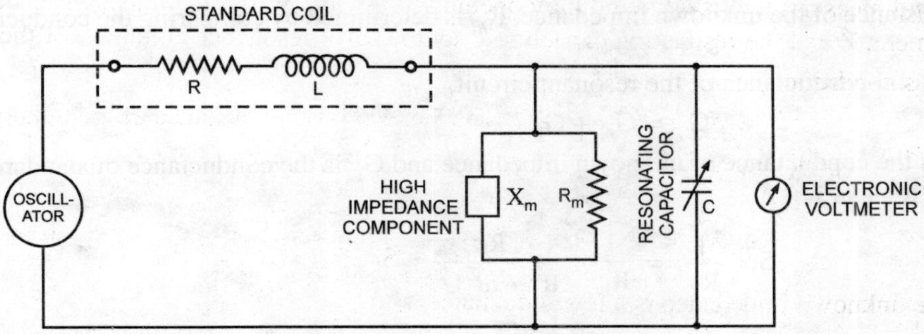

Fig. 7.34

The unknown impedance is measured in terms of its parallel components X_m and R_m. Firstly the measurement is carried out without connecting unknown impedance and reading of the resonating capacitor is noted with the standard coil in the circuit. Q-factor of the circuit is also calculated. Let their values be C_1 and Q_1 respectively.

Second measurement is carried out after inserting the unknown impedance in the circuit. New value of the resonating capacitor is obtained and the Q-factor of the circuit is determined. Let the corresponding values be C_2 and Q_2.

In first case *i.e.* when unknown impedance is not connected in the circuit, under resonant conditions

$$\omega L = \frac{1}{\omega C_1}$$

$$\text{and } Q_1 = \frac{\omega L}{R} = \frac{1}{\omega C_1 R}$$

In second case *i.e.* when unknown impedance is connected in the circuit, under resonant conditions the reactance of the standard coil will be equal to parallel reactances of the resonating capacitor X_{C2} and the reactance of unknown impedance X_m.

$$\text{So } X_L = \frac{X_{C2}\, X_m}{X_{C2} + X_m}$$

$$\text{or } X_L (X_{C2} + X_m) = X_{C2}\, X_m$$

$$\text{or } X_m = \frac{X_L\, X_{C2}}{X_{C2} - X_L} = \frac{\frac{1}{\omega C_1} \cdot \frac{1}{\omega C_2}}{\frac{1}{\omega C_2} - \frac{1}{\omega C_1}} = \frac{1}{\omega (C_1 - C_2)} \quad \text{...(7.62)}$$

If the reactance of the unknown impedance is capacitive then

$$X_m = \frac{1}{\omega C_m}$$

$$\text{or } C_m = \frac{1}{\omega X_m} = C_1 - C_2 \qquad \text{...(7.63)}$$

Total resistance of the circuit R_T will be equal to the product of the Q-factor of the total circuit and the reactance of standard coil under resonant condition

$$\text{So } R_T = Q_2 X_L = Q_2 X_{C1} = \frac{Q_2}{\omega C_1}$$

The resistance of the unknown impedance, R_m is determined by comparing the conductances in the circuit.

So the total conductance of the resonant circuit,

$$G_T = G_m + G_L$$

where G_m is the conductance of unknown impedance and G_L is the conductance of standard coil and is equal to $R/(R^2 + \omega^2 L^2)$

$$\text{So } \frac{1}{R_T} = \frac{1}{R_m} + \frac{R}{R^2 + \omega^2 L^2}$$

$$\text{or } \frac{1}{R_m} = \frac{\omega C_1}{Q_2} - \left(\frac{1}{R}\right)\left(\frac{1}{1 + \frac{\omega^2 L^2}{R^2}}\right) = \frac{\omega C_1}{Q_2} - \frac{1}{R}\left(\frac{1}{1+Q_1^2}\right)$$

Normally value of Q_1 is very high so $1 + Q_1^2$ can be taken equal to Q_1^2

$$\text{So } \frac{1}{R_m} = \frac{\omega C_1}{Q_2} - \frac{1}{RQ_1^2} = \frac{\omega C_1}{Q_2} - \frac{1}{RQ_1 \cdot \frac{1}{\omega C_1 R}}$$

$$\text{or } \frac{1}{R_m} = \frac{\omega C_1}{Q_2} - \frac{\omega C_1}{Q_1} = \frac{\omega C_1 Q_1 - \omega C_1 Q_2}{Q_1 Q_2} = \frac{\omega C_1 (Q_1 - Q_2)}{Q_1 Q_2}$$

$$\text{or } R_m = \frac{Q_1 Q_2}{\omega C_1 (Q_1 - Q_2)} \qquad \text{...(7.64)}$$

and the Q-factor of the unknown impedance,

$$Q_m = \frac{R_m}{X_m} = \frac{\dfrac{Q_1 Q_2}{\omega C_1 (Q_1 - Q_2)}}{\dfrac{1}{\omega (C_1 - C_2)}} = \frac{(C_1 - C_2) Q_1 Q_2}{C_1 (Q_1 - Q_2)} \qquad \text{...(7.65)}$$

7.5.3. Errors. Various errors occuring in measurements are discussed below:

***(i)* Error Due To Distributed Capacitance.** The distributed capacitance or self capacitance of the measuring circuit introduces error in the measurement and is the most important source of error. This source of error affects the actual value of Q and the inductance of the coil. Distributed or self capacitance C_d of a coil can be measured by making two measurements at different frequencies. The circuit diagram is shown in Fig. 7.35. The coil under test is connected directly to the test terminals of the measuring circuit, as shown in the Fig. Resonating capacitor is tuned for resonance at two different frequencies *i.e.* f_1 and f_2 (normally f_2 is taken twice of f_1 *i.e.* $f_2 = 2f_1$). Let the values of resonating capacitor be C_1 and C_2 at frequencies of f_1 and f_2 respectively. So

$$f_1 = \frac{1}{2\pi\sqrt{L(C_1 + C_d)}}$$

$$\text{and } f_2 = \frac{1}{2\pi\sqrt{L(C_2 + C_d)}}$$

$$\text{as } f_2 = 2f_1$$

$$\text{So } \frac{1}{2\pi\sqrt{L(C_2 + C_d)}} = \frac{2}{2\pi\sqrt{L(C_1 + C_d)}}$$

$$\text{or } \frac{1}{C_2 + C_d} = \frac{4}{C_1 + C_d}$$

$$\text{or } C_d = \frac{C_1 - 4C_2}{3} \quad ...(7.66)$$

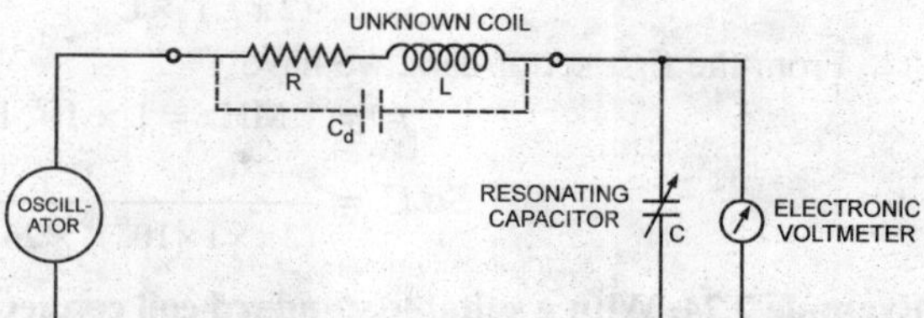

Fig. 7.35

Distributed capacitance reduces the effective value of Q of a coil and the difference between measured value and actual value depends on the value of the distributed capacitance and the resonating capacitor. The true value of Q is obtained from the following relation

$$\text{True value of Q} = Q_m\left(\frac{C + C_d}{C}\right) \quad ...(7.67)$$

where Q_m is the effective Q of the coil and is usually considered indicated Q, C is the resonating capacitor and C_d is the distributed capacitance.

***(ii)* Error Due To Residual Inductance of the Instrument.** The error due to residual inductance of the instrument is very small, of the order of 0.015 μH, and so affects the measurement of only very small inductors, say inductors of value lesser than 0.5 μH.

***(iii)* Error Due To Shunt Resistor.** The shunt resistor R_{sh} of the Q-meter circuit introduces error only when resistance of the test coil is comparable to the shunt resistor say for example a coil of resistance 0.2 Ω is measured with shunt resistor of 0.2 Ω. Otherwise error due to shunt resistor R_{sh} is very small and can be neglected as in case of measuring Q of a coil of the resistance 20 Ω with a shunt resistor of 0.02 Ω.

***(iv)* Error Due To Conductance of Voltmeter.** The error may be introduced due to shunting effect of conductance of the voltmeter connected across the resonating capacitor at higher frequencies. But this error is too small that it can be neglected.

Example 7.23. A coil of inductance L is tuned by a capacitance C to resonate at 1 MHz. A standard capacitor of 230 *pF* is tuned to resonate with the same coil at 2 MHz. If the resonance is indicated by a VTVM of capacitance 8 pF in parallel with the circuit, calculate the value of L and C.

Solution: For a simple series arrangement of tuned circuit, the resonant frequency is given by

$$\text{Resonant frequency, } f_r = \frac{1}{2\pi\sqrt{LC}} \quad ...(i)$$

$$\text{So inductance of coil, } L = \frac{1}{(2\pi f_r)^2 \times C}$$

From the second set of data, we have

$$f_r = 2 \text{ MHz} = 2 \times 10^6 \text{ Hz and } C = (230 + 8) \text{ pF} = 238 \times 10^{-12} \text{ F}$$

$$\text{So L} = \frac{1}{(2\pi \times 2\times 10^6)^2 \times 238\times 10^{-12}} = 26.6\ \mu\text{H}\ \textbf{Ans.}$$

From Eq. (*i*) we also have

$$C = \frac{1}{(2\pi f_r)^2 \times L}$$

From the first set of data, we have

$$f_r = 1\text{ MHz} = 1\times 10^6\text{ Hz}$$

$$\text{So C} = \frac{1}{(2\pi \times 1\times 10^6)^2 \times 26.6\times 10^{-6}} = 952\text{ pF}\ \textbf{Ans.}$$

Example 7.24. With a suitable standard coil connected to a Q meter resonance is obtained with a frequency f_1 with resonating capacitor set at C_1; the indicated Q-factor is Q_1. An unknown impedance is connected in series with the standard coil and resonance is re-established by resetting the resonating capacitor at C_2, the corresponding Q-factor being Q_2. Determine (*i*) the resistive and (*ii*) reactive components of the unknown impedance when C_1 = 208 pF, Q_1 = 80, C_2 = 184 pF and Q_2 = 50. The frequency is 165 kHz.

Solution : Given that $C_1 = 208\text{ pF} = 208\times 10^{-12}\text{ F}$

$C_2 = 184\text{ pF} = 184\times 10^{-12}\text{ F}$

Frequency, $f = 165\text{ kHz} = 165\times 10^3\text{ Hz}$

$Q_1 = 80$

$Q_2 = 50$

(*i*) Resistive component of unknown impedance is given by the equation

$$R_m = \frac{1}{\omega}\left(\frac{1}{C_2Q_2} - \frac{1}{C_1Q_1}\right)$$

$$= \frac{1}{2\pi\times 165\times 10^3}\left(\frac{1}{184\times 10^{-12}\times 50} - \frac{1}{208\times 10^{-12}\times 80}\right) = 46.88\ \Omega\ \textbf{Ans.}$$

(*ii*) Reactive component of unknown impedance,

$$X_m = \frac{1}{\omega}\left(\frac{1}{C_2} - \frac{1}{C_1}\right) = \frac{1}{2\pi\times 165\times 10^3}\left(\frac{1}{184\times 10^{-12}} - \frac{1}{208\times 10^{-12}}\right) = 605\ \Omega\ \textbf{Ans.}$$

Example 7.25. A coil is tuned to resonance at 500 kHz with a resonating capacitance of 36 μμF. At 250 kHz the resonance is obtained with a resonating capacitance of 160 μμF. Determine the self capacitance and inductance of the coil.

Solution. Given that $f_1 = 250\text{ kHz} = 250\times 10^3\text{ Hz}$

$C_1 = 160\ \mu\mu\text{F} = 160\times 10^{-12}\text{ F}$

$f_2 = 500\text{ kHz} = 500\times 10^3\text{ Hz}$

and $C_2 = 36\ \mu\mu\text{F} = 36\times 10^{-12}\text{ F}$

$$\text{Self capacitance of the coil, } C_d = \frac{C_1 - 4C_2}{3} = \frac{160\times 10^{-12} - 4\times 36\times 10^{-12}}{3} = 5.33\ \mu\mu\text{F}\ \textbf{Ans.}$$

$$\text{Since resonating frequency, } f_1 = \frac{1}{2\pi\sqrt{L(C_1 + C_d)}},$$

$$L = \frac{1}{(2\pi f_1)^2\ (C_1 + C_d)} = \frac{1}{(2\pi\times 250\times 10^3)^2\ (160 + 5.33)\times 10^{-12}}$$

$$= 2{,}451\ \mu\text{H}\ \textbf{Ans.}$$

Example 7.26. Describe the circuit and working of a Q meter. A coil was tested using a Q meter and following information was obtained.

At a frequency of 3 MHz, the value of capacitance to give maximum voltage across the capacitor was 251 pF while at 6 MHz frequency it was 50 pF. Calculate the self capacitance of the coil.

[U.P. Technical Univ. Elec. Measurements and Measuring Instruments 2005-06]

Solution. Given that $C_1 = 251$ pF at $f_1 = 3$ MHz

$C_2 = 50$ pF at $f_2 = 6$ MHz

$$\text{Self capacitance of coil, } C_d = \frac{C_1 - 4C_2}{3} = \frac{251 - 4\times 50}{3} = 17 \text{ pF} \quad \textbf{Ans.} \qquad \because f_2 = 2f_1$$

Example 7.27. Explain the working of Q meter. To find the self capacitance of a coil by Q meter, the resonance was obtained with

***(i)* tuning capacitor of 1,530 pF at 1.0 MHz and**

***(ii)* tuning capacitor of 162 pF at 3.0 MHz.**

What is the value of self capacitance? *[U.P.S.C. I.E.S. Electrical Engineering-I, 1998]*

Solution: From Art. 7.5.3 we have

$$f_1 = \frac{1}{2\pi\sqrt{L(C_2 + C_d)}} \qquad ...(i)$$

$$f_2 = \frac{1}{2\pi\sqrt{L(C_2 + C_d)}} \qquad ...(ii)$$

and $f_2 = 3f_1$ $\qquad \because f_2 = 3.0$ MHz and $f_1 = 1.0$ MHz

$$\text{So} \quad \frac{3}{2\pi\sqrt{L(C_1 + C_d)}} = \frac{1}{2\pi\sqrt{L(C_2 + C_d)}}$$

or $9\,C_2 + 9\,C_d = C_1 + C_d$

$$\text{or Self capacitance of the coil, } C_d = \frac{C_1 - 9C_2}{8} = \frac{1{,}530 - 9\times 162}{8} = 9 \text{ pF} \quad \textbf{Ans.}$$

Example 7.28. A circuit consisting of an unknown coil, a resistance and a variable capacitor connected in series is tuned to resonance using a Q meter. If the frequency is 450 kHz and the resonating capacitor is set at 250 pF, determine the effective inductance and resistance of unknown coil. The Q meter indicates 105 and the resistance is 0.75 Ω.

Solution: Resonating frequency, $f = 450$ kHz $= 450 \times 10^3$ Hz

Resonating capacitance, $C = 250$ pF $= 250 \times 10^{-12}$ F

Resistance, $R_{sh} = 0.75\ \Omega$.

$Q = 105$

Under resonant condition

$$\text{Inductance of coil, } L = \frac{1}{(2\pi f)^2 C} = \frac{1}{(2\pi\times 450\times 10^3)^2\times 250\times 10^{-12}} = 500\ \mu\text{H} \quad \textbf{Ans.}$$

$$\text{Resistance of the coil, } R = \frac{\omega L}{Q} - R_{sh} = \frac{2\pi\times 450\times 10^3\times 500\times 10^{-6}}{105} - 0.75 = 12.71\ \Omega \quad \textbf{Ans.}$$

Example 7.29. A coil of resistance 5 Ω is connected in the circuit shown in Fig. 7.32 and resonance occurs when oscillator frequency is 500 kHz and the resonating capacitor is set at 120 pF. Determine the percentage error introduced in the calculated value of Q if a resistance of 0.02 Ω is used across the oscillatory circuit.

Solution. The true or effective Q of the coil,

$$Q_{\text{True}} = \frac{1}{\omega CR} = \frac{1}{2\pi\times 500\times 10^3\times 120\times 10^{-12}\times 5} = 530.5$$

The indicated or calculated Q of the coil

$$Q_{Indicated} = \frac{1}{\omega C (R + 0.02)} = \frac{1}{2\pi \times 500 \times 10^3 \times 120 \times 10^{-12} \times 5.02} = 528.4$$

$$\text{Percentage error} = \frac{Q_{True} - Q_{Indicated}}{Q_{True}} \times 100 = \frac{530.5 - 528.4}{530.5} \times 100 = 0.4\% \textbf{ Ans.}$$

Example 7.30. Tests using a Q meter, on a radio-tuning coil to establish the self-capacitance, gave the following test results.

(*i*) With the radio coil connected normally resonance was obtained at 800 kHz with the tuning capacitor of the Q meter set at 95 pF.

(*ii*) With a standard inductor connected in place of radio coil, resonance was obtained at 2.5 MHz and this condition was not changed when the radio coil was connected in parallel with the standard inductor. Determine the self capacitance of the radio coil.

Solution. If the inductance and self capacitance of the radio coil are L and C_d respectively then from first test

$$L = \frac{1}{\omega_1^2 (C_1 + C_d)} \quad ...(i)$$

where ω_1 is the angular frequency and C_1 is the tuning capacitor setting to produce resonance in first setting.

In second test the reason for resonance condition of the standard inductor being unchanged is that a very high impedance has been placed in parallel, this being the resonant circuit of the radio coil and its own self capacitance *i.e.* the radio coil in parallel with its own self capacitance is resonant at 2.5 MHz. Thus we can write

$$L = \frac{1}{\omega_2^2 C_d} \quad ...(ii)$$

Substituting $L = \frac{1}{\omega_2^2 C_d}$ from Eq. (*ii*) in Eq. (*i*) we get

$$\frac{1}{\omega_2^2 C_d} = \frac{1}{\omega_1^2 (C_1 + C_d)}$$

$$\text{or } \omega_1^2 (C_1 + C_d) = \omega_2^2 C_d$$

$$\text{or } C_d = \frac{\omega_1^2 C_1}{\omega_2^2 - \omega_1^2} = \frac{f_1^2 C_1}{f_2^2 - f_1^2}$$

Substituting $C_1 = 95 \times 10^{-12}$ F, $f_1 = 800 \times 1{,}000$ Hz and $f_2 = 2.5 \times 10^6$ Hz, in above equation we get

$$C_d = \frac{(800 \times 1{,}000)^2 \times 95 \times 10^{-12}}{(2.5 \times 10^6)^2 - (800 \times 1{,}000)^2} = 10.84 \text{ pF } \textbf{Ans.}$$

Example 7.31. A coil is tuned to resonance at 1 MHz with a resonating capacitance of 480 pF. At 2 MHz the resonance is obtained with resonating capacitance of 90 pF. Determine (*i*) the self capacitance of the coil (*ii*) true and indicated values of Q when R = 10 Ω for both the measurements.

Solution : Given that $f_1 = 1 \text{ MHz} = 1 \times 10^6$ Hz

$C_1 = 480 \text{ pF} = 480 \times 10^{-12}$ F

$f_2 = 2 \text{ MHz} = 2 \times 10^6$ Hz

$C_2 = 90 \text{ pF} = 90 \times 10^{-12}$ F

Resistance of coil, R = 10 Ω

(*i*) Self capacitance of the coil, $C_d = \frac{C_1 - 4C_2}{3} = \frac{480 \times 10^{-12} - 4 \times 90 \times 10^{-12}}{3} = 40$ pF **Ans.**

(*ii*) Indicated or effective Q for first measurement

$$= \frac{1}{2\pi f_1 (C_1 + C_d) R} = \frac{1}{2\pi \times 1 \times 10^6 \times 520 \times 10^{-12} \times 10} = 30.607 \textbf{ Ans.}$$

$$\text{True Q for first measurement} = \frac{1}{2\pi f_1 C_1 R} = \frac{1}{2\pi \times 1 \times 10^6 \times 480 \times 10^{-12} \times 10} = 33.157 \textbf{ Ans.}$$

Indicated or effective Q for second measurement

$$= \frac{1}{2\pi \times 2 \times 10^6 \times 130 \times 10^{-12} \times 10} = 61.214 \textbf{ Ans.}$$

$$\text{True Q for second measurement} = \frac{1}{2\pi \times 2 \times 10^6 \times 90 \times 10^{-12} \times 10} = 88.42 \textbf{ Ans.}$$

EXERCISES

1. In the circuit shown in Fig. 1 when I = 0, show that

$$\frac{\omega L_1}{R_1} = \frac{1}{\omega C_4 R_4}$$ **[U.P.S.C. I.E.S. Electrical Engineering I, 2002]**

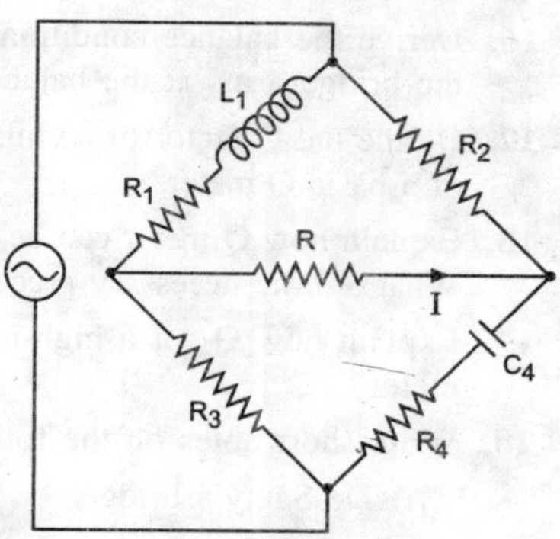

Fig. 1

2. Describe how an unknown capacitance can be measured with the help of De Sauty's bridge. What are the limitations of this bridge and how are they overcome by using modified De Sauty's bridge ? Draw phasor diagram to illustrate your answer.
[U.P. Technical Univ. Elec. Measurements and Measuring Instruments 2005-06]

3. Explain the working principle of Schering bridge and derive an expression for measurement of unknown capacitor. Draw the phasor diagram under null condition and explain how dissipation factor of the capacitor can be calculated.
[U.P. Technical Univ. Elec. Measurements and Measuring Instruments, 2001-02]

4. Draw circuit diagram of a high voltage Schering bridge. What are its main features ? Draw a phasor diagram of the bridge and derive the equations of balanced conditions.
[Pb. Univ. Elec. Measurements-I, June 1993]

5. Prove that Schering bridge can be used to measure the insulating properties and values of capacitance with high precision. **[U.P. Technical Univ. Electronics Measurements and Instrumentation, 2006-07]**

6. Describe the Wien bridge method for measuring unknown frequencies in the audio range. What are the other applications of this bridge ? **[U.P.S.C. I.E.S. Electrical Engineering-I, 1993]**

7. How can the frequency be determined using a bridge ? Draw this bridge and derive condition for balance. Why and how are two resistances and capacitances made equal ?**[U.P.S.C. I.E.S. Electrical Engineering-I, 2001]**

8. Explain Maxwell's bridge for measurement of unknown inductance. Determine condition for balance. Mention its applications. **[U.P. Technical Univ. Elec. Measurements and Measuring Instruments 2005-06]**

9. Explain how Maxwell's inductance-capacitance bridge is used for the measurement of inductance of a coil. Derive the mathematical relationship for unknown inductance and Q factor of a coil in terms of other known parameters of the bridge. What are the advantages and disadvantages of the bridge. ?

10. Draw a neat schematic diagram of the Anderson bridge. Deduce the equations when the bridge is under balance condition. Explain clearly how you may measure self inductance by using this bridge. Draw the phasor diagram of the voltages and currents of the bridge arms at balance.

11. For Anderson's bridge, derive the relation for unknown impedance. What are the advantages and limitations of this bridge ? **[U.P. Technical Univ. Elec. Measurements and Measuring Instruments, 2002-03]**

12. Describe the working of Hay's bridge. It is said that this bridge is suitable for measuring high-Q inductors. Give reasons fore such a statement.
[U.P. Technical Univ. Electronics Measurements and Instrumentation 2004-05]

13. Draw the circuit of that bridge which is useful in measuring high-Q coils. Derive the unknown components values in terms of circuit elements.

14. In a Hay bridge, the four arms are $R_1 - L_1$, R_2, $R_3 - C_3$, R_4 connected in clockwise order. Show that, under the phase null condition, Q of the coil is given by

$$Q = \frac{\text{Voltage across } C_3}{\text{Voltage across } R_3}$$

[U.P.S.C. I.E.S. Electrical Engineering I, 2003]

15. Describe the working of Hay's bridge for measurement of inductance. Derive the equations for balance condition and draw the phasor diagram under balance condition. Why is this bridge suited for measurement of inductance of high Q-coils ?
[U.P. Technical Univ. Elec. Measurements and Measuring Instruments 2005-06]

16. Derive the balance conditions of Owen's bridge. Draw the vector diagram of currents and voltages of the bridge arms at the balance condition. **[U.P. Technical Univ. Electronics Measurements and Instrumentation 2004-05]**

17. Define the Q-factor of a coil. Explain with a circuit diagram the construction and principle of operation of a basic Q-meter.

18. Explain how Q meter can be used for the measurement of resistance and inductance of an unknown coil. What are the necessary precautions to be taken while using a Q meter ?

19. Explain how Q_x of a high-impedance coil and low-impedance coil are measured by a commercial Q meter?

20. Write short notes on the following :
(*i*) De Sauty's bridge.
(*ii*) Schering bridge.
(*iii*) Wien bridge.
(*iv*) Maxwell bridge.
(*v*) Anderson bridge.
(*vi*) Q-meters **[PTU December 2005]**

SHORT ANSWER TYPE QUESTIONS WITH ANSWERS

Q. 1. Name the commonly used detectors for ac bridges.
Ans. Commonly used detectors for ac bridges are telephone receivers or head phones, vibration galvanometer and tunable amplifier detectors.

Q. 2. How do you modify the capacitor symbol to accommodate the loss in the capacitor ?
Ans. An imperfect capacitor (*i.e.* a capacitor having dielectric loss) is represented diagrammatically as a perfect capacitor C in parallel or in series with a resistance R.

Q. 3. What is meant by the term dielectric loss in a dielectric ?
Ans. There is a definite amount of dissipation of energy when a dielectric is subjected to alternating voltage. It is this dissipation of energy that is called the dielectric loss.

Q. 4. What is meant by 'dissipation factor' of a leakage capacitor ?
Ans. In a leakage capacitor, the leakage current does not lead applied voltage by exactly 90° *i.e.*, the

phase angle Φ is always less than 90°. The complementary angle δ (δ = 90°– Φ) is called the dielectric loss angle. Tangent of dielectric loss angle *i.e.*, tan δ is called the dissipation factor.

Q. 5. Capacitors with very high leakage resistance should be measured as a series R-C combination, not as a parallel R-C combination. Why ?

Ans. If a capacitor with a very high leakage resistance is measured as a parallel R-C combination, then resistance R_2 required for balancing the bridge would be inconveniently large for a precision adjustable resistor.

Q. 6. Why a spark is connected across resistance arms in a Schering bridge ?

Ans. A spark is connected across resistance arms in order to prevent dangerous rises of voltage across the capacitor arm of a Schering bridge.

Q. 7. What is meant by Q- factor of a coil ?

Ans. Q-factor of a coil, also known as quality factor or figure of merit of a coil, is the reciprocal of its power factor *i.e.*,

$$\text{Q-factor of a coil} = \frac{1}{\text{Cos}\phi} = \frac{Z}{R} = \frac{\omega L}{R} \text{ because R is very small in comparson to inductive reactance.}$$

Q. 8. What is meant by time constant of a coil ?

Ans. Time constant of a coil τ = L/R is defined as the time required for the current to rise to its final steady value if it is continued to rise at the initial rate or it may also be defined as the time required for the current to attain 0.632 of its final steady value in an inductive (R – L) circuit.

Q. 9. Give the application of Maxwell's inductance-capacitance bridge.

Ans. Maxwell's inductance-capacitance bridge is very convenient and useful bridge for the determination of inductance having medium Q factor ($1 < Q < 10$).

Q. 10. For what purpose the Anderson bridge is used ?

Ans. The Anderson bridge is used for precise measurement of inductance over a wide range.

Q. 11. Which bridge is preferred for measurement of high Q-coil ?

Ans. Hay bridge is preferred for measurement of inductance of high-Q coils.

Q. 12. Draw the circuit of that bridge which is useful in measuring high-Q coils.

Ans. Hay bridge is useful in measuring inductance of high-Q coils. The circuit diagram is given in Fig. 7.27(*a*).

Q. 13. Why Hay bridge is not suited for the measurement of low-Q factor of the inductors ?

Ans. Hay bridge is not suited for the measurement of low Q-factor of the inductors because in that case it is required to know the bridge frequency to a very accurate limit and also with low value of Q-factor, it gives poor convergence in balancing.

Q. 14. What is Q-meter ?

Ans. The Q-meter is an instrument designed for measurement of Q-factor of a coil as well as for the measurement of electrical properties of coils and capacitors.

Q. 15. On what principle does a Q-meter operate ?

Ans. The Q-meter operates on the principle of series resonance i.e., under resonant condition of an ac series circuit voltage across the capacitor is equal to the applied voltage times of Q of the circuit.

Q. 16. What is Q-factor of the coil ?

Ans. The ratio of the inductive reactance to the effective resistance of the coil is called the Q-factor of the coil.

Q. 17. What are the different parameters that can be measured using a Q-meter ?

Ans. The Q-meter is used for measuring Q-factor, inductance, effective resistance, self capacitance, bandwidth and capacitance.

Q. 18. What are the factors which affect the measurement accuracy of Q-meter ?

Ans. Factors affecting measurement accuracy are:

(*i*) Distributed capacitance or self capacitance of the coil.

(*ii*) Residual inductance of the instrument.

(*iii*) Conductance of voltmeter.

(*iv*) Shunt resistor of Q-meter.

Q. 19. Why the actual Q-factor of the coil is somewhat larger than the calculated Q-factor ?

Ans. Calculated value of Q-factor is somewhat smaller than its actual value because Q-factor measurement includes the losses of the resonating capacitor, voltmeter and the shunt resistance $R_{sh.}$

PROBLEMS

1. A dielectric sheet of thickness 1 mm is tested at 50 Hz between two electrodes of 10 cm diameter. The Schering bridge employed has a standard compressed air capacitor C_3 of 100 pF; a non-inductive resistor R_4 of 350 Ω in parallel with a variable capacitor C_4 and a non-inductance variable resistor R_2. At balance C_4 = 0.4 μF, R_2 = 250 Ω. Calculate the power factor and the permittivity of the sheet.

[Kanpur Univ. Elec. Measurements 1986]

[Ans. 0.04398; 2]

2. Fig. 7.9 (*a*) shows a Schering bridge circuit used for measuring the power loss in dielectrics. The specimens are in the form of discs 0.3 cm thick and have a dielectric constant of 2.3. The area of each electrode is 314 cm^2 and the loss angle is 9′ for a frequency of 50 Hz. $\varepsilon_0 = 8.855 \times 10^{-12}$ F/m. The fixed resistors of the network has a value of 1,000 Ω and the fixed capacitance is 50 μF. Find the values of variable resistor and capacitor required at balance. Derive the expressions used, if any.

[U.P.S.C. I.E.S. Electrical Engineering I, 1994]

[Ans. 0.0833 μF; 234.55 MΩ]

3. The four arms of a bridge are:
Arm AB: an imperfect capacitor C_1 with an equivalent series resistance r_1.
Arm BC: a non-inductive resistance R_3.
Arm CD: a non-inductive resistance R_4.
Arm DA: an imperfect capacitor C_2 with an equivalent resistance r_2 in series with a resistance R_2
Supply at 450 Hz is given between terminals A and C, and detector is connected between B and D. At balance R_2 = 5 Ω, R_3 = 2,000 Ω, R_4 = 2,950 Ω, C_2 = 0.5 μF and r_2 = 0.4 Ω. Calculate the values of C_1 and r_1 and also the dissipation factor for the capacitor. **[Ans.** 0.7375 μF; 3.66 Ω; 0.007632]

4. A capacitor is tested by Schering bridge. It forms one arm AB of the bridge. The other arms are: AD-a non-reactive resistance of 100 Ω; DC-a non-reactive resistance of 300 Ω shunted by a capacitor of 0.5 μF; BC—a standard loss free capacitor of 100 μμF, the supply frequency is 50 Hz. The bridge is at balance with the above components.
Deduce the balance conditions of the bridge and hence find out the capacitance and the power factor of the capacitor under test. **[Ans.** 296.7×10^{-12} F; 0.0471]

5. A sample of bakelite was tested by the Schering bridge method at 25 kV, 50 Hz. Balance was obtained with a standard condenser of 10 μF capacitance, a condenser of capacitance 0.5 μF in parallel with a non-reactive resistor of 309 Ω and a non-reactive resistor of 100 Ω. Determine the equivalent series resistance and the power factor of the specimen. **[Ans.** 0.458 MΩ; 0.048513]

6. Arm AB is an imperfect capacitor C_1 with a leakage resistance R_1 ; arm BC is a standard capacitor C_2 in series with non-reactive resistor of 700 Ω; arms CD and DA are non-reactive resistors of 5,000 Ω each. At balance C_2 = 0.0115 μF and frequency $f = 7{,}500/2\pi$.
Determine the values of C_1 and R_1. AC supply is connected across BD and a detector across AC. Draw the phasor diagram. **[Ans.** 0.0115 μF; 0.1913 MΩ]

7. A balanced 1,000 Hz bridge has the following configuration; arm AB, R_1 = 1,000 Ω in parallel with C_1 = 0.053 μF; BC, R_2 = 1,500 Ω in series with C_2 = 0.53 μF; CD, the unknown; DA pure capacitance C_4 = 0.265 μF. Find the R and L or C constants of the unknown. Draw the phasor diagram of the bridge at above frequency. **[Ans.** R = 119.46 Ω; 0.165625 μF]

8. The four arms of a Maxwell bridge are arranged as follows:
AB and BC are non-reactive resistances of 250 Ω each, DA is a standard variable reactor L of resistance 35.2 Ω and CD comprises a standard variable resistor R in series with a coil of unknown impedance. Balance is obtained when L = 25.6 mH and R = 25 Ω. Find the resistance and inductance of the coil. **[Ans.** 32.7 Ω; 26.6 mH]

9. The arms of an ac Maxwell bridge are arranged as follows: AB is a non-reactive resistor of 1,000 Ω in parallel with a condenser of capacitance 0.5 μF; BC is a non-reactive resistor of 600 Ω; CD is an unknown inductive resistor; and DA is a non-reactive resistor of 400 Ω. If balance is obtained under these conditions, find the value of the resistance and inductance in CD.
[Pb. Univ. Elec. Measurements-I, June 1993]
[Ans. 2,400 Ω; 0.12 H]

10. A coil of inductance L_1 and resistance R_1 is placed in the arm AB of an ac bridge. At balance other arms of the bridge have the following values :
R_3 in arm BC = 500 Ω (non-inductive resistance). R_2 in arm AD = 300 Ω (non-inductive resistance).
C_4 = 0.5 μF (air capacitor) in arm CD connected in parallel with a non-inductive resistance.
R_4 = 1,500 Ω. Calculate the Q-factor of the coil at 1,000 Hz. **[Ans.** 4.71]

11. In the diagram (Fig. 1) calculate the values of P and L under balanced conditions. **[Ans.** 1000 Ω, 50 H]

12. The four arms of a Maxwell's capacitance bridge at balance are: arm *ab*, an unknown inductance L_1 having an inherent resistance R_1; arm *bc*, a non-inductive resistance of 100 Ω; arm *cd* a capacitor of 0.5 μF in parallel with a resistance of 100 Ω; arm *da*, a resistance of 1,000 Ω. The source is connected to *a* and *c* and the detector is connected between *b* and *d*. Derive the equations for balance and find the values of R_1 and L_1. Draw the phasor diagram of the bridge under conditions of balance.
[Pb. Univ. Elec. Measurements-I, June 1992]
[Ans. 1,000 Ω; 50 m H]

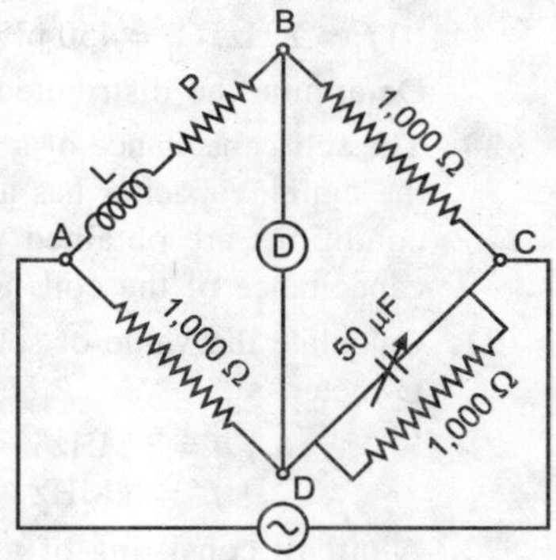

Fig. 1

13. In an Anderson bridge for measurement of inductance L_x and resistance R_x in the arm AB, the arms CD and DA have resistances of 600 Ω each and the arm CE has a capacitor of 1 μF capacitance. With ac supply at 100 Hz supplied across A and C balance is obtained with a resistance of 400 Ω in arm DE and 800 Ω in the arm BC. Calculate the value of L_x and R_x.
[Ans. 800 Ω; 1.12 H]

14. An ac bridge is connected as follows;
Branch AB is an inductive resistor, branches BC and ED are variable resistors, branches CD and DA are non-reactive resistors of 400 Ω each and branch CE is a condenser of 2 μF capacitance. The supply is connected to A and C and the detector to B and E. Balance is obtained when the resistance of BC is 400 Ω and that of ED is 500 Ω. Determine the resistance and inductance of AB.
[Kanpur Univ. Elec. Measurements 1986]
[Ans. 400 Ω, 1.12 H]

15. In an ac bridge the arrangement is as follows: AB is an inductive coil of inductance L and effective resistance R; BC is a non-reactive resistor of 100 Ω; CD and DA are non-reactive resistors of 200 Ω; each; CE is a loss-free capacitor of capacitance 1 μF; DE is a non-reactive resistance of 500 Ω under balanced condition. A supply of 10 V at 50 Hz is connected across AC and a vibration galvanometer between B and E. Calculate the value of R and L. Draw the phasor diagram for the balanced bridge.
[Ans. 100 Ω; 0.18 H]

16. The four arms of a Hay alternating current bridge are arranged as follows:
AB contains a coil of unknown impedance.
BC is a non-inductive resistor of 1,000 ohms.
CD is a non-reactive resistor of 883 ohms in series with a standard capacitor of 1.38 μF.
DA is a non-reactive resistor of 16.8 kΩ.
If the frequency of the applied voltage is 50 Hz and the bridge is balanced, determine the inductance and resistance of the coil. Derive the formula used. **[Ans.** 20.51 H; 2,357 Ω]

17. The resistance of an inductive coil is measured by an Owen bridge arranged as follows at balance. Arm AB contains the coil; arm BC contains a capacitor of 0.25 μF capacity in series with a resistance of 800 Ω, arm CD contains a loss-free capacitor of 1 μF capacity, arm DA contains a non-reactive resistance of 10 Ω. The supply frequency is 50 Hz.
(*a*) Deduce the balance condition of the bridge.
(*b*) Determine the values of the self-inductance, resistance and time constant of the coil.
(*c*) Draw the phasor diagram of the voltages and currents in the bridge arms at balance.
[Ans. 40 Ω; 8 mH; 0.2 ms]

18. A coil with a resistance of 10 Ω is connected in the circuit shown in Fig. 7.32 and resonance occurs when oscillator frequency of 0.5 MHz and the resonating capacitor is set at 120 pF. Determine the Q of the coil. **[Ans.** 265.26]

19. Two sets of measurements were carried out for measuring Q of a coil connected across the test terminals of the Q-meter.
(*i*) $f_1 = 2$ Hz, $C_1 = 450$ pF (*ii*) $f_2 = 4$ Hz, $C_2 = 90$ pF
Determine the distributed capacitance of the coil. **[Ans.** 30 pF]

20. The self capacitance of a coil is measured by a Q-meter. The circuit is set into resonance at 2 MHz and the tuning capacitor has a value of 460 pF. The frequency is now adjusted to 4 MHz and the resonance conditions are obtained with tuning capacitor set at a value of 100 pF. Calculate the value of self-capacitance of the coil. Prove the relationship used. **[Ans.** 20 pF]

21. Calculate the value of self capacitance of a coil when the following measurements are performed on a Q-meter.

$f = 2$ MHz C measured = 500 pF
$f' = 6$ MHz C′ measured = 50 pF **[Ans.** 6.25 pF]

22. A circuit consisting of a coil, a resistance and a variable capacitor connected in series is tuned to a resonance using a Q-meter. If the frequency is 500 kHz, the resistance is 0.5 Ω and the variable capacitor is set to 350 pF, calculate the effective inductance and resistance of the coil, if the Q-meter indicates 90. **[Ans.** 0.2895 mH; 9.61 Ω]

23. Determine the percentage error in Q measurement introduced by 0.02 Ω insertion resistance. The resonating capacitor is 135 pF and oscillator frequency at resonance is 3 MHz. The resistance of the coil is 10 Ω. **[Ans.** 0.2%]

24. The following observations were made while testing a high Q-factor coil be means of a circuit magnification meter : (*i*) with the test coil connected to the inductor terminals, resonance was obtained with resonating capacitance of 380 pF at a frequency of 1 MHz; (*ii*) with a standard coil connected to the inductor terminals, it was found that at a frequency of 9.6 MHz, resonance was not changed by connecting the test coil in parallel with the standard coil. Determine the distributed capacitance of the coil under test. **[Ans.** 4.168 pF]

UNIT-IV

Cathode Ray Oscilloscopes

CONTENTS IN THIS UNIT

CRO: CRT, Waveform display, Time base, Dual trace oscilloscope, Measurement of voltage, frequency and phase by CRO, Oscilloscope probes, Oscilloscope specifications and performance.

Delay time base oscilloscopes, Sampling oscilloscope, DSO, DSO applications.

CHAPTER 8: Cathode Ray Oscilloscopes.

CHAPTER 9: Special Oscilloscopes.

8

Cathode Ray Oscilloscopes

8.1 INTRODUCTION

The cathode ray oscilloscope is an extremely useful and the most versatile laboratory instrument used for studying wave shapes of alternating currents and voltages as well as for measurement of voltage, current, power and frequency, in fact, almost any quantity that involves amplitude and waveform. It allows the user to see the amplitude of electrical signals as a function of time on the screen.

Time recording devices, such as pen and strip chart recorders, have existed for a long time; however, the oscilloscope is capable of much fast operations, since it is completely electronic in nature. The oscilloscope is a kind of voltmeter which uses beam instead of a pointer, and kind of recorder which uses an electron beam instead of a pen. Rather than the recording of events over a period of few seconds, which is typical of the mechanical type recorder, the oscilloscope is capable of displaying events that take place over periods of microseconds and nanoseconds. Meter movements can only follow instantaneous variations up to a few cycles per second (Hz) while the high-speed oscillographs are limited to displaying signals below 500 Hz. But because of unique display mechanism, CRO are being made which can follow signals with frequencies up to 500 MHz.

The more expensive models can measure signals at frequencies up to 500 MHz and even the cheapest models can measure signals of frequencies up to 20 MHz. One particularly strong point that goes in favour of the oscilloscope is its high input impedance, typically 1 MΩ, which means that the instrument has a negligible loading effect in most measurement situations. Some drawbacks of oscilloscopes include their fragility (being built around a cathode ray tube) and their moderately high cost. Also, whilst the accuracy of the best instruments is about ± 1%, the accuracy of the cheapest one is only ± 10%. The most important aspects in the specification of an oscilloscope are its bandwidth,* its rise time and its accuracy.

In most oscilloscopes, the amplifier is direct-coupled, which means that it amplifies dc voltages by the same factor as low-frequency ac ones. For such instruments, the minimum frequency measurable is zero and the bandwidth (BW) can be interpreted as the maximum frequency where the sensitivity

* The bandwidth (BW) is defined as the range of frequencies over which the oscilloscope amplifier gain is within 3 dB of its peak value.

(deflection per volt) is within 3 dB of the peak value. The oscilloscope must be chosen such that the maximum frequency to be measured is well within the BW. The – 3 dB (the – 3 dB point is where the gain is 0.707 times its maximum value) specification means that an oscilloscope with a specified accuracy of ± 2% and BW of 100 MHz will have an accuracy of ± 5% when measuring 30 MHz signals and this accuracy will fall still further at higher frequencies. Thus when applied to signal amplitude measurement, the oscilloscope is only usable at frequencies up to about 0.3 times its specified BW.

The *rise time* is the transit time between the 10% and 90% levels of the response when a step input is applied to the oscilloscope. Oscilloscopes are normally designed such that

$$\text{BW} \times \text{rise time} = 0.35$$

$$\text{Thus, for BW of 100 MHz, rise time} = \frac{0.35}{100{,}000{,}000} = 3.5 \text{ ns}$$

Although most oscilloscopes are monochromatic, colour oscilloscopes are finding applications in computers and television.

The applications of oscilloscopes have been enhanced on account of many recent developments. Most modern oscilloscopes are capable of accepting two or more inputs displaying them simultaneously. This may be achieved through using a split beam or by using a multiple beam tube. Sampling oscilloscopes are available for high speed applications.

Storage oscilloscopes can be employed for capturing transient signal and then display them for periods varying from a few minutes to several years. An analog oscilloscope makes use of a modified form of a conventional cathode ray tube (CRT) to store the trace. The digital storage oscilloscope first converts the analog signal to a digital form and stores it in digital memory. The signal can then be recalled for display as and when needed.

Many oscilloscopes are now available with instrumentation interface (IEEE 488) bus capabilities, so that they can be employed as a part of measurement test bed, with the instrument's controls set at a remote location, and the readings digitized and retrieved for the purposes of recording and analysis.

An oscilloscope is a relatively complicated instrument which is constructed from a number of subsystems, and it is necessary to consider each of these in turn in order to understand how the complete instrument functions.

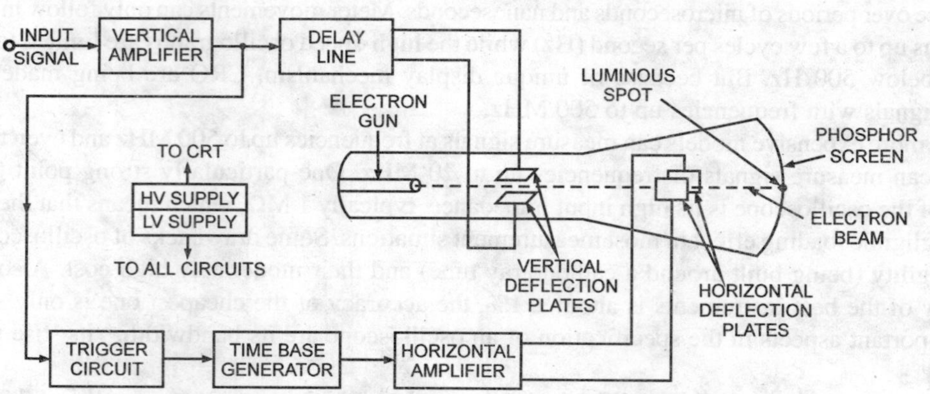

Fig. 8.1 *Block Diagram of a General Purpose CRO*

The instrument employs a cathode ray tube (usually abbreviated as CRT), which is the heart of the oscilloscope. It generates the electron beam, accelerates the beam to a high velocity, deflects the beam to create the image, and contains a phosphor screen where the electron beam eventually becomes visible.

For accomplishing these tasks various electrical signals and voltages are required, which are provided by the power supply circuit of the oscilloscope. Low voltage supply is required for the heater of the electron gun for generation of electron beam and high voltage, of the order of few thousand volts, is required for cathode ray tube to accelerate the beam. Normal voltage supply, say a few hundred volts, is required for other control circuits of the oscilloscope.

Horizontal and vertical deflection plates are fitted between electron gun and screen to deflect the beam according to input signal. Electron beam strikes the screen and creates a visible spot. This spot is deflected on the screen in horizontal direction (X-axis) with constant time dependent rate. This is accomplished by a time base circuit provided in the oscilloscope. The signal to be viewed is supplied to the vertical deflection plates through the vertical amplifier, which raises the potential of the input signal to a level that will provide usable deflection of the electron beam. Now electron beam deflects in two directions, horizontal on X-axis and vertical on Y-axis. A triggering circuit is provided for synchronizing two types of deflections so that horizontal deflection starts at the same point of the input vertical signal each time it sweeps. A basic block diagram of a general purpose oscilloscope is shown in Fig. 8.1. Cathode ray tube and its various components will be discussed in the following articles.

8.2 CATHODE RAY TUBE

The cathode ray tube (CRT) is the heart of a cathode ray oscilloscope (CRO). The remaining part of the CRO consists of circuit to operate the CRT.

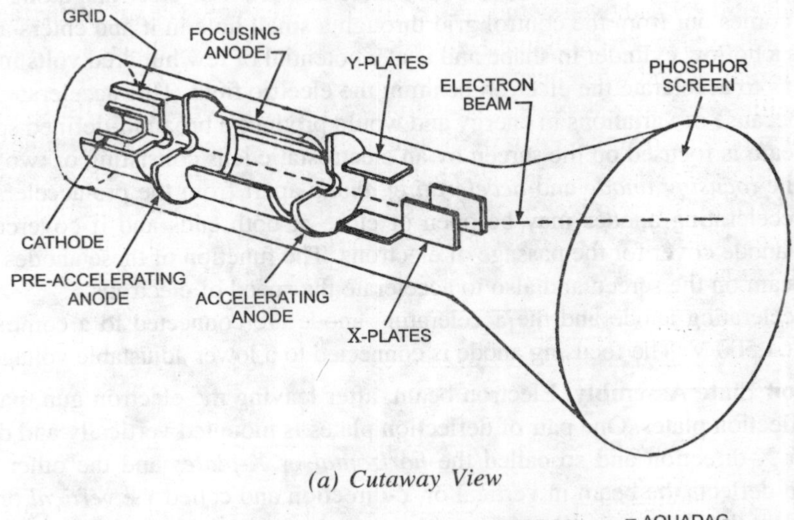

(a) Cutaway View

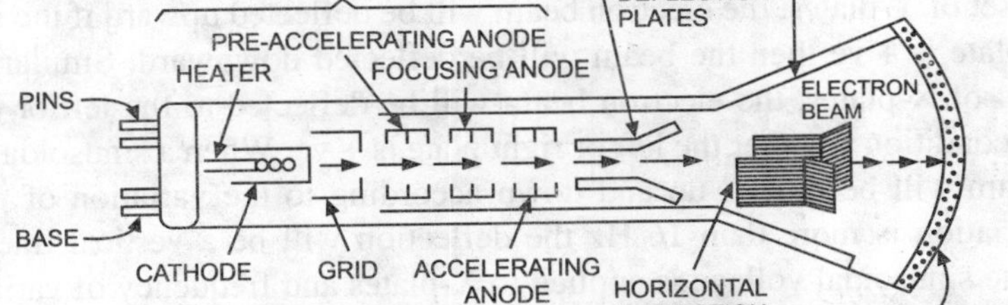

(b) Internal Structure

Fig. 8.2 *Cathode Ray Tube*

Cathode ray tube essentially consists of an electron gun for producing a stream of electrons, focusing and accelerating anodes for producing a narrow and sharply focused electron beam, horizontal and vertical deflection plates for controlling the beam path and an evacuated glass envelope with phosphor screen giving bright spot when struck by a high velocity electron beam, and base through which connections are made to different components.

1. Electron Gun Assembly. The electron gun assembly consists of an indirectly heated cathode, a control grid surrounding the cathode, a focusing anode and an accelerating anode. The sole function of the electron gun assembly is to provide a focused electron beam which is accelerated towards the phosphor screen. The cathode is a nickel cylinder coated with an oxide coating of barium and strontium and emits plenty of electrons, when heated. The emitting surface of the cathode should be as small as possible, theoretically a point. Rate of emission of electrons or say the intensity of electron beam depends on the magnitude of cathode current, which can be controlled by the control grid in a manner similar to a conventional vacuum tube. The typical values of current and voltage required by an indirectly heated cathode are 600 mA at 6.3 V (ac or dc). The special low power designs use 140 mA at 1.5 V. The *control grid* is usually a metal cylinder covered at one end but with a small hole in the cover. This is usually a metal cup of low permeability steel, about 15 mm in diameter and 15 mm long. An aperture of approximate 0.25 mm is drilled in the grid cap for the passage of electrons through it. The grid is kept at negative potential (variable) with respect to cathode and its function is to vary the electron emission and so the brilliancy of the spot on the phosphor screen. The hole in the grid is provided to allow passage for electrons through it and concentrate the beam of electrons along the axis of tube. Electron beam comes out from the control grid through a small hole in it and enters a pre-accelerating anode, which is a hollow cylinder in shape and is at a potential of few hundred volts more positive than the cathode so as to accelerate the electron beam in the electric field. This accelerated beam would be scattered now because of variations in energy and would produce a broad ill-defined spot on the screen. This electron beam is focused on the screen by an electrostatic lens consisting of two more cylindrical anodes called the *focusing anode* and *accelerating anode* apart from the pre-accelerating anode. The focusing and accelerating anodes may be open or close at both ends and if covered, holes must be provided in the anode cover for the passage of electrons. The function of these anodes is to concentrate and focus the beam on the screen and also to accelerate the speed of electrons.

The pre-accelerating anode and the accelerating anode are connected to a common positive high voltage of about 1,500 V. The focusing anode is connected to a lower adjustable voltage of 500 V.

2. Deflection Plate Assembly. Electron beam, after leaving the electron gun, passes through the two pairs of deflection plates. One pair of deflection plates is mounted vertically and deflects the beam in horizontal or X-direction and so called the *horizontal* or X-*plates* and the other pair is mounted horizontally and deflects the beam in vertical or Y-direction and called the *vertical* or *Y-plates*. These plates are to deflect the beam according to the voltage applied across them. For example if a constant *pd* is applied to the set of Y-plates, the electron beam will be deflected upward if the upper plate is + ve. In case the lower plate is + ve then the beam will be deflected downward. Similarly if a constant *pd* is applied to the set of X-plates, the electron beam will be deflected to the left or right of the tube axis according to the condition whether the left or right plate is + ve. When a sinusoidal voltage is applied to Y-plates, the beam will be moved up and down according to the variation of plate potential. If the frequency of variation is more than 16 Hz the deflection will be a vertical line in the centre of the screen. In case the sinusoidal voltage is applied to X-plates and frequency of variation is more than 16 Hz the deflection will be a horizontal line. If potentials are applied to both sets of plates simultaneously, the deflection will be an oblique line. The amount of deflection is in proportion to the voltage applied to the pair of plates.

3. Screen For CRT. As we know that some crystalline materials, such as phosphor, have property of emitting light when exposed to radiation. This is called the *fluorescence* characteristic of the materials. These fluorescent materials continue to emit light even after radiation exposure is cutoff. This is called the *phosphorescence* characteristic of the materials. The length of time during which phosphorescence occurs is called the *persistence* of the phosphor.

Persistence is usually measured in terms of the time required for the CRT image to decay to a certain percentage (usually 10 percent) of the original light output. It is expressed as short, medium and long. Persistence is called short if it lasts for microseconds, medium if it lasts for milliseconds and long in case it lasts for seconds to several minutes. *Short persistence* is required for extremely high speed phenomena, to avoid smearing and interference caused when one image persists and overlaps with the next one. *Medium persistence* traces are mostly used for general purpose applications. Long persistence traces are used for transients as they keep the fast transient on the screen for observation after the transient has disappeared.

The end wall of the CRT, called the *screen,* is coated with phosphor. When electron beam strikes the CRT screen, a spot of light is produced on the screen. The phosphor absorbs the kinetic energy of the bombarding electrons and emits energy at a lower frequency in a visual spectrum. Among the fluorescent materials used are zinc orthosilicate giving a green trace very suitable for visual observations; calcium tungstate giving blue and ultraviolet radiations very suitable for photography and zinc sulphide with other materials giving a white light suitable for TV. Zinc phosphate gives a pronounced after glow and is useful when studying transient phenomena because the trace persists for short while after the transient has disappeared.

The intensity of light emitted from the screen of CRT, known as *luminance,* depends upon several factors such as (*i*) on the number of bombarding electrons striking the screen per second (*ii*) on the energy with which the bombarding electrons strike the screen which is determined by the accelerating potential (more the accelerating potential more the luminance) (*iii*) on the time the beam strikes a given area of the phosphor *i.e.* sweep speed and (*iv*) on the physical characteristics of the phosphor itself. Almost all manufacturers provide their customers with a choice of phosphor materials. Table 8.1 summarizes the characteristics of the some of the commonly used phosphors.

Table 8.1 *Phosphor Data Chart*

Phosphor Type	*Fluorescence*	*Phosphorescence*	*Relative Luminance**	*Decay To 0.1% (ms)*	*Remarks*
P_1	Yellow-green	Yellow-green	50%	95	General purpose; replaced by P_{31} in most applications.
P_2	Blue-green	Yellow-green	55%	120	Good compromise for high- and low-speed applications.
P_4	White	White	50%	20	Television displays.
P_7	Blue	Yellow-green	35%	1,500	Long decay; Observation of low-speed phenomena.
P_{11}	Purple-blue	Purple-blue	15%	20	Photographic applications.
P_{31}	Yellow-green	Yellow-green	100%	32	General purpose-brightest available phosphor.

* Luminance is the photometric equivalent of brightness and is based on measurements made with a sensor having spectral sensitivity approximating the human eye. P_{31} is the reference phasor.

4. Glass Body and Base. The whole assembly is protected in a conical highly evacuated glass housing through suitable supports. The inner walls of CRT between neck and screen are usually coated with a conducting material known as *aquadag* and this coating is electrically connected to the accelerating anode. The coating is provided in order to accelerate the electron beam after passing between the deflecting plates and to collect the electrons produced by secondary emission when electron beam strikes the screen. Thus the coating prevents the formation of – ve charge on the screen and state of equilibrium of screen is maintained.

In some tubes, particularly CRTs with magnetic focusing (such as TV picture tubes), the accelerating anode is dispensed with entirely and the conductive coating is used as the final accelerating anode.

In small CRTs, connections to various electrodes are brought out through pins in the base of the tube, as shown in Fig. 8.2 (*b*). Larger type and medium sized high performance tubes operate at very high voltages and these leads are brought out through the sides of the glass envelope.

Horizontal and vertical marks are marked on the screen of the CRT to provide user a correct measurement. These marks, usually in rectangular form, are called *graticule*.

Size. Size refers to the screen diameter. CRTs for oscilloscopes are available in sizes 1, 2, 3, 5 and 7 inches. 3 inch is most common for portable instruments. For example a CRT having a number 5 GP_1 is 5 inch tube with a medium persistence green trace. The first number indicates that it is a 5 inch tube and P_1 indicates a medium persistence green trace. Both round and rectangular CRTs are found in oscilloscopes. The vertical viewing size is 8 cm and horizontal is 10 cm.

Comparison Between CRT and TV Picture Tubes. (*i*) The CRT used in a CRO uses electrostatic method of focusing whereas the TV picture tube always uses electromagnetic focusing.

(*ii*) In a CRT used in CRO, connection to the accelerating anode is brought out through pins in the base with other electrode connections, as illustrated in Fig. 8.2 (*b*). But in a TV picture tube the lead in from accelerating anode is brought out through the sides of the glass envelope.

(*iii*) In a CRT used in CRO, extra high tension (EHT) supply to accelerating anode is provided from mains supply through EHT transformer whereas in TV picture tube EHT supplied to accelerating anode is generated in the line output stage during flyback. During flyback the anode current of line output stage is suddenly cutoff and the anode voltage generates a spike of 2 to 3 kV amplitude. By increasing the number of turns on the primary winding to form an auto-transformer of say 1:5 turns, pulses of 15 kV or more are obtained. After rectification and smoothing, EHT is obtained and applied to accelerating anode. This method of applying EHT has an advantage because if input to the line and tube fails, no EHT is produced and thus the tube is saved.

8.3 FOCUSING DEVICE

Because of the mutual repulsion between the electrons, the convergent beam coming out from the accelerating anode has a tendency to spread from the axis. To aviod this and to bring the beam to a sharp focus at the screen, some focusing device is required. The focusing device may be either electrostatic or magnetic. Electrostatic focusing is used in a CRO whereas magnetic focusing is used in a TV picture tube.

8.3.1. Electrostatic Focusing. An electrostatic focusing system is shown in Fig. 8.3. Electrostatic lens consists of three anodes, with the middle anode at a lower potential than the other two electrodes.

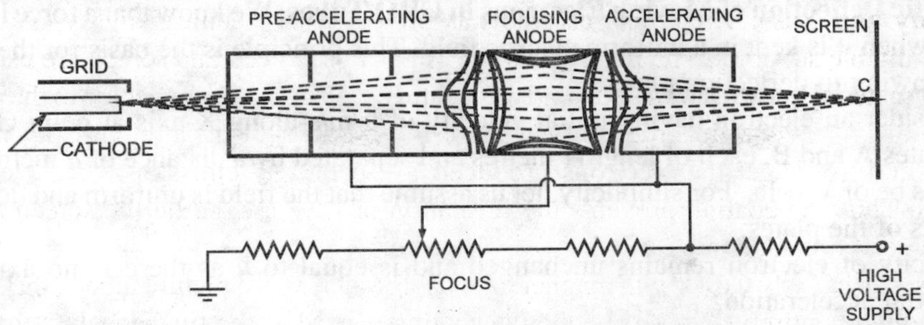

Fig. 8.3 *Electrostatic Focusing System of a CRT*

In Fig. 8.4 two anodes and its electrostatic lines and equipotential surfaces are shown. A pd is kept between these two electrodes so that an electric field is generated between them. Spreading of electric field is caused because of repulsion between electric lines. If equipotential lines are drawn, as shown in Fig. 8.4, they would bulge at the centre of the two anodes.

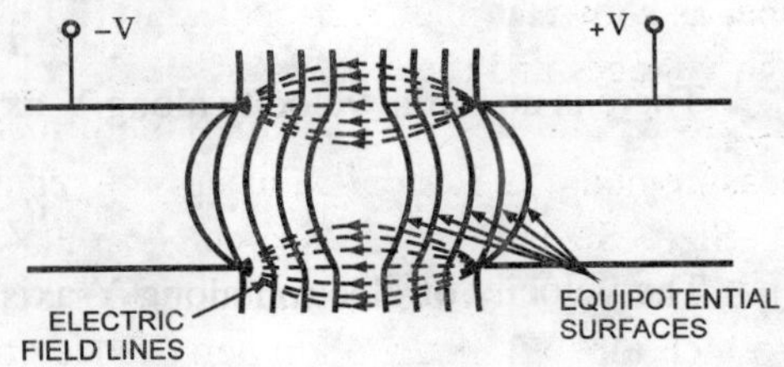

Fig. 8.4

As we know that electrons move in a direction opposite to that of electric field lines and equipotential surfaces are perpendicular to the electric field lines so force on the electron is exerted in the direction normal to the equipotential surface.

Electrons entering at the centre line of the two anodes experience no force but electrons displaced from the centre line experience a force normal to the direction of equipotential surface and deflects, as shown in Fig. 8.5. In Fig. 8.5 an equipotential surface is shown, in which an electron with velocity v_1 and at an angle θ_1 to the normal of equipotential surface enters and experiences a force in a direction normal to the equipotential surface. Thus the velocity of the electron increases to v_2. This force on the electron is exerted in the direction normal to equipotential surface so only the normal component of electron velocity v_{1N} increases to v_{2N} and the tangential component of velocity v_{1T} remains the same.

From Fig. 8.5

$$v_{1T} = v_1 \sin \theta_1$$

$$v_{2T} = v_2 \sin \theta_2$$

$$\text{but } v_{1T} = v_{2T}$$

$$\text{so } v_1 \sin \theta_1 = v_2 \sin \theta_2$$

$$\text{or } \frac{v_2}{v_1} = \frac{\text{Sin } \theta_1}{\text{Sin } \theta_2}$$

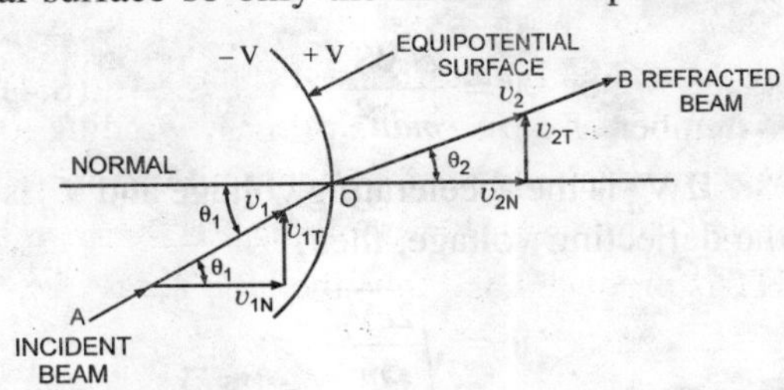

Fig. 8.5 *Refraction of an Electron Ray at an Equipotential Surface*

From the above equation it is obvious that equipotential surface acts as a concave lens in geometrical optics. That is why, this focusing system is named as an *electrostatic lens.*

Now if we go back and refer to Fig. 8.3, it can be seen that because of middle anode at a lower potential, electron beam coming from the cathode and passing through the first concave electrostatic lens tends to become more aligned with the axis of CRT and when it enters at the second concave electrostatic lens, formed between two anodes at different potentials, it is focused at the phosphor screen. Focal length of the electrostatic lens can be adjusted by varying potential of middle anode with respect to other two anodes. Thus the electron beam can be made to focus at the screen very precisely.

Electrostatic Deflection of Moving Electrons in CRO Tubes. We know that a force is experienced by an electron when it is kept in a uniform electric field. This principle is the basis for the deflection of electron beam owing to deflection plates.

Let us consider an electron having initial velocity of u m/s along X-axis at point O in the space between the plates A and B, each of length *l* metres and separated by a distance of *d* metres. Let the *pd* across the plates be of V volts. For simplicity, let us assume that the field is uniform and does not extend beyond the ends of the plates.

Axial velocity of electron remains unchanged and is equal to *u* as there is no axial force and, therefore, no axial acceleration.

The period during which an electron remains in the region between the two plates is given by the equation.

$$t = l/u \qquad \text{...(8.1)}$$

There is no initial velocity along Y-axis but has an acceleration along Y-axis given by the equation.

$$a_y = \frac{eE}{m} = \frac{e}{m} \cdot \frac{V}{d} \text{ m/s}^2 \qquad \text{...(8.2)}$$

The velocity of electron along Y-axis after time *t* is given by the equation.

$$v_y = 0 + a_y t = \frac{e}{m} \cdot \frac{V}{d} \cdot \frac{l}{u} \qquad \text{...(8.3)}$$

After the electron leaves the region between the deflection plates, it travels in a straight line, because there is no field acting on it. A simple extension of this development will show that if this line is extended backward, it intercepts the X-axis at the centre of the plates *i.e.* at $x = l/2$ (approximately). Let S be distance along the X-axis from this point to the screen. Then the deflection *y* can be determined by similar triangles. Thus

$$\frac{y}{S} = \frac{v_y}{v_x} = \frac{\frac{e}{m} \cdot \frac{V}{d} \cdot \frac{l}{u}}{u}$$

$$\text{or} \quad y = \frac{elVS}{mdu^2} \qquad \text{...(8.4)}$$

If V_a is the accelerating voltage and V_d is the deflecting voltage, then,

$$u = \sqrt{\frac{2eV_a}{m}}$$

$$\text{or} \quad u^2 = \frac{2eV_a}{m} \qquad \text{...(8.5)}$$

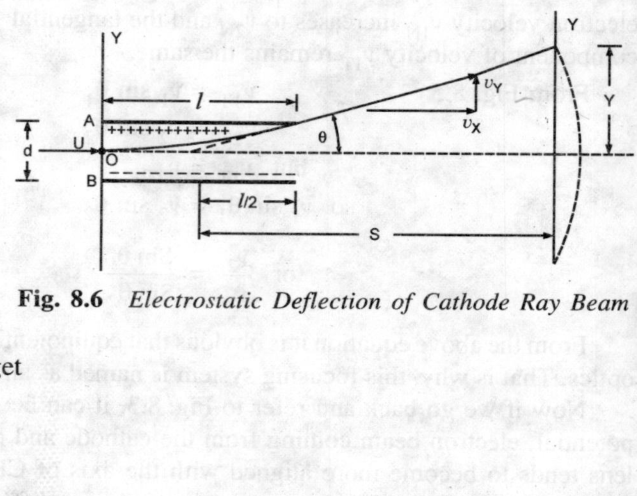

Fig. 8.6 *Electrostatic Deflection of Cathode Ray Beam*

Substituting $u^2 = \frac{2eV_a}{m}$ in Eq. (8.4) we get

$$y = \frac{elSV_d}{md \frac{2eV_a}{m}} = \frac{lSV_d}{2dV_a}$$

From the above equation we can conclude that for a fixed accelerating voltage V_a and dimensions of CRT, the deflection of electron beam on the screen is directly proportional to the deflection voltage V_d. So CRT can be employed as a *linear voltage indicating device.* Image on the screen of CRT follows the variation of deflection voltage V_d in a linear fashion.

The *deflection sensitivity* of a cathode ray tube, which is defined as the vertical deflection of the beam on the screen per unit deflecting voltage, is given by the equation

$$\frac{y}{V_d} = \frac{lS}{2dV_a} \text{ m/V} \qquad ...(8.6)$$

The deflection factor, the reciprocal of deflection sensitivity, is given by the equation

$$\frac{V_d}{y} = \frac{2dV_a}{lS} \text{ V/m} \qquad ...(8.7)$$

$$\text{Length of the trace obtained on the screen} = 2y = \frac{lSV_d}{dV_a} \qquad ...(8.8)$$

From the above equations for deflection sensitivity and deflection factor it can be said that deflection factor for a CRT varies linearly with the accelerating voltage. A high accelerated electron beam possesses more kinetic energy and so produces a brighter image on the screen of the CRT but a high deflection voltage is required for deflecting the beam.

Typical values of deflection sensitivity are 0.1 mm/volt to 1.00 mm/volt.

Example 8.1. The X-deflecting plates in a cathode ray tube are 5 mm apart and 25 mm long. The centre of the plates is 20 cm from the screen. The accelerating voltage is 3,000 V. Find the rms value of the sinusoidal voltage applied to X-deflecting plates if the length of the trace obtained on the screen is 10 cm. Find also the electrostatic deflection sensitivity of the above CRT.

Solution : Length of plates, l = 25 mm = 0.025 m

Distance between plates, d = 5 mm = 0.005 m

The distance between the screen and centre of plates,

S = 20 cm = 0.2 m

Accelerating voltage, V_a = 3,000 volts

Trace length = 2y = 10 cm = 0.1 m

$$\text{Deflecting voltage, } V_d = \frac{2dV_a y}{lS} = \frac{2 \times 0.005 \times 3{,}000 \times 0.05}{0.025 \times 0.2} = 300 \text{ volts} \qquad \because y = \frac{lSV_d}{2dV_a}$$

RMS value of the sinusoidal voltage applied to the X-deflecting plates,

$$V_{rms} = \frac{300}{\sqrt{2}} = 212 \text{ V } \textbf{Ans.}$$

$$\text{Deflection sensitivity} = \frac{lS}{2dV_a} = \frac{0.025 \times 0.2}{2 \times 0.005 \times 3{,}000}$$

$$= 0.1667 \times 10^{-3} \text{ m/V} = 0.167 \text{ mm/V } \textbf{Ans.}$$

Example 8.2. Calculate the maximum velocity of electrons in a CRT having a cathode anode voltage of 1,000 V. Assume the electrons to leave the cathode with zero velocity. Charge of electron = 1.6 ×1^{-19} C and mass of electron = 9.1 × 10^{-31} kg. [*U.P.S.C. I.E.S. Electrical Engineering I, 1995; A.M.I.E. Winter 1998; Summer 1999*]

Solution : Accelerating voltage, V_a = 1,000 V

$e = 1.6 \times 10^{-19}$ C

$m = 9.1 \times 10^{-31}$ kg

$$\text{Maximum velocity of electrons} = \sqrt{2 \times V_a \times \frac{e}{m}} = \sqrt{\frac{2 \times 1{,}000 \times 1.6 \times 10^{-19}}{9.1 \times 10^{-31}}} = 18.75 \times 10^6 \text{ m/s } \textbf{Ans.}$$

Example 8.3. The horizontal deflection plates in a CRT are 20 mm long and 5 mm apart. The centre of the plates is 20 cm from the screen. Accelerating voltage is 2,500 V. Determine the deflection sensitivity.

[*U.P. Technical Univ. Measurements & Instrumentation 2003-04*]

Solution : Length of plates, $l = 20 \text{ mm} = 0.02 \text{ m}$

Distance between plates, $d = 5 \text{ mm} = 0.005 \text{ m}$

The distance between the screen and centre of plates,

$$S = 0.20 \text{ m}$$

Accelerating voltage, $V_a = 2{,}500 \text{ V}$

$$\text{Deflection sensitivity} = \frac{lS}{2dV_a} = \frac{0.02 \times 0.20}{2 \times 0.005 \times 2{,}500} = 0.16 \text{ mm/V} \quad \textbf{Ans.}$$

Example 8.4. How much voltage is required across two deflection plates separated by 1 cm to deflect an electron beam 1° if the effective length of the deflection plates is 2.5 cm and the accelerating potential 1,000 V ? Deduce the formulae used.

Solution : From Fig. 8.6, $\text{Tan } \theta = \frac{y}{S}$

Substituting from Eq. (8.8), $\frac{y}{S} = \frac{lV_d}{2dV_a}$ we have

$$\text{Tan } \theta = \frac{lV_d}{2dV_a} \quad \text{...}(i)$$

where l is the effective length of the deflection plates and is equal to 2.5 cm; d is the separation between deflection plates and is equal to 1 cm; V_a is the accelerating voltage and is equal to 1,000 V and V_d is the voltage required across the deflection plates to deflect the electron beam through an angular deflection $\theta = 1°$.

$$\text{So } V_d = \frac{2dV_a}{l} \tan \theta = \frac{2 \times 1 \times 1{,}000}{2.5} \tan 1° = 13.97 \text{ V } \textbf{Ans.}$$

Example 8.5. An electrostatically deflected cathode ray tube has plane parallel deflecting plates which are 2.5 cm long and 0.5 cm apart, and the distance from their centre to the screen is 20 cm. The electron beam is accelerated by a potential difference of 2,500 V and is projected centrally between the plates. Calculate the deflecting voltage required to cause the beam to strike the screen and find the corresponding deflection of the screen.

[*Pb. Technical Univ. Electronics Measurements and Instrumentation December 2005*]

Solution : Length of plates, $l = 2.5 \text{ cm} = 0.025 \text{ m}$

Distance between plates, $d = 0.5 \text{ cm} = 0.005 \text{ m}$

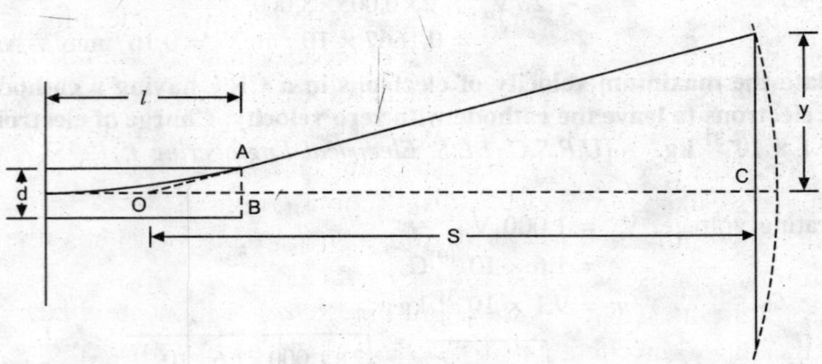

Fig. 8.7

Distance from the centre of plates to the screen,

$$S = 20 \text{ cm} = 0.2 \text{ m}$$

$$\text{Accelerating voltage, } V_a = 2{,}500 \text{ V}$$

$$\text{Deflection, } y = \frac{OC}{OB} \times AB = \frac{S \times d/2}{l/2}$$

$$= \frac{S \times d}{l} = \frac{0.2 \times 0.005}{0.025} = 0.04 \text{ m or } 4 \text{ cm} \quad \textbf{Ans.}$$

8.4 POST DEFLECTION ACCELERATION

The amount of light given off by the screen is dependent on the amount of energy transferred to the phosphor by the electron beam. In case the electron beam is to be deflected at a faster rate, permitting the oscilloscope to respond to fast occurring events, the velocity of the electron beam must be great, otherwise, the light output will decrease. It means that for a fast oscilloscope it is desirable to accelerate electron beam to the greatest amount possible, while on the other hand the greater electron beam velocity will make it more difficult to deflect the beam.

It can be seen that the greater the accelerating voltage the more difficult it is to deflect the electron beam. This would need higher deflection voltages, but, more important, the time change of voltage *i.e.*, $\frac{dv}{dt}$ is also greater owing to higher voltage. Because of this higher currents to charge the deflection plate capacitance, in addition to higher deflection voltage, would be required. This becomes a very significant problem for high-frequency oscilloscopes with frequency responses exceeding 100 MHz. Modern CRTs employ a two-step acceleration to overcome this problem. First, the electron beam is accelerated to a relatively low velocity through a few thousand volt potential. The beam is then deflected and, after deflection, is further accelerated to the desired final velocity. In this way, *post deflection acceleration* (acceleration after the deflection) does not affect the deflection sensitivity. The CRT making use of postacceleration deflection is called the *post deflection acceleration tube.*

Figure 8.8(*a*) illustrates a post deflection acceleration CRT using a mesh that further increases the amount of the electron beam scan. In this case the electron beam is accelerated and deflected in a fashion similar to previous case of the simple tube. However, the beam is further accelerated through a high potential of 10 kV or more, after the deflection. Thus the deflection sensitivity remains unaffected. A metallic mesh, suspended in the electron beam, acts as a magnifying lens that further increases the deflection and thus improves the deflection sensitivity. With this technique, deflection sensitivity can remain of the order of 0.5 to 5 V/mm even though the total electrons beam acceleration exceeds 10 kV.

There are several drawbacks of the mesh type post deflection acceleration CRT. First, the mesh tends to defocus the electron beam and make the spot broader than it would be in the absence of mesh interfering with the beam. Second, the mesh conducts some of the electron beam away from the screen. This reduces the beam current causing reduction in spot intensity. Another problem with the post-deflection acceleration CRT, is that the electron beam tends to be defocused in the vicinity of the deflection plates due to repulsion from charge distributions within the tube. Several recent advances in the design of CRT have eliminated the mesh and reduced these problems, thus providing a high-performance electron gun for use in high-frequency CRTs.

Electron gun for a meshless CRT is shown in Fig. 8.8(*b*). The electron beam is produced from a conventional heated cathode surrounded by the control grid. The first accelerating anode and two focus electrodes follow and provide focus, as well as the first accelerating potential. These focus electrodes are different from the cylindrical elements used in a conventional tube—they are constructed from individual metal wafers with noncylindrical holes in the centre. This allows for a different focusing characteristic in the horizontal plane and the vertical plane, typically divergent in one plane while being convergent in the other. The holes in the centre of the metal wafers can be formed with greater precision than in a formed cylinder, and thus greater tolerances can be achieved at a lower cost.

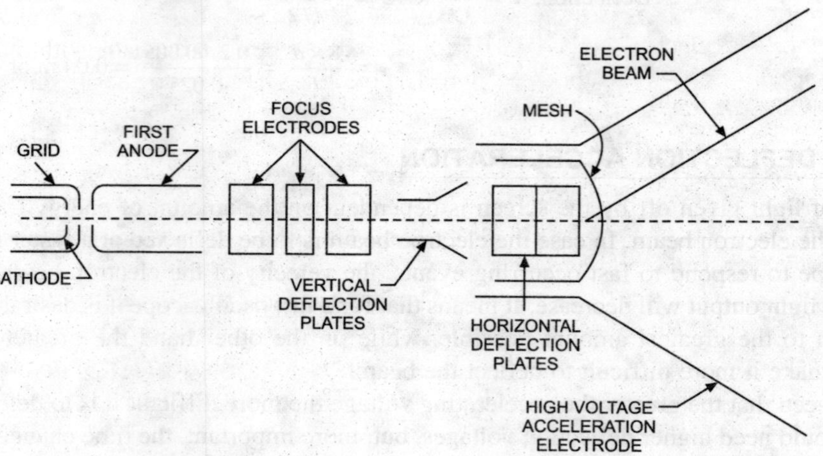

(a) Post deflection Acceleration Oscilloscope Tube Using a Scan Expansion Mesh

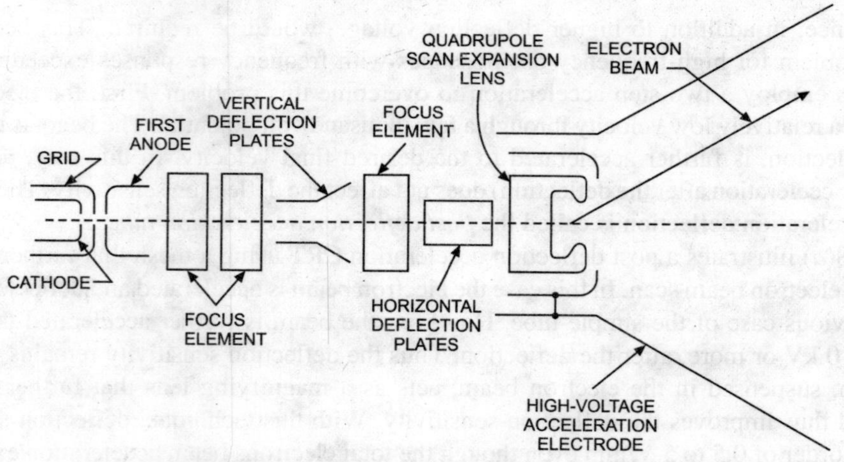

(b) Meshless Scan Expansion Post deflection Acceleration CRT

Fig. 8.8

After the two focusing electrodes, the beam passes through the vertical deflection plates. The beam at this point is not completely focused, which reduces the magnitude of beam distortion owing to the internal charge distributions. The beam will be further focused after deflection to give a fine spot.

After vertical deflection plates, the beam passes through a scan expansion lens that increases the magnitude of beam bending in the vertical plane. The beam is then deflected in the horizontal direction and passed through another electron lens, which provides additional focusing.

The beam is accelerated to the final velocity by a *quadrupole lens,* which provides not only an increase in electron velocity, but adds to the scan angle (scan expansion, which is similar to the mesh in the previous case) without defocusing or distorting the electron beam.

This design results in an increased sensitivity, typically 2.3 V/cm for the vertical deflection and 3.7 V/cm in the horizontal direction. The difference between the vertical and horizontal deflection sensitivities is owing to the fact that the vertical deflection occurs at a lower beam velocity. Because the horizontal deflection of the oscilloscope involves only a time linear sweep, while the vertical deflection needs complex waveforms, the more sensitive deflection should be reserved for the vertical direction.

Using the meshless electron gun, 100 MHz plus oscilloscopes can be made with integrated circuits using only 40 or 50 V or even less for deflection. The meshless tube, being considerably shorter, results in smaller and lighter oscilloscopes for laboratory and portable applications.

8.5 BASIC CONTROLS

Number of controls are required to be provided on a panel of CRO to facilitate its proper functioning. Intensity control is provided for adjustment of brightness of the spot on the screen. It is accomplished by varying the voltage between the first and second anodes. The horizontal and vertical position controls are provided for moving the beam on any part of the screen. It is accomplished by applying a dc voltage to horizontal or vertical deflection plates. Similarly there are other numerous controls in a CRO, which will be discussed, in detail, here.

A block diagram of a basic CRO is shown in Fig. 8.9.

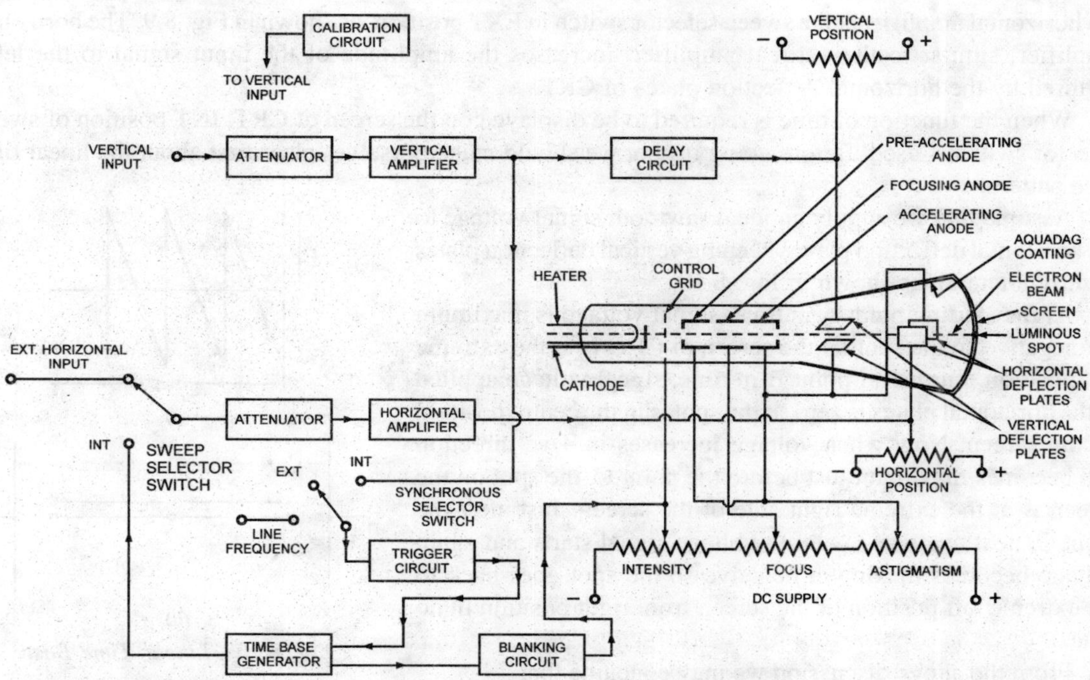

Fig. 8.9 *Detailed Block Diagram of a CRO*

1. Vertical Deflection System. The function of vertical deflection system is to provide an amplified signal of the proper level to drive the vertical deflection plates without introducing any appreciable distortion into the system.

The input sensitivity of many CROs is of the order of a few millivolts per division and the voltage required for deflecting the electron beam varies from approximately 100 V (peak to peak) to 500 V depending on the accelerating voltage and the construction of the tube. Thus the vertical amplifier is required to provide this desired gain from millivolt input to several hundred volt (peak-to-peak) output. Also the vertical amplifier should not distort the input waveform and should have good response for entire band of frequencies to be measured.

The deflection plates of CRO act as plates of a capacitor and when the input signal frequency exceeds over 1 MHz, the current required for charging and discharging of the capacitor formed by the deflection plates increases. So the vertical amplifier should be capable of supplying current enough to charge and discharge the deflection plate capacitor.

As we know that electrical signal is delayed by a certain amount of time when transmitted through an electronic circuitry. In CRO, output signal voltage of the vertical amplifier is fed to the vertical plates of CRT and some of its portion is used for triggering the time base generator circuit, whose output is supplied to the horizontal deflection plates through horizontal amplifier. The whole process, which includes generating and shaping of a trigger pulse and starting of a time-base generator and then its amplification, takes time of the order of 100 ns or so. So the input signal of the vertical deflection plates of a CRT is to be delayed by at least the same or little more amount of time to allow the operator to see the leading edge of the signal waveform under study on the screen. For this purpose, *delay line circuit* is introduced between vertical amplifier and the plates of CRT, as shown in Fig. 8.9.

2. Horizontal Deflection System. External signal is applied to horizontal deflection plates through the horizontal amplifier at the sweep selector switch in EXT position, as shown in Fig. 8.9. The horizontal amplifier, similar to the vertical amplifier, increases the amplitude of the input signal to the level required by the horizontal deflection plates of CRT.

When the function of time is required to be displayed on the screen of CRT, INT position of sweep selector switch is used. Before going further we should make ourselves clear first about the linear time base pattern.

Assume that we supply an ideal sawtooth signal voltage to the horizontal deflection plates, keeping vertical deflection plates at zero potential, as shown in Fig. 8.10.

At the starting point A in time, signal voltage is maximum but negative so the spot on the screen of CRO is at the extreme left position. Further at point B in time, signal voltage applied to the horizontal plates is zero so the spot is in the centre position on the screen. Now when voltage increases in + ve direction and becomes maximum just before the point C, the spot on the screen is at the extreme right side of the screen. Just after the point C, next cycle of sawtooth voltage signal starts and again voltage becomes maximum negative so the spot goes back to the extreme left position of the screen from right position in no time.

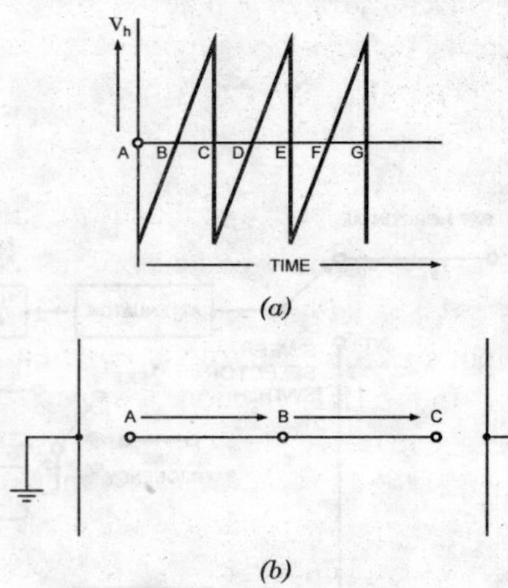

Fig. 8.10 *Linear Time Base*

From the above discussion we may conclude that

(*i*) The spot moves from left to right over the same path again for every cycle of sawtooth voltage applied to the horizontal deflection plates, so a horizontal line appears on the screen of CRO.

(*ii*) The spot moves from left to right on the screen with the uniform speed. Thus it produces a linear time base to display function of time on the screen of CRO.

For making idea of time base more clear let us discuss an application. Suppose a sine-wave voltage signal V_v of time period T is applied to the vertical deflection plates and a sawtooth voltage signal V_h of time period T is applied to horizontal deflection plates, as shown in Fig. 8.11.

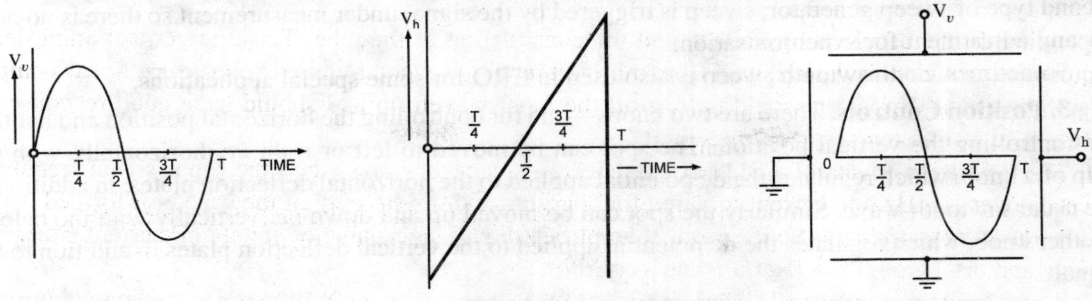

(a) Voltage Signal Applied To Vertical Deflection Plates

(b) Sawtooth Wave Voltage Signal Applied To Horizontal Deflection Plates

(c) Pattern of Signal on Screen of CRO

Fig. 8.11

At zero time, the spot is at extreme left vertically central position on the screen because of zero value of V_v and maximum negative value of V_h. At time $\frac{T}{4}$, the spot is at one-fourth way on the screen in horizontal direction and at maximum positive deflection above the centre line in vertical direction because of maximum positive value of V_v. At time $\frac{T}{2}$, values of both V_v and V_h are zero, so the spot is at the central position of the screen. At time $\frac{3T}{4}$, the spot is the three-fourth way on the screen in horizontal direction and at the maximum negative deflection in vertical direction. Finally, at the end of time T, the spot is at extreme right vertically central position of the screen and then it moves back to begin a new trace. In this way, sine-wave voltage applied to the vertical deflection system appears on the screen. If the period of sine-wave is reduced to half then two sine-wave cycle appears on the screen.

From the above discussion, it is obvious that the following conditions are to be satisfied in order to have a waveform of the input signal applied to vertical deflection system as a stationary pattern on the screen of CRO.

(*i*) Both horizontal and vertical signals must start at the same instant.

(*ii*) Ratio of frequency of horizontal and vertical signals should be a rational or fractional number.

For satisfying the above conditions, sawtooth wave is generated and synchronised with the vertical input signal by the trigger circuit and time base generator, as shown in Fig. 8.9 and explained above.

In the INT position of sweep selector switch, horizontal amplifier receives an input from the time base generator, which provides a time base, and controls the rate at which the beam is scanned across the face of the CRT. Time base generator is triggered or initiated by a trigger circuit which ensures that the horizontal sweep starts at the same point of the vertical input signal.

As explained earlier, it is necessary to synchronise the sweep with the signal under measurement in order to obtain stationary pattern. Ratio of frequency of the time base and of the signal under measurement should be a rational number, otherwise pattern on the screen will not be stationary. A synchronous selector switch is used, as shown in Fig. 8.9, to select the type of synchronisation. In the internal mode of switch the trigger is obtained from the vertical amplifier, input of which is signal under measurement.

In the external position of switch, the trigger is obtained from the external source. In the third position of switch *i.e.* line, trigger is obtained from the power supply *i.e.* 230 V and 50 Hz.

Two types of sweep generators are usually used. In the first one sawtooth signal of constant frequency is generated whether there is any input signal for vertical signal or not. That is why it is called *free running type*. In this it is essential to adjust the frequency of the sawtooth to get stationary pattern. In the second type of sweep generator, sweep is triggered by the signal under measurement so there is no need for any adjustment for synchronisation.

Sometimes, non-sawtooth sweep is also used in CRO for some special applications.

3. Position Controls. There are two knobs—one for controlling the horizontal position and another for controlling the vertical position. The spot can be moved to left or right *i.e.* horizontally with the help of a knob, which regulates the dc potential applied to the horizontal deflection plates, in addition to the usual sawtooth-wave. Similarly the spot can be moved up and down *i.e.* vertically with the help of another knob, which regulates the dc potential applied to the vertical deflection plates in addition to the signal.

4. Intensity Control. The potential of the control grid with respect to cathode is controlled with the help of potentiometer in order to control the intensity of brightness of the spot. The grid potential determines the amount of electrons leaving the cathode and thus controls the intensity of beam. A larger number of electrons in the beam causes a brighter spot to appear on the screen.

Caution. Care must be taken to prevent the electron beam from burning spots on the screen. A stationary spot should be kept on very low intensity. If the intensity is kept high, the spot must be kept moving. If a "halo" appears around the spot, the intensity is too high. Before turning the oscilloscope on, turn down the intensity.

5. Focus Control. In the electron gun of a CRT, middle anode is kept at lower potential with respect to other two anodes and it acts like an electrostatic lens and focal length of this lens can be varied by varying the potential of the middle anode with respect to other two anodes. So focusing of an electron beam is done by varying the potential of middle anode with the help of a potentiometer, as shown in Fig. 8.9. By increasing the + ve potential applied to the focusing anode the electron beam can be narrowed and the spot on the screen can be made a pin point.

6. Astigmatism. This is an additional focusing control and is analogous to astigmatism in optical lenses. A beam that is focused at the centre of the screen would be defocused at the edges of the screen because the lengths of the electron paths are different for the centre and the edges. Adjustment of this control gives a sharp focus over the entire screen. This control is affected by varying the potential of deflection plates and accelerating anodes.

7. Blanking Circuit. Sawtooth sweep voltage is applied to horizontal deflection plates of CRT which moves the spot on the screen following a straight horizontal line from left to right during the sweep period. When the spot moves slowly so that its rate of movement exceeds the threshold of persistence vision, the spot appears as a solid line. Below this threshold limit, only spot or some portion of line after the spot appears. If the movement of the spot is fast, it appears as thin and dim horizontal line or may be invisible.

In Fig. 8.10 (*a*) a sawtooth voltage waveform is shown, which is an ideal one. In this waveform retrace time is zero. But in practice it is not possible to achieve this, so there is some retrace time of waveform, as shown in Fig. 8.12. During this retrace time, the spot moves from right to left and leads to confusion. So this trace is blanked out by applying a high – ve voltage to the grid during the retrace time. This blanking voltage is usually triggered by time base generator.

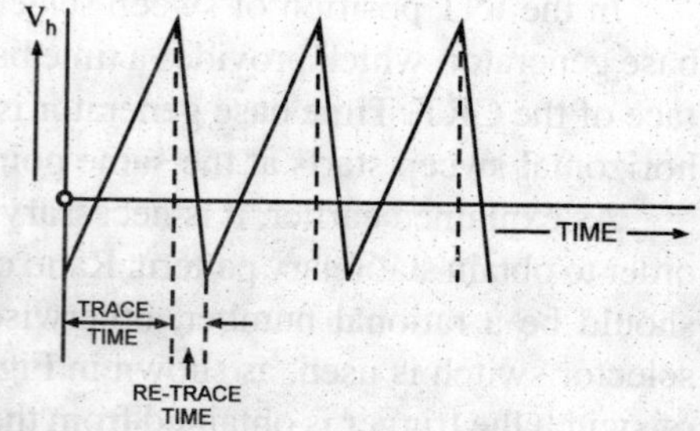

Fig. 8.12 *Sawtooth Voltage Waveform*

8. Calibration Circuit. Normally an oscillator, which generates a known and fixed voltage in square waveform, is fitted in the CRO for calibration purpose.

8.6 OSCILLOSCOPE TIME BASE

Time base is an axis, normally *x*-axis, that is used to represent time so that variations of quantities such as voltages, currents etc., or their waveforms can be plotted w.r.t. time.

Most of the CRO applications involve measurement or display of a quantity which varies with respect to time. This requires that the CRT spot moves across the screen with a constant velocity. For this a voltage which varies linearly with time has to be applied to one set of deflection plates. This voltage is used to sweep the electron beam across the screen, so it is called a *sweep voltage*. Because of its shape, it is also sometimes called a *sawtooth* or *ramp voltage.* The circuits which develop these linearly varying voltages are called *time base generators* or *sweep circuits*.

The purpose of a time base is to apply a voltage to the horizontal deflection plates such that the horizontal movement of the spot is proportional to time. This voltage, known as a *sweep waveform* (in the form of a ramp), must be applied repetitively, such that the motion of the spot across the screen appears as a straight line when a dc level is applied to the input channel. Furthermore, this *time base voltage* must be synchronised with the input signal in the general case of a time-varying signal, so that a steady picture is obtained on the oscilloscope screen. The length of time taken for the spot to traverse the screen is controlled by a *Time/division switch*, which sets the duration of time taken by the spot to travel between two marked divisions on the screen, thereby allowing signals at a wide range of frequencies under measurement.

Each cycle of the sweep waveform is initiated by a pulse from a pulse generator. The input to the pulse generator is a sinusoidal signal known as a *triggering signal*, with a pulse being generated every time the triggering signal crosses a predetermined slope and the voltage level condition. This condition is defined by the 'trigger level' and 'trigger slope' switches. The former selects the voltage level on the trigger signal, usually zero, at which the pulse is produced, whilest the latter selects whether pulsing occurs on a positive or negative going part of the triggering waveform.

Synchronisation of the sweep waveform with the measured signal is most easily achieved by deriving the trigger signal from the measured signal, a procedure which is known as 'internal triggering'. Alternatively, 'external triggering' can be applied if the frequencies of the triggering signal and measured signals are related by an integer constant such that the display is stationary. External triggering is necessary when the amplitude of the measured signal is too small to drive the pulse generator, and it is also employed in applications where the phase difference between two sinusoidal signals of same frequency is to be measured. It is very convenient to use the 50 Hz line voltage for external triggering when measuring signals at mains frequency, and this is often given the name 'line triggering'.

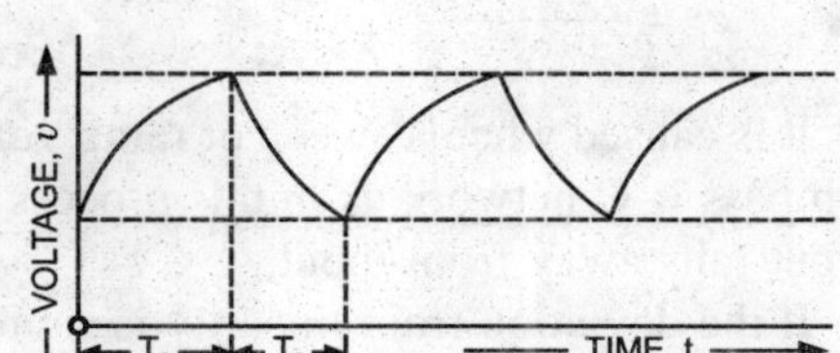

Fig. 8.13 *A General Sweep Voltage Waveform*

Typical form of a time base voltage (or sweep voltage) is shown in Fig. 8.13. The voltage starts from some initial value, also may be zero, increases linearly with time to a maximum or peak value, and then returns to its initial value. The time required to return to its initial value is called *restoration time* or *flyback time* (T_r) while the time occupied by the linear portion of the waveform is called *sweep time* T_s. Sometimes, the shape of the waveform during the restoration or retrace time T_r and restoration time itself are of little significance.

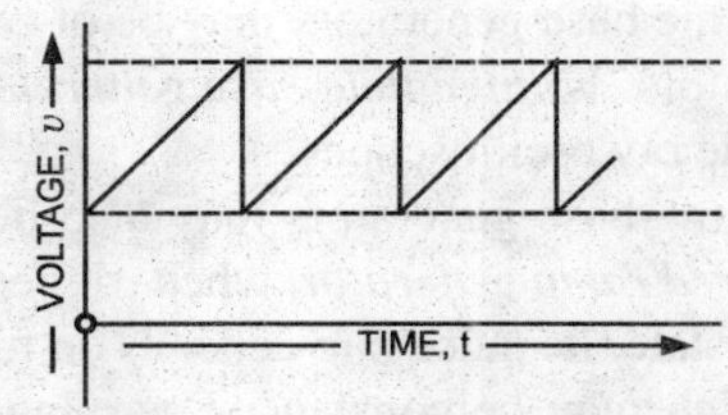

Fig. 8.14 *An Idealized Sweep Voltage Waveform*

However, in some cases it is required that T_r should be very small as compared to T_s. In such a case, when T_r is negligibly small, the waveform will appear as shown in Fig. 8.14. The waveform shown in Fig. 8.14 is an idealized waveform of sweep voltage, where the voltage is exactly linear during the sweep period and the flyback time is zero.

Time base generators do not ordinarily provide sweep voltages that are exactly linear, although every effort is made to obtain reasonable linearity in rise of voltage. Moreover, a linear sweep may get distorted in the course of transmission through a coupling network. The deviation from linearity may be expressed in the following ways.

Slope or *sweep speed error* e_s is defined as the ratio of difference in slope at the beginning and end of sweep to the initial value of slope *i.e.*

$$e_s = \frac{\text{Difference in slope at the beginning and end of slope}}{\text{Initial value of slope}} \quad \text{...(8.9)}$$

Such error is significant where constant sweep (rate of change of sweep voltage with time) is an essential requirement as in case of a general purpose CRO.

Displacement error, e_d is defined as the ratio of maximum difference between the actual sweep voltage and the linear sweep voltage passing through the beginning and end points of the actual sweep to maximum value attained by sweep voltage, as illustrated in Fig. 8.15 *i.e.*

$$e_d = \frac{(v_s - v_s')_{\max}}{V_S} \quad \text{...(8.10)}$$

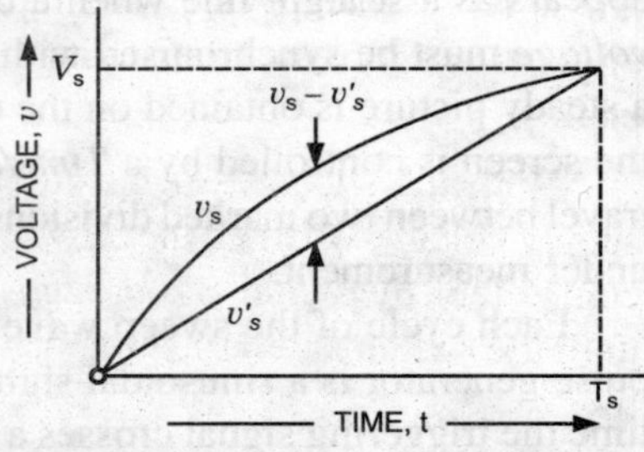

Fig. 8.15 *Displacement Error*

It is used to define nonlinearity of time base signal and is significant in some timing applications.

Transmission error, e_t is defined as the ratio of the difference between maximum amplitude of input and output signals to the maximum amplitude of signal, as shown in Fig. 8.16 *i.e.*

$$e_t = \frac{(V_s' - V_s')}{V_s} \quad \text{...(8.11)}$$

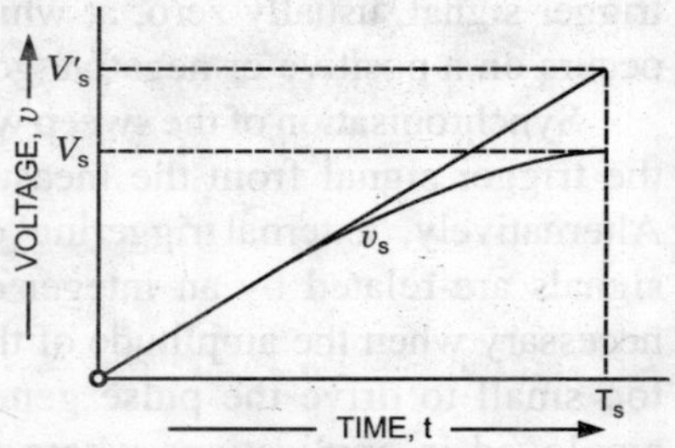

Fig. 8.16 *Transmission Error*

It is caused when a sweep or ramp voltage is passed through a high pass R-C network as in this process maximum amplitude of output falls away from input.

If the deviation from linearity is small, sweep voltage can be approximated by the sum of a linear and a quadratic terms in *t*, then

$$e_d = \frac{1}{8}e_s = \frac{1}{4}e_t \quad \text{...(8.12)}$$

Time base generators may be of two types viz. voltage time base generators and current time base generators. *Voltage time base generators* generate a voltage linearly varying with time and are used in cathode ray oscilloscopes.

Time-base generators may also be classified as triggered free-running and synchronised type. A *triggered ramp generator,* when triggered by a ramp signal, generates a cycle and then comes back to its stable state. Its pulse rate depends on frequency or pulse repetition rate of applied trigger, the frequency of which may be constant or varying. A *free-running time-base generator* switches forth and back between two levels and its frequency of oscillation depends on the parameters of the circuit. In a *synchronised time base generator*, frequency of oscillation is determined by a clock signal or synchronising voltage.

All of the above sweep circuits are capable of generating waves of either polarity (*i.e.* positive or negative).

8.6.1. Exponential Sweep Circuit. Figure 8.17 represents a basic sawtooth voltage generator. Switch S periodically changes its position from point 1 to point 2 (although a mechanical switch is shown in the circuit, the switching action is actually performed electronically by means of a separate circuit not shown here). The switching action causes the capacitor C to be alternately charged and discharged. The resulting waveform of output voltage v_{out} is shown in Fig. 8.18. As already explained, the time T_1 is called the *sweep time* and is represented by T_s and time T_2 is called the *retrace* or *return time* and is represented by T_r. The return or retrace time is required in display systems to return the beam to its original starting point before the next cycle begins.

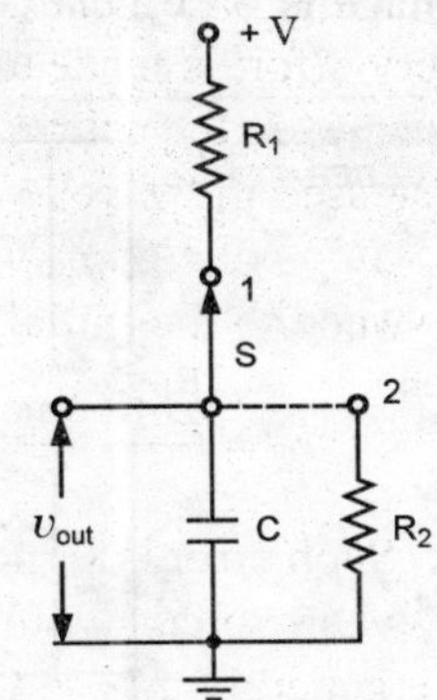

Fig. 8.17 *Circuit For a Sawtooth Voltage Generator*

With the switch S in position 1, the capacitor tends to charge to the supply voltage V, called the *aiming potential*. Before the capacitor can charge completely, the switch S is brought to position 2, and the capacitor discharges until the switch is returned to position 1, at which time the cycle is repeated. The output voltage is given by

$$v_{out}(t) = V - (V - V_1)\, e^{-t/R_1C} \quad \text{for } 0 \le t \le T_1 \qquad ...(8.13)$$

$$\text{and } v_{out}(t) = V_2\, e^{-(t-T_1)/R_2C} \quad \text{for } T_1 \le t \le T_2 \qquad ...(8.14)$$

When $t = T_1$, $v_{out}(T_1) = V_2$ so that

$$T_1 = R_1C \log_e \frac{V - V_1}{V - V_2} \qquad ...(8.15)$$

and at $t = T_2$, $v_{out}(T_2) = V_1$, so that

$$T_2 = R_2C \log_e \frac{V_2}{V_1} \qquad ...(8.16)$$

The sweep amplitude V_s is given as

$$V_s = V_2 - V_1 \qquad ...(8.17)$$

The *sweep linearity* is a measure of how close to the ideal the exponential waveform comes. From Fig. 8.18 it is seen that for smaller values of V_2, the straight line and exponential charges are nearly identical. In general, the sweep voltage should be a small fraction of the aiming potential or supply voltage if good linearity is to be achieved.

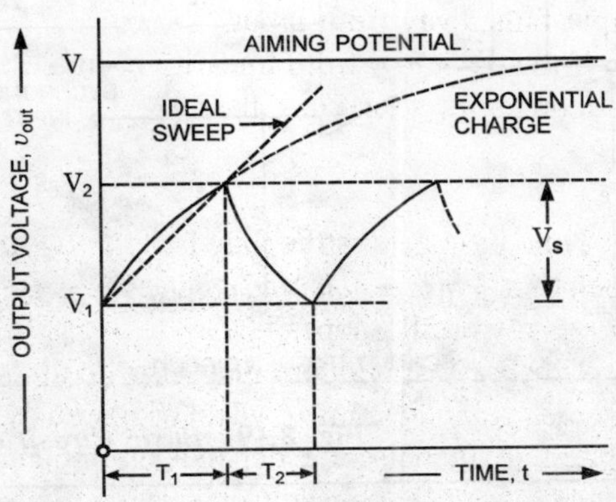

Fig. 8.18 *Output Waveform For Circuit Given in figure*

Basic circuit of an oscilloscope time base or sweep generator using Schmitt trigger is shown in Fig. 8.19. It consists of two main components: a ramp generator and a noninverting Schmitt trigger circuit.* Output voltage (V_1) of the ramp generator is applied to the Schmitt trigger through resistor R_5. Because of very large voltage gain (typically 200 000), a very small difference between the inverting and noninverting terminals makes the Schmitt trigger output to be *saturated*. It means that the output voltage is very close to either the positive or negative supply voltages. Typically, the saturated output voltage is $+(V_{CC} - 1\text{ V})$ or $-(V_{EE} - 1\text{ V})$.

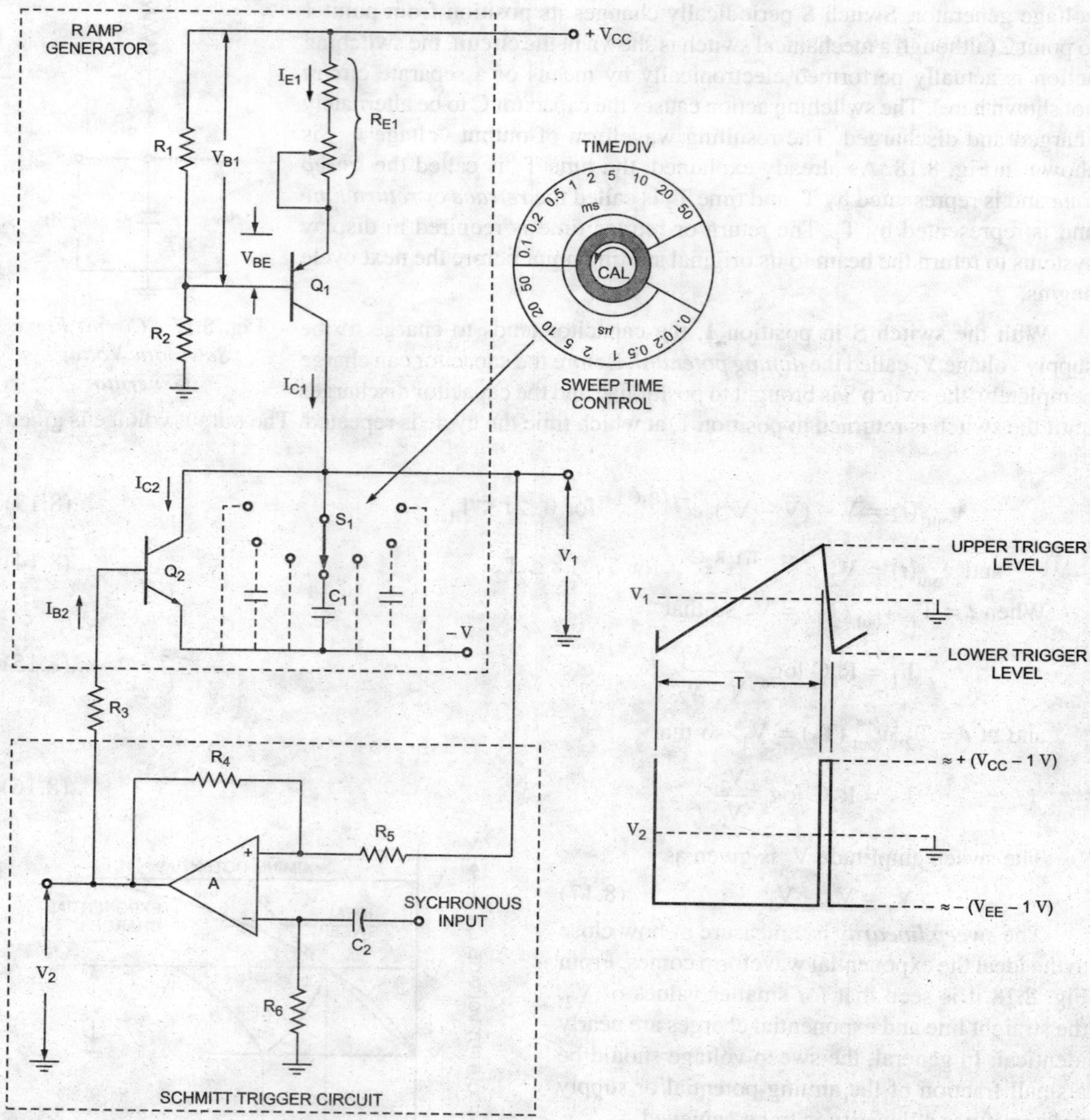

Fig. 8.19 *Basic Circuit of an Oscilloscope Sweep Generator*

* A Schmitt trigger circuit is a fast operating voltage level detector. When the input voltage arrives at the upper or lower trigger levels the output changes rapidly. The circuit operates with almost any type of input waveform, and it provides a pulse-type output.

Assume that the output of the Schmitt trigger is negative, and that the ramp input to the Schmitt trigger is at its minimum level. The voltages at both ends of potential divider $R_4 - R_5$ are negative and therefore the junction of resistors R_4 and R_5 must also be negative. Thus, the noninverting terminal voltage of the op-amp is below the level of the (grounded) inverting terminal, and ouput of the op-amp remains saturated in a negative direction. This keeps transistor Q_2 biased off.

With the growth of the ramp voltage, the voltage at the junction of resistors R_4 and R_5 rises toward ground. When the ramp voltage attains a high enough positive level, the noninverting input terminal is eventually raised slightly above ground. This makes the output of the op-amp to switch rapidly from the negative saturated level to saturation in the positive direction.

The level of the input voltage (to R_5)which makes the output to switch in positive direction is known as *upper trigger point* (UTP). For the output to go negative once more, the input to resistor R_6 must go down to a negative level, which will pull the noninverting terminal below the groud. The negative input voltage level required to trigger the output to negative saturation is known as *lower trigger point* (LTP). UTP and LTP are numerically equal.

The trigger voltage levels may be given as

$$\text{UTP, LTP} = \pm (V_{CC} - 1\text{ V}) \times \frac{R_5}{R_4} \quad \text{...(8.18)}$$

Potential divider $R_1 - R_2$ provides a constant bias voltage V_{B1} to the base of PNP transistor Q_1. This makes the voltage drop across emitter resistor R_{E1} a constant quantity, and maintains emitter current I_{E1} constant. Transistor collector current I_{C1}, being approximately equal to its emitter current I_{E1}, is a constant current given by

$$I_{C1} \simeq \frac{V_{B1} - V_B}{R_{E_1}} \quad \text{...(8.19)}$$

Thus, in the ramp generator section of the time base circuit shown in Fig. 8.19 transistor Q_1 with its associated components acts as a *constant current source*. Current I_{C1} can be adjusted, if required, by adjusting the variable portion of emitter resistor R_{E_1}.

When the output of the Schmitt trigger circuit is positive, a base current, I_{B2} flows into the NPN transistor Q_2 through resistor R_3. This switches NPN transistor Q_2 into saturation and short-circuits the capacitor C_1. When the trigger circuit output goes negative, NPN transistor Q_2 *switches off* and current I_{C1} flows into capacitor C_1 charging it linearly and generating a ramp output. Ramp voltage is given by equation

$$\Delta V_1 = \frac{I_{C1}T}{C_1} \quad \text{...(8.20)}$$

where ΔV_1 is the capacitor voltage change during time T and C_1 is the capacitance.

The capacitor voltage continues to increase linearly until it attains the upper trigger level for the Schmitt trigger. Then the Schmitt trigger output goes positive once again, turning transistor Q_2 *on* and rapidly discharging capacitor C_1. When the voltage across capacitor C_1 goes down to the lower trigger level of the Schmitt trigger, the trigger output switches to negative again. Transistor Q_2 gets once again switched *off*, and the voltage across capacitor C_1 begins to increase linearly once more.

The amplitude of the ramp waveform obviously depends upon the UTP and LTP of the Schmitt trigger circuit. The time period τ of the ramp depends on charging current I_{C1} and the capacitance of capacitor C_1. I_{C1} can be adjusted by varying the emitter resistance R_{E_1}. As illustrated, any one of the several capacitors can be switched into the circuit and thus time period τ can be changed. Individual capacitors can be selected to produce ramps that sweep the electron beam across the CRT screen at 1 div/10 ms, or 1 div/20 ms, or at any one of several other rates. The capacitor selection switch on the oscilloscope front panel is identified as TIME/DIV.

Example 8.6. The sweep generator circuit shown in Fig. 8.19 has R_{E_1} = 5.6 kΩ, C_1 = 0.2 μF, V_{B_1} = 6.3 V, and the trigger levels for the Schmitt trigger are ± 2.5 V. Determine the peak-to-peak amplitude and time period of the ramp waveform.

Solution: Capacitor voltage variation,

$$\Delta V_1 = 2 \times (\text{UTP, LTP}) = 2 \times 2.5 = 5 \text{ V}$$

Charging current,

$$I_{C1} = \frac{V_{B_1} - V_{BE}}{R_{E_1}} = \frac{6.3 \text{ V} - 0.7 \text{ V}}{5.6 \text{ k}\Omega} = 1 \text{ mA} \quad \textbf{Ans.}$$

Time period,

$$\tau = \frac{\Delta V_1 \times C_1}{I_{C1}} = \frac{5 \text{ V} \times 0.2 \,\mu\text{F}}{1 \text{ mA}} = 1 \text{ ms} \quad \textbf{Ans.}$$

Ramp waveform is shown in Fig. 8.20.

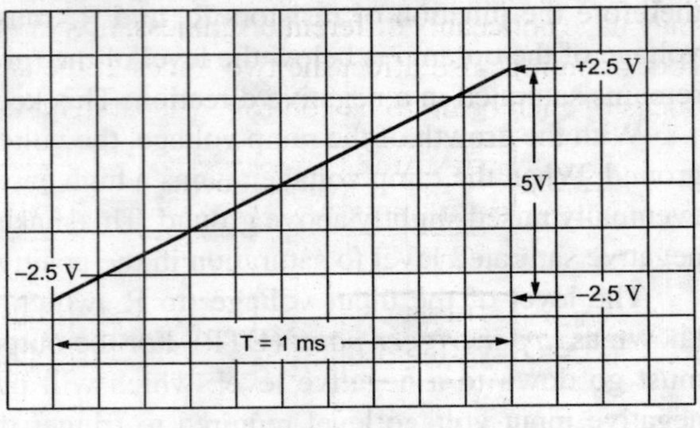

Fig. 8.20

8.7 MULTIPLE TRACE OSCILLOSCOPE

In the analysis of electronic circuits and systems, its is quite useful to be able to view the behaviour of two or more voltages simultaneously. No doubt, it can be done by using two oscilloscopes, but in addition to extra cost it is difficult to trigger the sweeps of each oscilloscope at precisely the same time and to ensure that the sweep generators operate precisely the same. Even if this should be achieved, the two traces to be viewed will be on two different oscilloscope tubes, and not one atop the other where comparison can be made easily.

There is an elegant solution to this problem, which requires a special CRO that uses two separate electron guns and deflection systems mounted in the same CRT to display two different electron beams simultaneously. This feature should not be confused with the *dual trace feature* of other oscilloscopes which alternately display two signals on one screen by time-sharing one electron gun and deflection system (this is a more common and cheaper solution to the problem mentioned above, as opposed to dual beam solution). By being able to display and trigger each beam independently, dual beam oscilloscopes have the capability of displaying unusual combinations of signal characteristics. For example, one beam may be used to show the phase shift between two sinusoidal signals by displaying their Lissajous pattern. Meanwhile the other beam can be displaying the time variation of one of the signals through the use of the triggered-sweep mode.

A **dual beam** oscilloscope can also display two simultaneous non-current signals of short duration. This cannot be done on a dual trace oscilloscope. An example of this application is illustrated in Fig. 8.21, where the input and output waveforms of a pulse-shaping circuit are both exhibited on the same screen.

The block diagram of a dual beam CRO is depicted in Fig. 8.22. The two channels may have a common time base system, as shown in Fig. 8.22, or they may have independent time base circuits.

In case of common time base, only one beam can be synchronised at a time, so the input signals must have the same frequency or must be related harmonically so that both beams are locked on the screen of the CRT. An independent time base allows different sweep rates for the two beams but size and weight of the oscilloscope is increased.

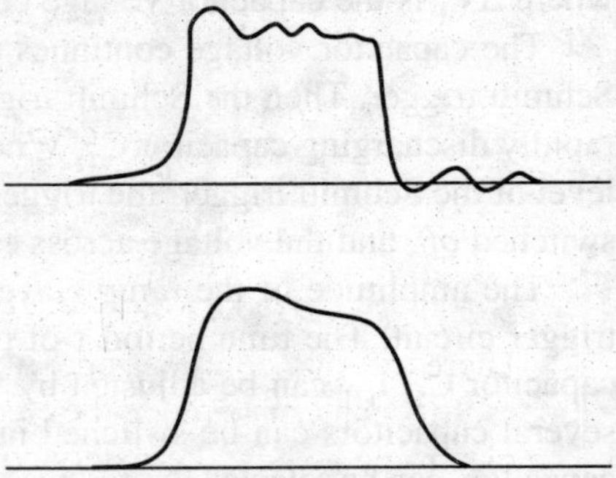

Fig. 8.21

There are two methods of obtaining separate beams. In some oscilloscopes the output of a single electron gun is mechanically split into two separate beams. This is called split-beam technique. Other oscilloscopes use a CRT which has two separate electron guns. The drawback of the split-beam construction is that the two displays may have noticeably different brightness, if operated at widely spaced sweep speeds. The brightness and focus controls also affect the two traces at the same time. Double gun tube allows the brightness and focus of each beam to be controlled separately but it is bulkier than a split-beam tube.

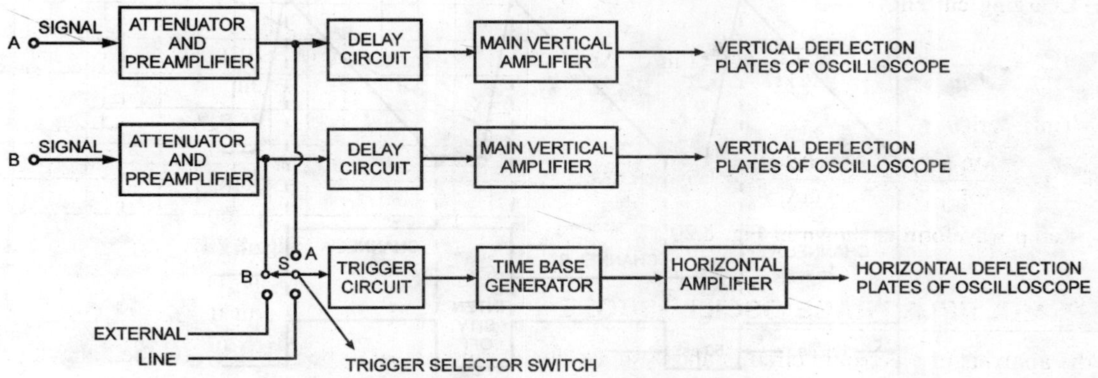

Fig. 8.22 *Block Diagram of a Dual Beam CRO*

Dual-beam CROs are not as versatile as the dual-trace CROs because the later can operate in a number of modes. Dual beam CROs are only used in systems where absolute independence of the vertical channels is required.

The block diagram of a **dual trace oscilloscope** is given in Fig. 8.23. It employs a single electron gun and produces a dual trace display by means of electronic switching of two separate input signals. As illustrated in block diagram, there are two separate vertical input channels (or circuits) A and B and they employ separate attenuators and preamplifier stages. So the amplitude of each signal can be independently controlled. The outputs of the preamplifiers are supplied to the electronic switch, which is capable of passing one signal at a time into the main vertical amplifier of the oscilloscope. The display of the input is continuous and simultaneous to the eyes although it is sampled display.

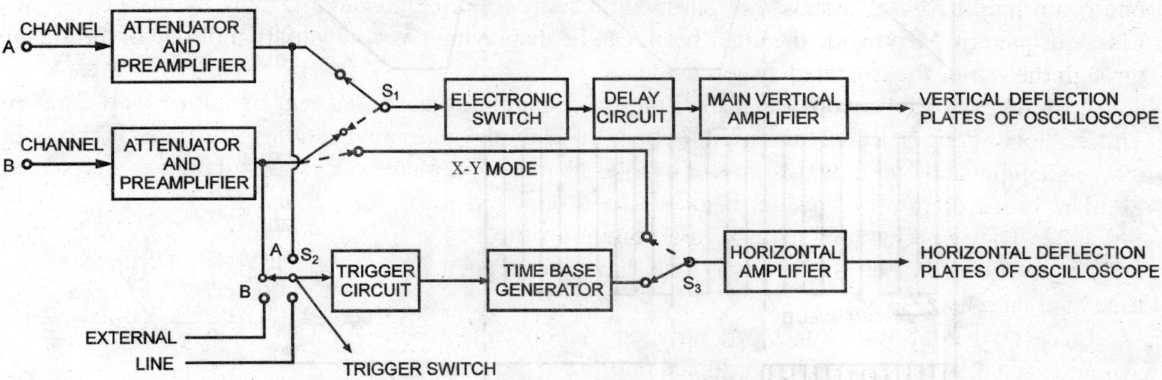

Fig. 8.23 *Block Diagram of a Dual Trace Oscilloscope*

There are two common operating modes for the electronic switch, called the *alternate* and *chop* modes. In the alternate mode the electronic switch feeds each signal alternately to the vertical amplifier. The electronic switch alternately connects the main vertical amplifier to channels A and B and adds a different dc component to each signal and this dc component directs the beam alternately to the upper or lower half of the screen. The switching takes place at the start of each new sweep of the sweep or time-base generator. The switching rate of the electronic switch is synchronised to the sweep rate, so that the cathode ray tube spot traces the channel A signal on one sweep and the channel B signal on the succeeding sweep as illustrated in Fig. 8.24. The alternate mode cannot be used for displaying very low frequency signals.

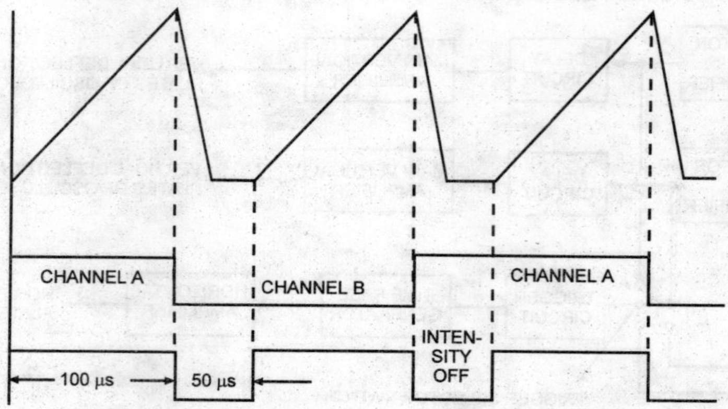

Fig. 8.24 *Time Relation of a Dual Channel Vertical Amplifier in Alternate Mode*

The sweep trigger signal is available from channels A or B and it is picked off before the electronic switch. This arrangement maintains the correct phase relationship between the two input signals.

In the X-Y mode of operation, the time-base generator is disconnected and the channel B is connected to the horizontal amplifier. Since both preamplifiers are identical and have the same delay time, it is possible to make accurate X-Y measurements.

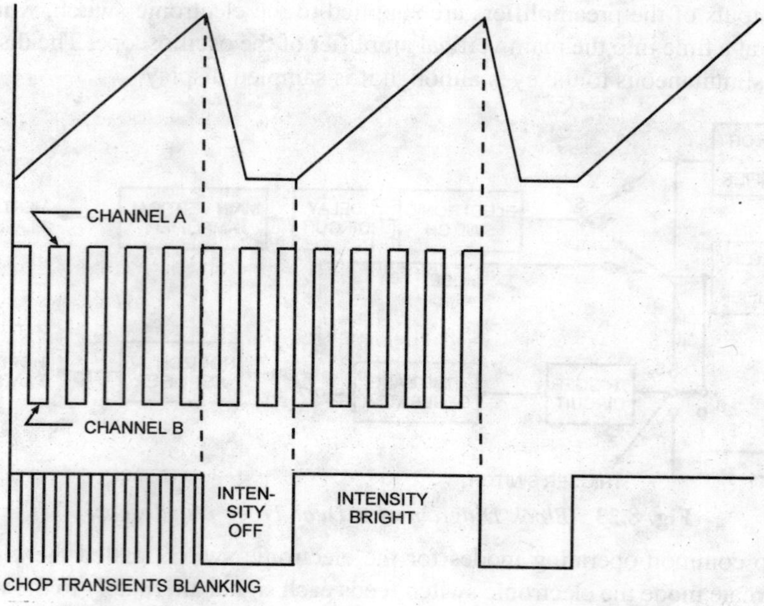

Fig. 8.25 *Time Relation of a Dual-Channel Vertical Amplifier in Chop Mode*

In the chop mode, the electronic switch free runs at a high frequency of the order of 100-500 kHz, entirely independent of the frequency of the time-base generator. The result is that small segments from channels A and B are connected alternately to the main vertical amplifier at a relatively fast chopping rate of 500 kHz and displayed on the screen. In case of the chopping rate much faster than horizontal sweep rate, the display will be a continuous line for each channel. If the chopping rate is slow, the continuity of the display is lost and it is better to use the alternate mode of operation.

In the *added mode of operation* a single image can be displayed by the addition of signals from channels A and B *i.e.*, (A + B) etc. With polarity inversion switches in both channels, it is possible to display A + B , A – B, B – A, – A – B.

Dual trace oscilloscopes are widely used in industries and research laboratories.

Specifications of Dual Trace Oscilloscope

Display. 8 × 10 cm rectangular mono-accelerator internal graticule CRT 2 kV EHT. Trace rotation by front panel preset.

Vertical Deflection

Input Channels	:	Two CH_1 and CH_2 (identical)
Bandwidth (– 3 dB)	:	DC to 20 MHz (2 Hz to 20 MHz on ac)
Sensitivity	:	2 mV/cm to 10 V/cm in 1-2-5 sequence
Accuracy	:	± 3%
Variable Sensitivity	:	> 2, 5 : 1 range allowing continuous adjustment of sensitivity from 2 m V/cm to 25 V/cm
Input Impedance	:	1 MΩ/28 pF approx.
Input Coupling	:	AC/DC
Input Protection	:	400 V dc or pk, ac

Display Modes

Single Trace	:	CH_1 or CH_2
Dual Trace	:	Chopped or Alternate Modes automatically selected by the TB switch. Between 0.5 µs/cm and 0.2 ms/cm selecting the ALTERNATE MODE. Whilst on ranges slower than 0.2 ms/cm selecting the CHOP MODE. The CHOP frequency – 500 kHz.
Add	:	CH_1 and CH_2 added to give the algebraic sum of the two channels.
Invert CH_2	:	CH_2 may be inverted and when used in conjunction with add mode giving the algebraic difference of the two channels. *X-Y* CH_2 input giving *Y* deflection and CH_1 input giving X deflection.
Bandwidth (– 3 dB)	:	DC to 1 MHz with less than 3° phase shift at 50 kHz.

Horizontal Deflection

Time base	:	0.5 µs/cm to 0.2 s/cm, 18 ranges in 1-2-5 sequence.
Accuracy	:	± 3%, (to 200 ns/cm)
X Expansion	:	× 10 push button giving fastest speed of 50 ns/cm.
Variable Sweep	:	> 2.5 : 1 allowing continuous coverage from 0.5 µs/cm to 0.5 s/cm

Trigger

Variable level control with Bright Line ON/OFF facility.

With Bright Line On, the time base free-runs when insufficient signal (20 Hz – 20 MHz) is present or when the selected level is outside the range of the input signal.

Source : Internal CH_1 or CH_2 External or Line

Slope : + or –

Coupling : DC, AC or TV (active sync. separator with line frame selected by TB switch between 50 and 100 μs/cm)

Sensitivity : Internal — DC coupled 2 mm to 2 MHz, 5 mm to 20 MHz
AC coupled 2 mm, 10 Hz – 2 MHz 5 mm, 4 Hz – 20 MHz

External — DC coupled 100 mV to 2 MHz, 400 mV to 20 MHz
AC coupled 100 mV, 10 Hz to 2 MHz 400 mV, 4 Hz to 20 MHz

External Input Impedance : 100 kΩ/10 pF approx.

External Input Protection : 250 V dc or pk, ac

Additional Facilities

Calibrator : 1 V, 2% square wave at approx. 1 kHz

Ramp Output : Approx. + 3.5 V ramp from 5 kΩ

Z Mod Input : DC coupled, 2V visible mod sensitivity, + 40 V cut-off sensitivity, input impedance 10 kΩ/10 pF approx. Maximum input 100 V dc or pk ac.

Supply : 220/240 V ± 10%; 45 to 65 Hz approx. 40 VA

Operating Temperature Range : 0 to + 50°C (+ 15 to + 35°C for full accuracy)

Dimensions : 140 × 305 × 460 mm

Weight : 6 kg approx.

8.8 MEASUREMENT OF VOLTAGE, FREQUENCY AND PHASE

8.8.1. Measurement of Voltage. Cathode ray oscilloscope can be used for measurement of voltage of any electrical signal as the deflection of the electrostatic beam is directly proportional to the deflection plate voltage.

For measurement of direct voltage, firstly the spot is centred on the screen without applying any voltage signal to the deflection plates. Then direct voltage to be measured is applied between a pair of deflection plates and deflection of the spot is observed on the screen. The magnitude of the deflection multiplied by the deflection factor gives the value of direct voltage applied. Usually the screen is calibrated for fixed operating condition, so by reading the scale, voltage can be measured directly by CRO.

In case of measurement of alternating voltage of sinusoidal waveform, it is applied between a pair of deflection plates and the length of the straight line is measured. Knowing the deflection sensitivity, the peak-to-peak value of applied ac voltage can be determined. The rms value of ac voltage applied will be equal to this peak value divided by $2\sqrt{2}$ for sinusoidal waveform.

Figure 8.26 illustrates two sinusoidal waves with different amplitudes and time periods. Waveform A has peak-to-peak amplitude of 4.8 vertical divisions on the graticule, while peak-to-peak amplitude of wave B is 2 vertical divisions. It is very important to keep central vernier knob on the VOLTS/DIV control in its calibrated (CAL) position before measuring the amplitudes of waveform.

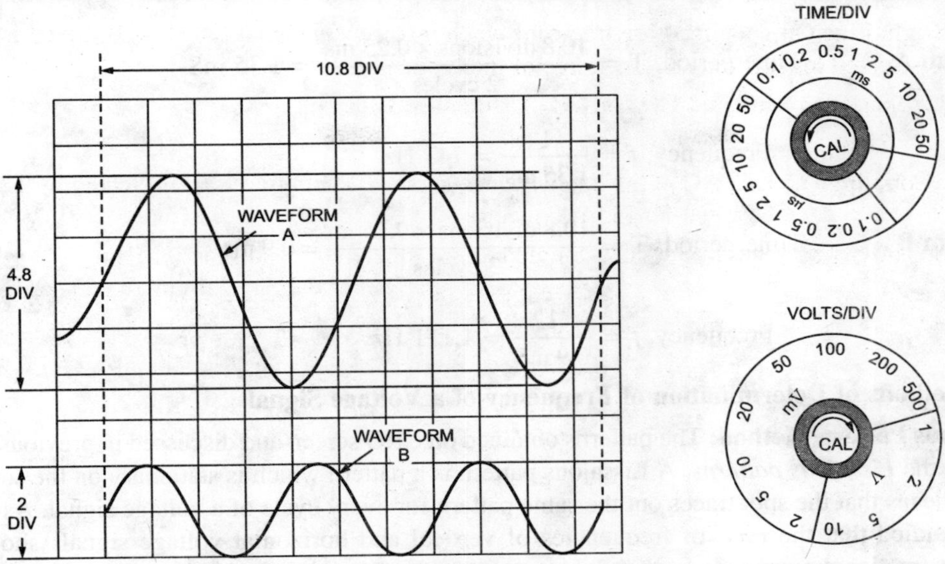

Fig. 8.26

Peak-to-peak voltage of waveform = Vertical peak-to-peak divisions × VOLTS/DIV

With the VOLTS/DIV control at 50 mV, the peak-to-peak voltage of

Waveform A = 4.8 divisions × 50 mV/division = 240 mV

Waveform B = 2 divisions × 50 mV/division = 100 mV

$$\text{RMS value of waveform A} = \frac{240}{2\sqrt{2}} \text{ mV} = 84.25 \text{ mV}$$

$$\text{RMS value of waveform B} = \frac{100}{2\sqrt{2}} \text{ mV} = 35.36 \text{ mV}$$

For measurement of current, the current under measurement is passed through a known non-inductive resistance and the voltage drop across it is measured by CRO, as mentioned above. The current can be determined simply by dividing the voltage drop measured by the value of non-inductive resistance. When the current to be measured is of very small magnitude, the voltage drop across non-inductive resistance (small value) is usually amplified by a calibrated amplifier.

The current and voltage can be measured simultaneously by using double beam cathode ray oscilloscope.

8.8.2. Determination of Frequency. The frequency of any waveform is reciprocal of time period, which may be determined by measuring the time for one cycle in horizontal divisions and multiplying by the setting of the TIME/DIV control.

$$\text{Time period, T} = \text{Horizontal divisions per cycle} \times \text{Time/division} \quad \ldots(8.21)$$

$$\text{and Frequency, } f = \frac{1}{T} = \frac{1}{\text{Horizontal divisions per cycle} \times \text{Time/division}} \quad \ldots(8.22)$$

Here also, it is necessary to set the central vernier knob on the TIME/DIV control in its calibrated position.

Assuming that the TIME/DIV control is set 0.25 ms, the time period and frequency of waveform shown in Fig. 8.26 are:

Waveform A Time period, $T = \frac{10.8 \text{ divisions} \times 0.25 \text{ ms}}{2 \text{ cycles}} = 1.35 \text{ mS}$

Frequency, $f = \frac{1}{1.35 \text{ ms}} = 741 \text{ Hz}$

Waveform B Time period, $T = \frac{10.8 \text{ divisions} \times 0.25 \text{ ms}}{3 \text{ cycles}} = 0.9 \text{ ms}$

Frequency, $f = \frac{1}{0.9 \text{ mS}} = 1{,}111 \text{ Hz.}$

Other Methods of Determination of Frequency of a Voltage Signal

1. Lissajous Pattern Method: The patterns obtained on CRO screen and discussed in previous sections are called the *Lissajous patterns.* A Lissajous pattern is a pattern which is stationary on the screen of a CRO. It means that the spot traces out the same pattern for every cycle of a voltage signal. As we have already studied that the ratio of frequencies of vertical and horizontal voltage signals should be a rational or fractional number to have steady pattern, so the condition for having a Lissajous pattern on the CRO screen is

$$\frac{f_y}{f_x} = \frac{A}{B} \quad \text{where A and B are integers.}$$

Lissajous patterns are usually of two types. First one is closed Lissajous pattern and has no free end. Second one is open Lissajous pattern and has free ends. Both types of Lissajous patterns are shown in Fig. 8.27.

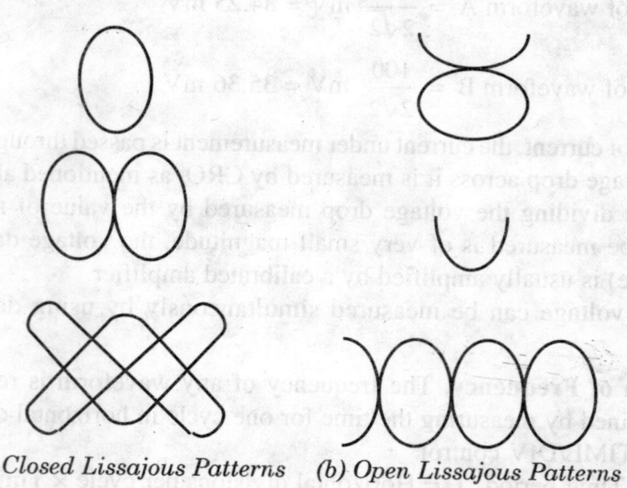

(a) Closed Lissajous Patterns (b) Open Lissajous Patterns

Fig. 8.27

In a Lissajous pattern ratio of frequency of vertical signal to the frequency of horizontal signal is equal to the ratio of positive Y-peaks to positive X-peaks in that pattern.

$$\text{So} \quad \frac{f_y}{f_x} = \frac{\text{Positive Y-peaks in pattern}}{\text{Positive X-peaks in pattern}} \qquad \text{...(8.23)}$$

Thus by counting the positive Y-peaks and X-peaks on a Lissajous pattern, ratio of frequencies of two voltage signals can be determined. In case of an open Lissajous pattern, free end is treated as half peak. This will be clear from Examples 8.8 and 8.9.

Voltage signal of unknown frequency is applied to the vertical deflection plates and horizontal deflection plates are supplied by an accurately calibrated variable frequency source. Frequency of the variable frequency source is adjusted until a stationary pattern appears on the screen. Now by reading the value of frequency of horizontal signal, with the help of calibrated scale, frequency of voltage signal applied to vertical deflection plates can be known.

In case single loop stationary pattern is obtained the frequency of the sinusoidal voltage applied to vertical deflection plates is the same as that of the voltage applied to horizontal deflection plates. In case complex Lissajous figure is obtained, the frequency of alternating voltage may be determined using the relation

$$\frac{\text{Points of tangency to a vertical line}}{\text{Points of tangency to a horizontal line}} = \frac{\omega_x}{\omega_y} = \frac{f_x}{f_y} \quad ...(8.24)$$

One point, very interesting, to know is that sometimes we may have different types of patterns on the CRO screen for the voltage signals of the given frequencies, as shown in Fig. 8.28.

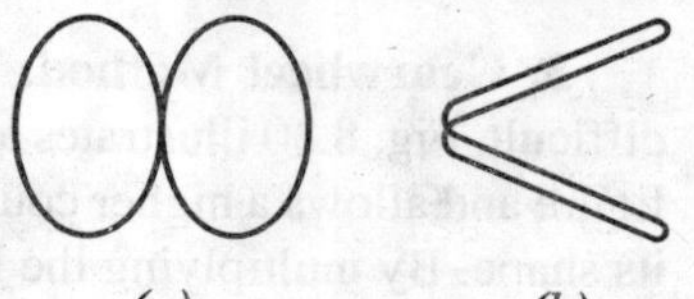

(a) (b)

Fig. 8.28

In both of the figures, Figs. 8.28 (*a*) and 8.28 (*b*), the ratio of Y-peaks to X-peaks is equal, so in both cases, ratios of frequencies of vertical and horizontal signals are same. But appearance of Lissajous patterns is different owing to different phase difference of voltage signals applied to vertical and horizontal deflection plates.

CRO is not a precision instrument for measuring frequency of an alternating voltage because the accuracy depends directly on the accuracy of calibrated scale of variable frequency source, which is usually a few per cent. It is used for rough estimate of frequency or when voltage waveform is so complex that a frequency counter would not operate reliably.

2. Spot Wheel Method of Display: With the increase in the ratio of the two frequencies under comparison, Lissajous pattern becomes more and more complicated and interpretation becomes difficult. In such cases, spot wheel method of display, as shown in Fig. 8.29 (*a*), can be advantageously applied. In this method, an intensity modulated circular or elliptical display is produced on the face of the CRT. The lesser of the two frequencies under comparison is applied to the CRT deflection plates through the resistance-capacitance (R–C) phase shifter, as shown, thereby generating the circular or elliptical sweep. The higher frequency is applied to the control grid of the CRT (biased slightly below cut-off), as depicted in Fig. 8.29 (*b*), thus modulating the beam intensity. Consequently, the pattern appears as a series of alternate bright and dark spots, as depicted in Fig. 8.29 (*c*), where the ratio of the high to low frequencies is given by the number of bright spots.

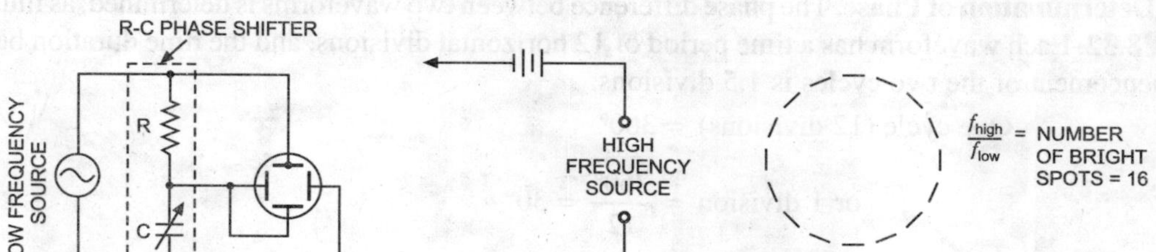

(a) Basic Circuit *(b) Input Circuit To The Control Grid* *(c) Spot Wheel Pattern*

Fig. 8.29 *Spot Wheel Method of Frequency Measurement*

If two sinusoidal signals, equal in magnitude but 90º out of phase are applied to the vertical and horizontal deflecting plates, a circle is formed. In case a single sinusoidal signal is applied to a phase shifting circuit, it is possible to have two outputs, equal in magnitude and 90º out of phase.

If the voltages across resistor R and capacitor C are equal, the pattern obtainable is a circle. When the applied voltages are unequal, the pattern obtainable is an ellipitcal.

Variation in grid-cathode potential causes the change in the density of the electron stream within the CRT and determines the intensity. When the CRT grid bias attains cut-off potential, there is no electron stream and fluorescence at the screen cannot occur. This property is used in comparing two frequencies, provided their ratio is an integer. Grid bias must be sufficient so that the negative peaks of the sinusoidal wave cut off the electron beam. It is not necessary that high-frequency signal be sinusoidal, it can be a square wave.

3. Gearwheel Method: Large number of loops in the Lissajous pattern makes their counting difficult. Fig. 8.30 illustrates a test method that makes use of a modulated ring pattern in place of looped figure and allows a higher count. This pattern is also known as a *gearwheel* or *toothed wheel*, because of its shape. By multiplying the known frequency by the number of teeth in the pattern we determine the unknown frequency. Test set up in shown in Fig. 8.31. A phase shift of 90º between the horizontal and vertical channels of the oscilloscope, required to give a ring or circle pattern with the known frequency f_v, is introduced by R-C phase shift network.

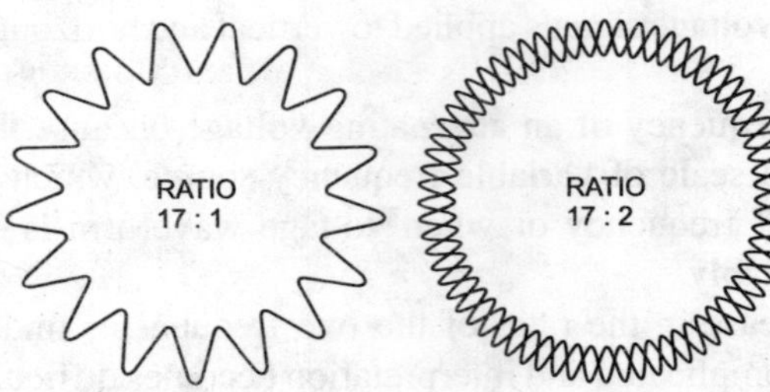

Fig. 8.30 *Gearwheel Pattern*

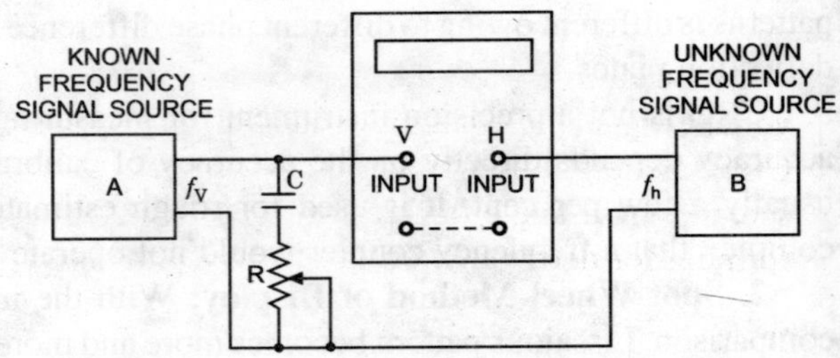

Fig. 8.31 *Gearwheel Method of Frequency Measurement*

A voltage from unknown frequency modulates the ring. With equal voltages across R and C, pattern obtainable will be circle while with unequal voltages it will be an ellipse. The unknown frequency must be larger than known frequency and the amplitude of the unknown frequency must be smaller than that of known frequency in order to avoid distortion.

8.8.3. Determination of Phase. The phase difference between two waveforms is determined, as illustrated in Fig. 8.32. Each waveform has a time period of 12 horizontal divisions, and the time duration between commencement of the two cycles is 1.5 divisions.

$$\text{One cycle (12 divisions)} = 360^\circ$$

$$\text{or 1 division} = \frac{360^\circ}{12} = 30^\circ$$

$$\text{Phase difference} = \text{Phase difference in divisions} \times \text{degrees/division}$$

$$= 1.5 \text{ divisions} \times 30^\circ = 45^\circ$$

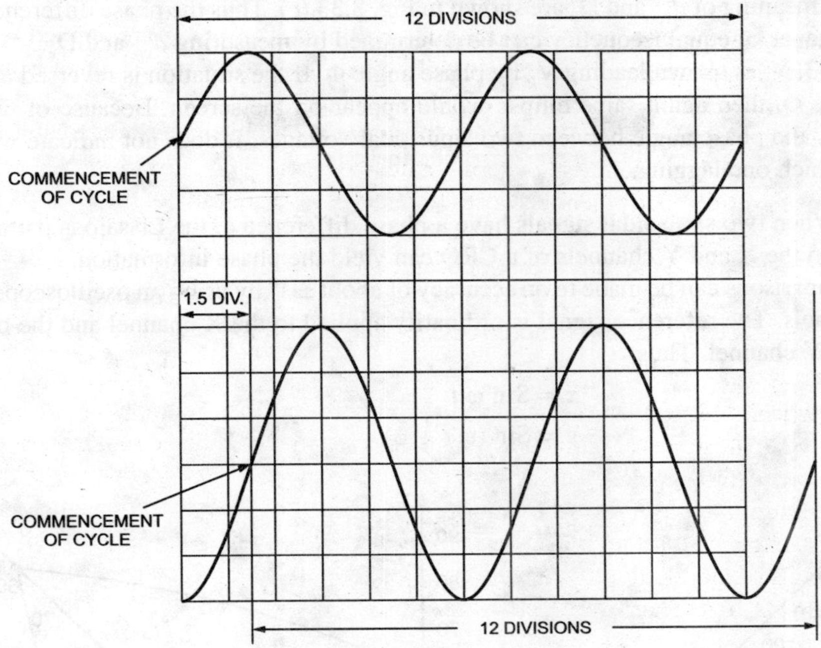

Fig. 8.32

Alternative Method of Determination of Phase: We have discussed that when two sinusoidal voltage signals of equal frequency having some phase difference are applied to the deflection plates of CRO, a straight line or an ellipse appears on the screen. In the case of straight line appearing on the screen, phase angle difference would be zero or 180º but in case of an ellipse we will have to use a formula for determination of phase difference.

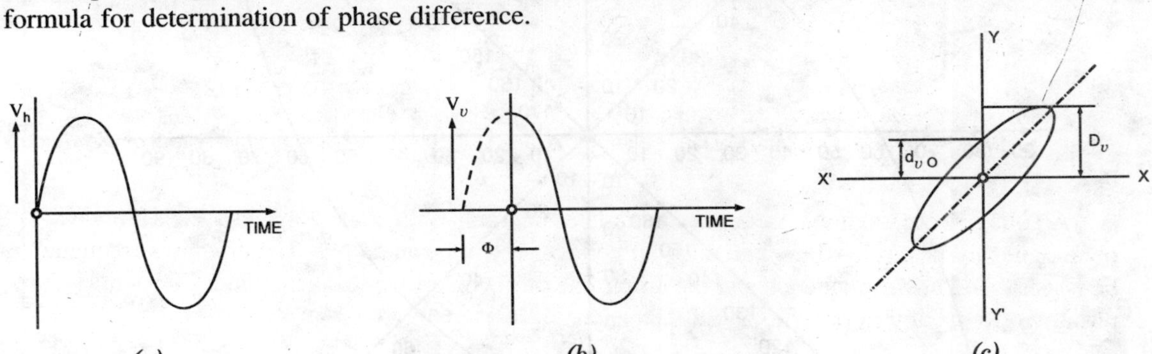

Fig. 8.33

Let there be two sinusoidal voltage signals given by equations

$$v_h = V_h \sin \omega t$$

$$v_v = V_v \sin (\omega t + \Phi) \text{ where } \Phi \text{ is the phase difference.}$$

Since deflection is directly proportional to amplitude of voltage,

$$d_h = D_h \sin \omega t$$

$$\text{and } d_v = D_v \sin (\omega t + \Phi)$$

At time $t = 0$, values of d_h and d_v are $d_{ho} = 0$ and $d_{vo} = D_v \sin \Phi$

$$\text{So } \sin \Phi = \frac{d_{vo}}{D_v}$$

Graphical meaning of d_{vo} and D_v are shown in Fig. 8.33 (*c*). Thus the phase difference between two sinusoidal voltages of equal frequency can be determined by measuring d_{vo} and D_v.

In Fig. 8.33, V_v is shown leading V_h by phase angle Φ. If the situation is reversed and V_h leads V_v by phase angle Φ, then again same ellipse would appear on the screen. Because of this fact we can determine only the phase angle between two sinusoidal voltages. It does not indicate which voltage is leading and which one lagging.

Procedure. When two sinusoidal signals have a phase difference ϕ, the Lissajous pattern obtained by feeding them to the X and Y channels of a CRO can yield the phase information.

Phase comparisons can be made to an accuracy of about ± 1° by using an oscilloscope with identical X and Y channels. The reference signal is ordinarily applied to the X channel and the phase-shifted is applied to the Y channel. Thus,

$$x = \text{Sin } \omega t$$

$$y = \text{Sin } (\omega t + \phi)$$

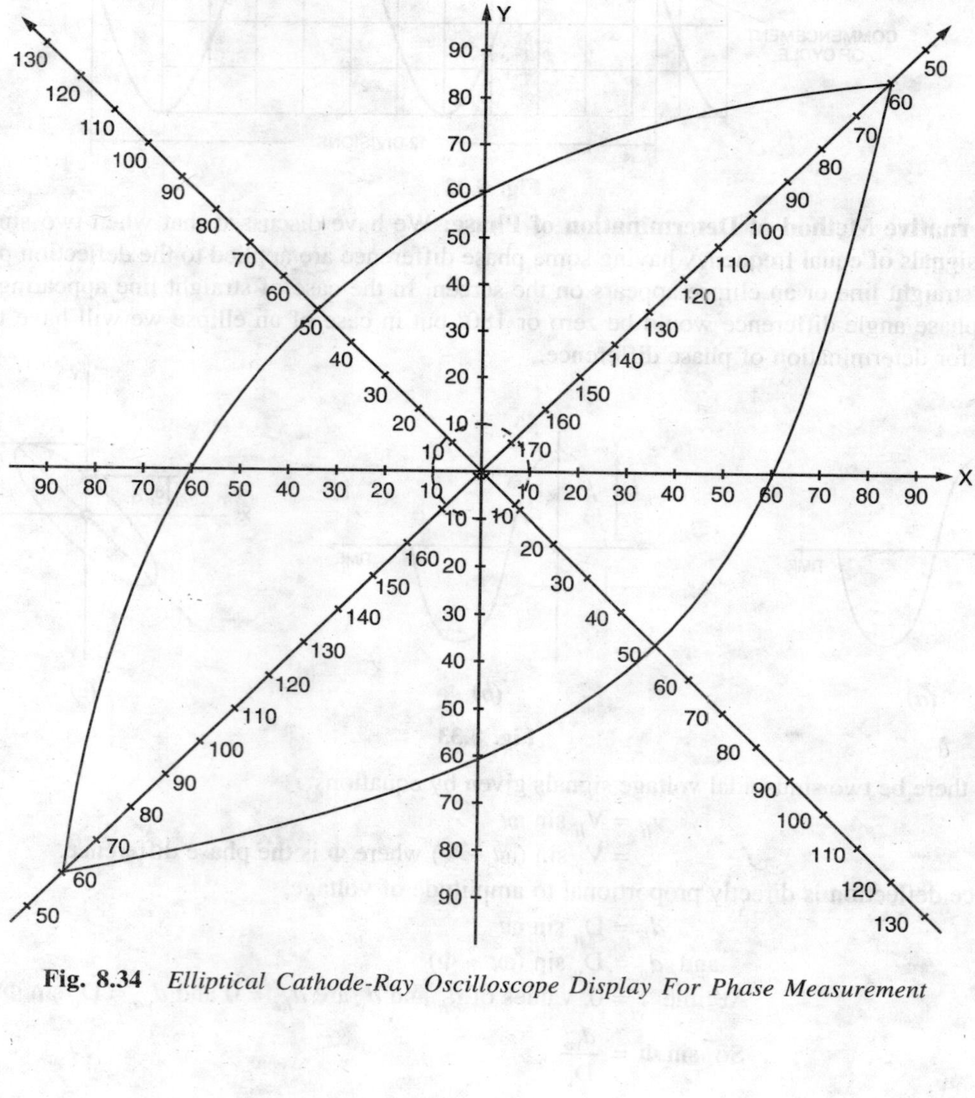

Fig. 8.34 *Elliptical Cathode-Ray Oscilloscope Display For Phase Measurement*

If $\phi = 0°$, a 45° straight line should be displayed. It is good idea to apply the same signal to both channels simultaneously and adjust the gains (and if necessary, the phase shift in one channel) to get a 45° straight line before making a measurement.

With either signal removed, 10 divisions of peak-to-peak deflection should be produced in the direction of the remaining signals. Then, with both signals applied, the phase can be estimated from the y or x intercepts of the resulting ellipse.

When $x = 0$, then $y = \sin \phi$ and when $y = 0$, then $x = \sin(-\phi)$; so all intercepts should be equal and the phase is $\pm \sin^{-1} y$ or $\sin^{-1} x$.

When $\phi \approx (2n - 1)\, 90°$, the x and y intercepts become very insensitive to small changes in ϕ, and so the accuracy is very poor.

A better method is to measure the minor axis of the ellipse. For $|\phi| < 90°$, this will lie on a line at – 45° as shown in Fig. 8.34, while for $|\phi| > 90°$, the minor axis will be along a + 45° line. If we call the peak x and y deflections unity, it is easy to show that the intercepts on the line at – 45° and + 45° are respectively.

$$u = \sqrt{2} \sin \frac{\phi}{2}$$

$$v = \sqrt{2} \cos \frac{\phi}{2}$$

Thus scale like those shown in Fig. 8.34 can be constructed to read the minor axis. The ellipse always crosses these lines at right angles, which removes one source of reading errors and the minor axis is a sensitive function of ϕ for all ϕ.

The use of oscilloscopes to measure phase is neither rapid nor very accurate, but has the great virtue that no special equipment is needed. When a great many or accurate phase measurements are to be made, other methods having higher inherent accuracy, faster read-out, and great immunity to noise and waveform distortion should be used.

Example 8.7. The vertical gain control of a CRO is set at a deflection sensitivity of 5 V/cm. An unknown ac voltage (sinusoidal) is applied to the Y-input. A 10 cm long straight line trace is observed on the screen. Determine the ac voltage.

Solution : The peak-to-peak value of unknown ac voltage, *i.e.*

$$V_{\text{peak-to-peak}} = \text{Length of trace} \times \text{deflection sensitivity} = 10 \times 5 = 50 \text{ V}$$

$$\text{Peak value of unknown ac voltage, } V_{\text{peak}} = \frac{V_{\text{peak-to-peak}}}{2} = \frac{50}{2} = 25 \text{ V}$$

$$\text{RMS value of unknown ac voltage} = \frac{V_{\text{peak}}}{\sqrt{2}} = \frac{25}{\sqrt{2}} = 17.667 \text{ V} \quad \textbf{Ans.}$$

Example 8.8. A closed Lissajous pattern is shown in Fig. 8.35. Find the ratio of frequencies of vertical and horizontal signals.

Solution : Positive Y-peaks in pattern shown in Fig. 8.35 = 3

Positive X-peaks in pattern shown in Fig. 1.49 = 2

$$\text{So } \frac{f_y}{f_x} = \frac{\omega_y}{\omega_x} = \frac{\text{Positive Y-peaks in pattern}}{\text{Positive X-peaks in pattern}} = \frac{3}{2} \quad \textbf{Ans.}$$

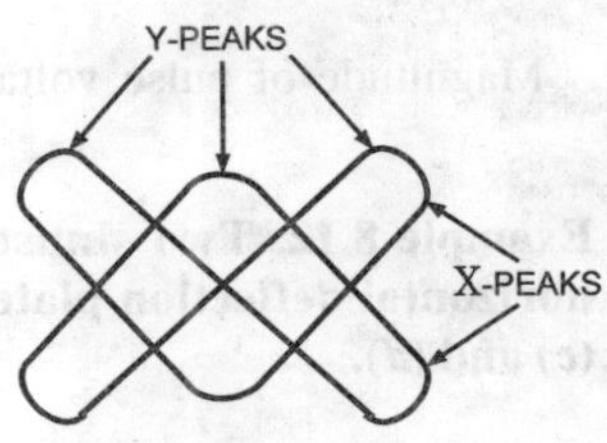

Fig. 8.35

Example 8.9. For a Lissajous pattern shown in Fig. 8.36, find out the frequency of vertical signal if the frequency of horizontal signal is 3 kHz.

Solution : In Fig. 8.36, positive Y-peaks in pattern $= 2 + \frac{1}{2} = 2\frac{1}{2}$

$$\text{Positive X-peaks in pattern} = \frac{1}{2} + \frac{1}{2} = 1$$

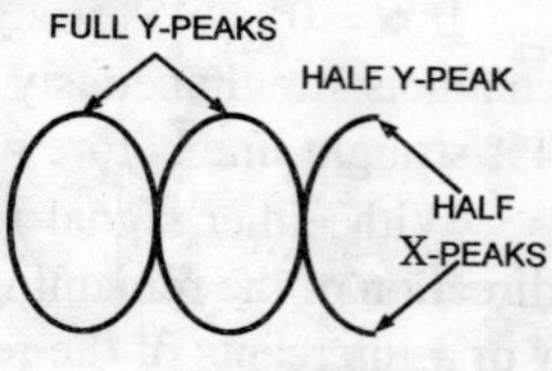

Fig. 8.36

$$\text{So } \frac{f_y}{f_x} = \frac{2\frac{1}{2}}{1} = 2\frac{1}{2}$$

$$\text{So frequency of vertical voltage signal} = 2\frac{1}{2} \times \text{frequency of horizontal voltage signal}$$

$$= 2\frac{1}{2} \times 3 = 7.5 \text{ kHz } \textbf{Ans.}$$

Example 8.10. A Lissajous pattern on an oscilloscope is stationary and has 5 horizontal tangencies and 2 vertical tangencies. The frequency of horizontal input is 1,000 Hz. Determine the frequency of vertical input.

[U.P.S.C. I.E.S. Electrical Engineering I, 1994]

Solution : Frequency of horizontal input, f_x = 1,000 Hz

Points of tangency to vertical line = 2

Points of tangency to horizontal line = 5

$$\text{Frequency of vertical input, } f_y = f_x \times \frac{\text{Horizontal tangencies}}{\text{Vertical tangencies}} = 1{,}000 \times \frac{5}{2} = 2{,}500 \text{ Hz } \textbf{Ans.}$$

Example 8.11. An oscilloscope has a display shown in Fig. 8.37. The signal amplitude control is set to 0.2 V/div and the time base control to 10 μs/div. Determine the mark-to-space ratio of the pulse waveform and the pulse frequency. Also determine the magnitude of the pulse voltage.

Solution : Taking standard practice of

1div = 1 cm for CRO wave displays

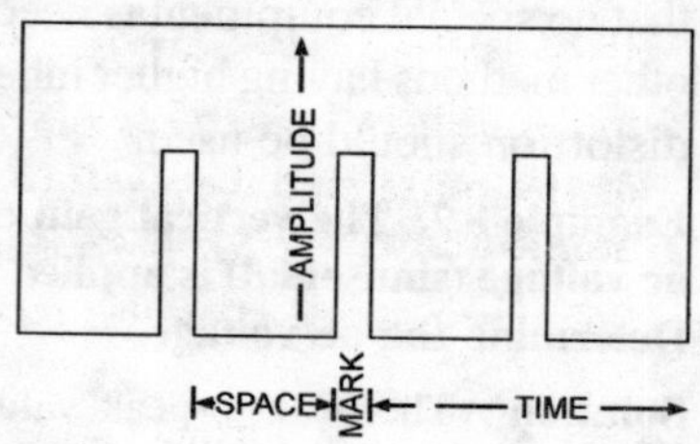

Fig. 8.37

By measuring to scale

Mark = 0.4 cm = 0.4 division

Space = 1.6 cm = 1.6 divisions

Amplitude = 2.15 cm = 2.15 divisions

$$\text{Mark-to-space ratio} = \frac{\text{Mark}}{\text{Space}} = \frac{0.4}{1.6} = \frac{1}{4} \textbf{ Ans.}$$

Pulse time period, T = Time for (space + mark) = 0.4 + 1.6 = 2 divisions

= 2 × time-base control setting = 2 × 10 μs = 20 μs

$$\text{Pulse frequency, } f = \frac{1}{T} = \frac{1}{20 \times 10^{-6}} = 50 \text{ kHz } \textbf{Ans.}$$

Magnitude of pulse voltage = Pulse amplitude in divisions × amplitude control setting

= 2.15 × 0.2 = 0.43 V **Ans.**

Example 8.12. Two sinusoidal voltage signals of equal frequency are applied to the vertical and horizontal deflection plates of CRO. Find the phase angle for each trace shown in Figs. 8.38 (*a*), (*b*), (*c*) and (*d*).

Solution : As we know that phase angle, $\Phi = \text{Sin}^{-1} \frac{d_{vo}}{D_v}$

(*a*) Pattern shown in Fig. 8.38 (*a*) is a straight line so value of d_{vo} = 0 and, therefore,

$$\text{Phase angle } \Phi = \text{Sin}^{-1} \frac{d_{vo}}{D_v} = \text{Sin}^{-1} \frac{0}{6} = 0° \textbf{ Ans.}$$

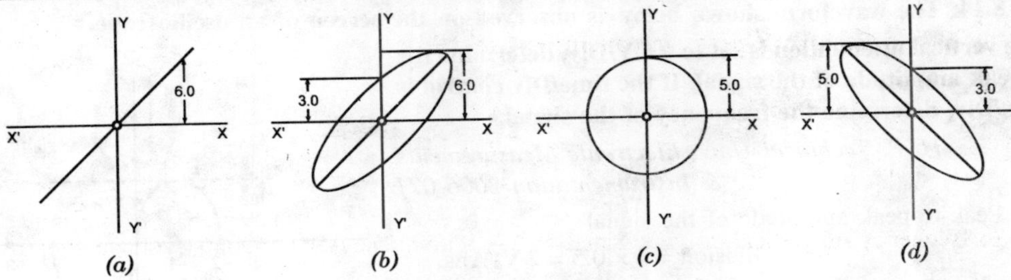

Fig. 8.38

(*b*) Pattern shown in Fig. 8.38 (*b*) is an ellipse. Values of d_{vo} and D_v are 3.0 and 6.0 respectively for this pattern.

$$\text{So phase angle for this pattern, } \Phi = \text{Sin}^{-1}\frac{3.0}{6.0} = \text{Sin}^{-1}\, 0.5 = 30^\circ \text{ or } 150^\circ$$

But referring to Fig. 8.38 (*b*) ellipse is in first and third quarters so valid value of phase angle $\phi = 30^\circ$ **Ans.**

(*c*) Pattern shown in Fig. 8.38 (*c*) is a circle in which values of d_{vo} and D_v are equal, so

$$\text{Phase angle } \Phi = \text{Sin}^{-1}\, 1.0 = 90^\circ \textbf{ Ans.}$$

(*d*) Pattern shown in Fig. 8.38 (*d*) is again an ellipse. Values of d_{vo} and D_v are 3.0 and 5.0 respectively.

$$\text{So phase angle, } \Phi = \text{Sin}^{-1}\frac{3.0}{5.0} = \text{Sin}^{-1}\, 0.6 = 36.9^\circ \text{ or } 143.1^\circ$$

But referring to Fig. 8.39 (*d*) ellipse is in 2nd and fourth quarters so the valid value of phase angle Φ is 143.1° **Ans.**

Example 8.13. A CRO is set to a time base of 0.1 ms/cm with a 2 V/cm amplitude. Sketch the display of the pulse signal waveform with a pulse repetition rate of 2,000 Hz and a duty cycle of 20%. Its amplitude is 6 V peak.

Solution : Pulse repetition rate, $f = 2{,}000$ Hz

$$\text{Time period of a pulse, } T = \frac{1}{f} = \frac{1}{2{,}000} = 0.5 \text{ ms}$$

$$\text{Duty cycle, } D = 20\% = 0.2$$

$$\text{Pulse duration} = D \times T = 0.2 \times 0.5 = 0.1 \text{ ms}$$

Choosing scale on axis of X, 1 cm = 0.1 ms and on axis of Y, 1 cm = 2 V, display of the given pulse signal waveform is sketched and is shown in Fig. 8.39.

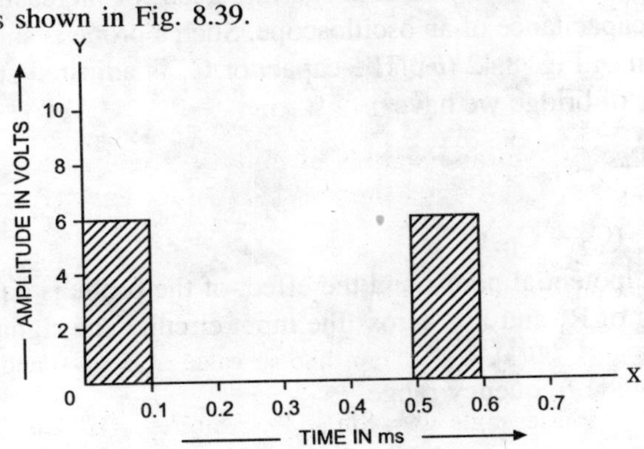

Fig. 8.39

Example 8.14. The waveform shown below is observed on the screen of an oscilloscope. If the vertical attenuation is set to 0.5 V/Div, determine the peak to peak amplitude of the signal. If the time/Div control is set to 2 μs/Div, determine the frequency of the signal.

[U.P. Technical Univ. Electronic Measurements & Instrumentation 2006-07]

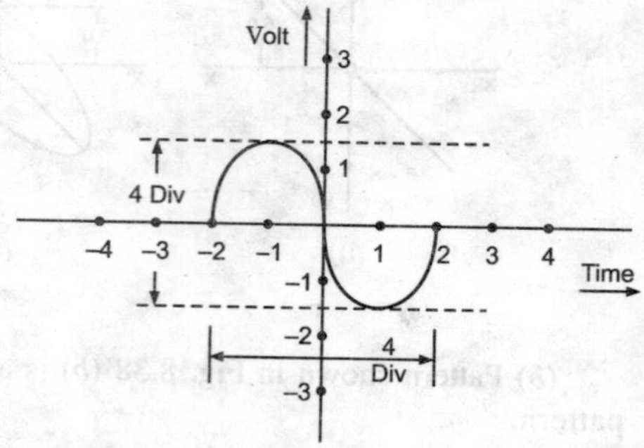

Fig. 8.40

Solution: Peak-to-peak amplitude of the signal

$$= 4 \text{ Division} = 4 \times 0.5 = 2 \text{ V} \textbf{ Ans.}$$

Pulse time period, T = 4 Division

$$= 4 \times 2\ \mu s = 8\ \mu s = 8 \times 10^{-6} \text{ s}$$

Signal frequency, $f = \frac{1}{T} = \frac{1}{8 \times 10^{-6}} = 125 \text{ kHz}$ **Ans.**

8.9 OSCILLOSCOPE PROBES

Probe is a means that connects the test circuit to the oscilloscope without altering, loading, or disturbing the test circuit. Oscilloscope probes of different types are described below.

1. *Direct Probe.* This is simplest of all the probes and uses a shielded coaxial cable. It avoids stray pick-ups which may cause troubles in measurement of low level signals. It is usually employed for low frequency and low impedance circuits. However, in using the probe of this type shunt capacitance of probe and cable is added to the input impedance and capacity of the scope and tends to lower the oscilloscope response to high impedance and high frequency circuits.

2. *Isolation Probe.* Such a probe is employed to avoid the undesirable circuit loading effects of the shielded probe. Isolation probe is made by placing a carbon resistor in series with the test lead, as illustrated in Fig. 8.41.

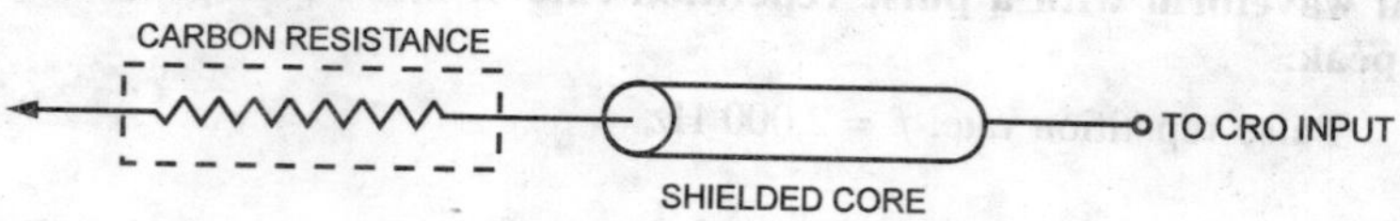

Fig. 8.41 *Isolation Probe*

Such a probe causes a slight change in the amplitude of the waveform and a slight change in the waveshape. To avoid this possibility, a high impedance compensated probe, called a *low capacitance* or a 10 : 1 probe is employed.

3. *High Impedance or* 10 : 1 *Probe.* Such probes are used for increasing the input resistance and reduce the effective input capacitance of an oscilloscope. Such a probe is shown in Fig. 8.42 (*a*) and its equivalent circuit is shown in Fig. 8.42 (*b*). The capacitor C_1 is adjusted so as to balance the bridge. Under balanced conditions of bridge we have

$$\frac{R_1}{\omega(C_2 + C_{IN})} = \frac{R_{IN}}{\omega C_1}$$

$$R_1 C_1 = R_{IN} (C_2 + C_{IN})$$

Thus, X and Y are equipotential points and the effect of the probe is equivalent to placement of a potential divider consisting of R_1 and R_{IN} across the input circuit. The signal attenuation is 10 : 1 *i.e.*,

$$\frac{R_1 + R_{IN}}{R_{IN}} = 10 : 1 \text{ over a wide frequency range.}$$

Capacitor C_1 is adjusted by connecting the probe tip to a square-wave of 1 kHz and observing the CRT display. For optimum response of CRT display, the value of C_1 is deemed to be appropriate.

$$\text{So } V_{out} = 0.1\ V_{IN} = \frac{V_{IN} \times R_{IN}}{R_1 + R_{IN}}$$

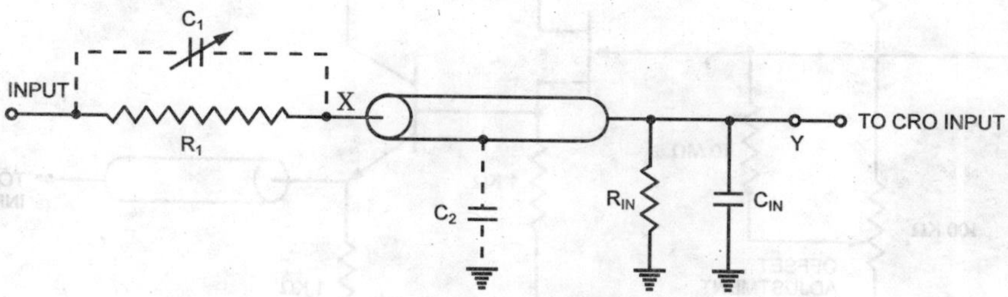

(a) 10 : 1 Probe

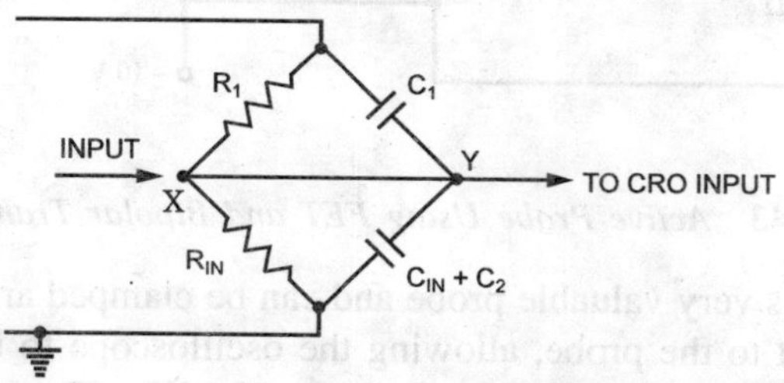

(b) Equivalent Circuit of 10 : 1 Probe

Fig. 8.42

4. *Active Probes.* The probe, described above, is also called the passive probe. Active probes are designed to provide efficient method of coupling high frequency, fast rise time signals to the CRO input. Usually active probes have very high input impedance, with less attenuation than passive probes described above. Active probes are bulky and more expensive as compared to passive probes, but they are useful for measurement of small signals, owing to their small attenuation.

A schematic diagram of an active probe using an FET and bipolar transistors is given in Fig. 8.43. Although the voltage gain of the FET follower circuit shown in Fig. 8.43 is unity, the follower circuit provides a power gain so that the input impedance can be increased. For eliminating the capacitance of the interconnecting cable and make the probe more effective, the FET must be mounted directly in the voltage probe tip. This requires that the power for the FET be supplied from the oscilloscope to the FET in the probe tip. The FET voltage follower drives a coaxiable cable, but instead of the cable connecting directly to the high input impedance of the oscilloscope, it is terminated in its characteristic impedance.

Since there is no signal attenuation between the FET amplifier and the probe tip, the range of the signals that can be handled by the FET probe is limited to the dynamic range of the FET amplifier, and is typically less than a few volts. This is a significant drawback of the FET probe. So for handling a large dynamic range, external attenuators are added at the probe tip. Active probes have limited use because the FET probe effectively becomes an attenuator probe. So, oscilloscopes are typically used with 10 : 1 attenuator probe.

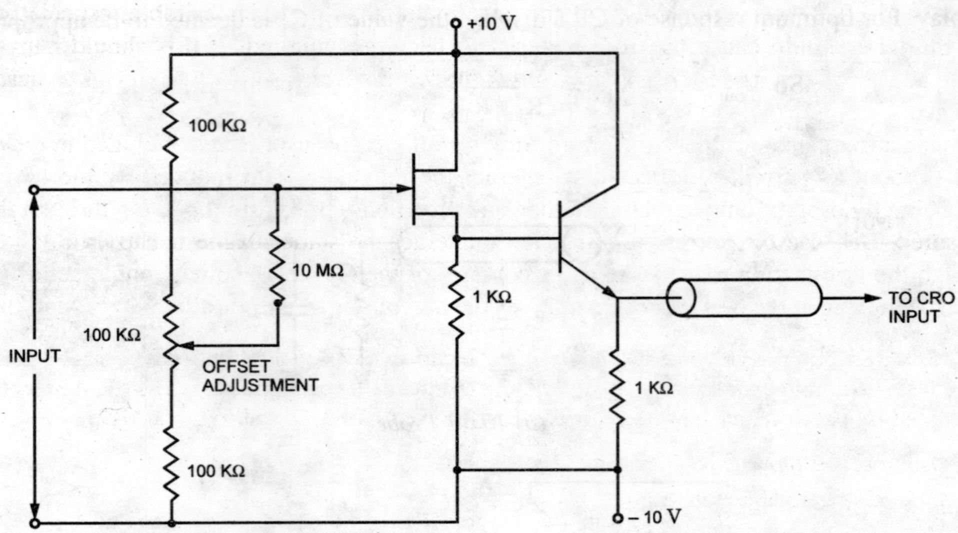

Fig. 8.43 *Active Probe Using FET and Bipolar Transistors*

5. *Current Probe.* This is very valuable probe and can be clamped around a wire carrying current without any physical contact to the probe, allowing the oscilloscope to be employed to measure the magnitude of the current with a frequency from 0 (dc) to 50 MHz. The current sensor consists of two parts, a conventional transformer for transforming alternating current to voltage, and a Hall effect device for converting direct current to a voltage, as illustrated in Fig. 8.44.

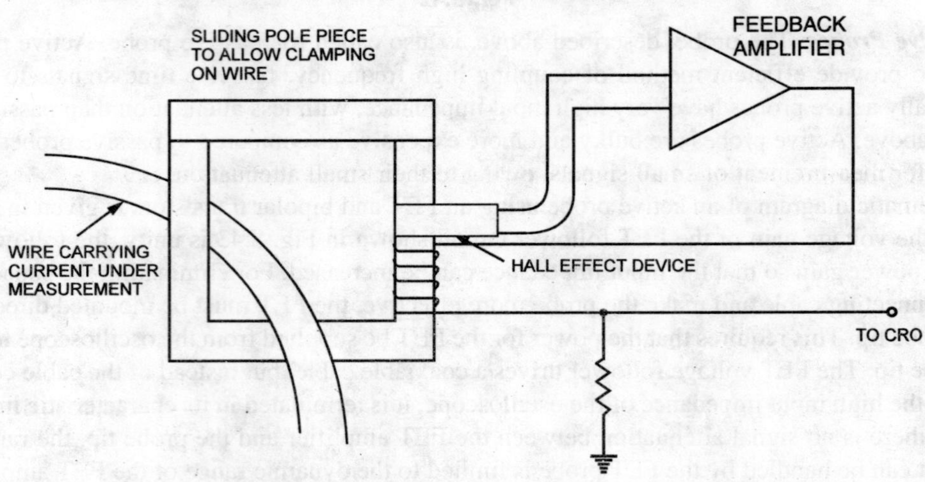

Fig. 8.44 *Current Probe Capable of Measuring From Zero (or DC) to Several MHz*

A magnetic core with a removable piece is used as the coupling element for the current probe. The wire carrying the current to be measured is inserted in the centre of the magnetic core and is similar to a primary of transformer. Alternating current in the wire will induce voltage in the secondary winding by transformer action. Any direct currents will not appear at the secondary of the current transformer.

In addition, the direct current passing through the wire will cause the magnetic flux in the core to increase and will affect the permeability of the core material. This is undesirable, especially if the current in the wire should cause the core material to become saturated. If this should happen, the transformer action of the current transformer will become severely affected and give inaccurate measurements.

To provide a frequency response to zero (or dc), a Hall effect sensor is also included in the current probe. The amount of current required to counteract the magnetic field induced by the wire being measured is directly proportional to the magnitude and direction of the current passing through the wire being measured. Because the current required to counteract the static magnetic flux in the core also flows through the terminating resistor for the secondary of the CT, this applied counteracting-current appears as a dc voltage at the oscilloscope and gives the magnitude of current flowing through the wire.

Example 8.15. Determine the value of C_1 required for compensation of high impedance probe (Fig. 8.42) when the oscillscope input capacitance is 36 pF and the coaxial cable capacitance is 150 pF. Also determine the input capacitance seen from the source. It is given that $R_1 = 10\ M\Omega$ and $R_{IN} = 1\ M\Omega$.

Solution: Oscilloscope input capacitance, $C_{IN} = 36$ pF

Coaxial cable capacitance, $C_2 = 150$ pF

Under balanced conditions of bridge,

$$\frac{R_1}{\omega(C_2 + C_{IN})} = \frac{R_{IN}}{\omega C_1}$$

$$\text{or } C_1 = \frac{R_{IN}(C_2 + C_{IN})}{R_1} = \frac{1 M\Omega\,(150 + 36)\text{ pF}}{10\ M\Omega} = 18.6 \text{ pF} \quad \textbf{Ans.}$$

Probe input capacitance C_T is C_1 in series with $(C_2 + C_{IN})$, so

$$\frac{1}{C_T} = \frac{1}{C_1} + \frac{1}{C_2 + C_{IN}} = \frac{1}{18.6} + \frac{1}{186} = \frac{11}{186}$$

$$\text{or } C_T = \frac{186}{11} \text{ pF} = 16.91 \text{ pF} \quad \textbf{Ans.}$$

Example 8.16. Determine the signal frequency at which the probe of Example 8.15 causes a 3 dB reduction in the signal from a 500 Ω source.

Solution: Probe will cause a 3 dB reduction when

$$X_{CT} = R_{source} = \frac{1}{2\pi f C_T}$$

$$\text{or Signal frequency, } f = \frac{1}{2\pi C_T R_{source}} = \frac{1}{2\pi \times 16.91 \times 10^{-12} \times 500} = 18.82 \text{ MHz} \quad \textbf{Ans.}$$

8.10 OSCILLOSCOPE SPECIFICATIONS AND PERFORMANCE

Sensitivity. The sensitivity is defined as the amplitude of input voltage that may be displayed on the CRO screen. It may also be defined as the ratio of voltage applied to the deflection in cm or division produced. Typically it is measured in volts/cm or volts per division. Sensitivity typically ranges from 2 mV/division to 10 V/division. For example if a sinusoidal wave with 10 mV peak-to-peak amplitude occupies only one vertical division on the screen, then the sensitivity of the CRO would be 10 mV/division. A higher V/division setting provides a smaller display. For careful examination, a displayed wave should typically occupy half the screen of oscilloscope (four vertical divisions). It means that the smallest wave that can be displayed must have a peak-to-peak value of 40 mV when the sensitivity is 10 mV per division. With a 10 : 1 probe, the input voltage must be 400 mV peak-to-peak to provide a four division display with a sensitivity of 10 mV/division.

Input Impedance. It is defined as the impedance that is offered by CRO to the signal under measurement. Typical value is 1 MΩ shunted by 25 pF.

Maximum Input. This specifies the maximum value that can be applied to a given CRO safely. DC and ac both values are specified.

Accuracy. The accuracy is expressed in terms of percentage. Typical value of accuracy for a CRO is ± 3%. That is an accuracy of 3 percent means that the actual full scale period of any sweep should not be more than 3% or less than the indicated one. The measuring accuracy of any waveform depends on the amplitude of the display and the specified accuracy. Because the trace thickness is about 5% of one vertical division, the best reading accuracy is ± 5% of one division. For a waveform with a peak-to-peak amplitude of 4 screen divisions this provides a reading accuracy of $\frac{\pm 5\%}{4}$, or ± 1.25%. Adding the ± 3% V/division accuracy provides an overall measuring accuracy of ± 4.25%. For wave that occupies only one vertical division, the measurement accuracy becomes ± (5% + 3%) *i.e.* ± 8%.

Frequency Response. The frequency response of an oscilloscope is defined as the highest and lowest waveform frequency that may be displayed with no more than 3 dB of attenuation. This is usually specified as dc (zero) to 20/50/100 MHz. For ac coupling, a low 3 dB frequency of 2 Hz is usually listed. The upper cutoff frequency f_H largely depends on the frequency response of the vertical deflection amplifiers, whereas the lower (ac) cutoff frequency f_L depends upon the coupling capacitor being switched into the input circuit.

For the oscilloscope upper cutoff frequency f_H to have a negligible effect on the displayed waveform, the signal frequency should not be more than $f_H/10$. For a negligible effect on the displayed waveform, the signal frequency should not be less than 10 times the lower cutoff frequency f_L of the oscilloscope.

Time Base Measurements. The number of horizontal divisions for one complete cycle of the displayed waveform represents its time period and it is given by

Time period, τ = Time base setting × horizontal divisions.

Consider an oscilloscope with a BW of 20 MHz. The time period of one cycle of 20 MHz waveform is $\frac{1}{20 \times 10^6}$ second *i.e.* 50 ns. The time base specification for such an oscilloscope is typically 0.25 s/division to 100 ns/division. The number of horizontal divisions occupied by any waveform can be determined from the relation

$$\text{Horizontal divisions} = \frac{\text{Time period, } \tau}{\text{Time/division}} = \frac{1}{f \times \text{time/division}} \quad ...(8.25)$$

Thus at 50 ns/division, one cycle of 20 MHz waveform would occupy 1 division.

The highest frequency that can be displayed is determined from the relation

$$f_{max} = \frac{1}{\text{Horizontal divisions} \times \text{time/division}}$$

For four horizontal divisions and a time base setting of 50 ns/division f_{max} would be 5 MHz.

The time base magnifier may be employed for expansion of the horizontal divisions occupied by one cycle of displayed waveform by a typical factor of 10. Thus, the 50 ns/division setting becomes 5 ns/division, and the maximum displayed frequency computed above becomes 50 MHz for one cycle occupying four horizontal divisions.

The measuring accuracy of the time base is typically mentioned as ± 5%. This becomes ± 7% with maximum setting of time base magnifier. At other magnifier settings between zero and maximum, the time base is not calibrated.

The lowest-frequency waveform that can be displayed at a setting of 0.2 s/division can be determined by first assuming that the time period occupies the full screen *i.e.*, 10 horizontal divisions. So,

$$\text{Frequency, } f = \frac{1}{\tau} = \frac{1}{10 \times 0.2} = 0.5 \text{ Hz}$$

Rise Time Measurements. The *rise time* is the transit time between the 10% and 90% levels of the response when a step input is applied to the oscilloscope. Oscilloscopes are normally designed such that

$$\text{BW} \times \text{rise time} = 0.35$$

$$\text{Thus, for BW of 100 MHz, rise time} = \frac{0.35}{100{,}000{,}000} = 3.5 \text{ ns}$$

Example 8.17. Compute the minimum time division sensitivity for an oscilloscope that is to be used to investigate a 20 MHz waveform. Assume that the time base magnifier expands the horizontal display by a factor of 10.

Solution: Time period of 20 MHz waveform,

$$\tau = \frac{1}{f} = \frac{1}{20 \times 10^6} = 50 \text{ ns}$$

For one cycle occupying 4 horizontal divisions,

$$\text{Minimum time/division} = \frac{\tau}{4} = \frac{50}{4} = 12.5 \text{ ns/division}$$

Using the 10 times magnifier to provide 12.5 ns/division

$$\text{Minimum time/division setting} = 10 \times 12.5 = 125 \text{ ns/division} \quad \textbf{Ans.}$$

EXERCISES

1. Explain with neat sketches working principle of a CRO.
2. What is the difference between a CRT and CRO ? Draw a neat block diagram of a general purpose CRO and explain function of each block. **[Pb. Univ. July 1992]**
3. What are the functions of the following components in a CRO ?
   (*i*) Fluorescent screen in a CRT.
   (*ii*) Vertical deflection plates in a CRT.
   (*iii*) Electron gun. **[U.P. Technical Univ. June 2001]**
4. Derive the expression for vertical deflection of an electron beam in a CRT and sensitivity of CRT. **[U.P. Technical Univ. Electronic Measurements & Instrumentation 2007-08]**
5. Explain the principle of electrostatic focusing of an electron beam in a CRO. **[U.P.S.C. I.E.S. Electrical Engineering-I, 1995]**

6. Draw the functional diagram of an oscilloscope. Show how the term synchronization is used in order to explain the functioning of an oscilloscope. If a 50 Hz line frequency is considered to be standard approximately what upper and lower frequency limits may be measured by using Lissajous figure ? What are limiting factors ?

   Draw the Lissajous figure in the case where the voltages applied to the vertical and horizontal plates are in the ratio of 1:2 with the initial phase difference of zero value.
7. Explain with block diagram the various parts of a CRT. What extra components are needed to make it a CRO ? Explain how would you measure frequency using a CRO. **[Pb. Univ. Elec. Measurements II, June 1993]**
8. Discuss different basic controls present in a cathode ray oscilloscope.

   How is the sensitivity of a cathode ray tube rated ?

   Explain how a CRO is used to measure the frequency of an alternating current flowing in a circuit.
9. What are the major components of a CRT ? Why are the operating voltages in a CRT arranged so that the deflection plates are nearly at ground potential ? **[U.P. Technical Univ. Measurements & Instrumentation 2003-04]**
10. What is the function of a time base circuit in a CRO ?

    Draw the basic circuit of an oscilloscope time base generator and explain its operation.
11. Differentiate between double beam CRO and dual trace CRO in detail.

    [U.P. Technical Univ. Electronic Measurements and Instrumentation 2006-07]
12. Explain the operation of the single-beam dual trace CRO. Also explain the use of this CRO in chop mode and in alternate mode. Sketch waveforms for illustration of the two modes of operation.
13. Explain why you need a sawtooth voltage to display a waveform on a CRO. Explain how Lissajous patterns are used in a CRO to measure an unknown frequency.
14. Explain how phase difference of two sinusoidal signals can be measured using a CRO.

    [U.P. Technical Univ. February 2001]
15. Explain how frequency, phase difference of two waveforms and dc voltage are measured using CRO.

    [U.P. Technical Univ. Basic Electronics September 2001]
16. Explain how from the Lissajous pattern appearing on the screen of a CRO, the phase and frequency of an unknown signal can be determined.
17. Explain how would you measure frequency of a signal on a CRO. **[U.P.S.C. I.E.S. Electrical Engineering I, 1994]**
18. What is a probe ? List the various types of probes with their applications.
19. State the function and explain the working of a 10 : 1 probe for a CRO.

    [U.P. Technical Univ. Electronic Measurements and Instrumentation 2006-07]
20. Write a brief note on CRO probes. **[U.P. Technical Univ. Electronic Measurements and Instrumentation 2005-06]**
21. Explain the following terms in reference to CROs.

    (*i*) Sensitivity (*ii*) Accuracy (*iii*) Frequency response (*iv*) Time base.
22. Discuss the accuracy of voltage amplitude and time measurements on an oscilloscope.

SHORT ANSWER TYPE QUESTIONS WITH ANSWERS

Q. 1. What is cathode-ray oscilloscope (CRO) ?

Ans. A CRO is an electronic device with a CRT as its main component and other associated circuits consisting of a power supply unit, a sawtooth wave generator, horizontal and vertical amplifiers.

Q. 2. How is CRO superior to ordinary measuring instruments ?

Ans. CRO is an electronic device that gives graphical representation of alternating quantities under examination.

The CRO gives very accurate measurements as it is free from the errors introduced by the moving parts. It is also free from damping mechanism and other inertia containing parts.

Q. 3. For what vertical and horizontal plates are provided in a CRO ?

Ans. Horizontal and vertical plates are provided between electron gun and screen to deflect the beam according to the input signal.

Q. 4. For what a triggering circuit is provided in a CRO ?

Ans. In a CRO, a triggering circuit is provided for synchronising two types of deflections so that horizontal deflection starts at the same point of the input vertical signal each time it sweeps.

Q. 5. What are the essential components of a CRT ?

Ans. The essential components of a CRT are electron gun, focusing and accelerating anodes, horizontal and vertical deflection plates, and evacuated glass envelope with phosphorescent screen.

Q. 6. For what electron gun assembly is provided in a CRT ?

Ans. The sole function of an electron gun assembly in a CRO is to provide a narrow and sharply focused electron beam which is accelerated towards the phosphor screen.

Q. 7. Why is the grid in a CRO provided with a hole in it ?

Ans. The hole in a grid of a CRO is provided to allow passage for electrons through it and concentrate the beam of electrons along the axis of the tube.

Q. 8. What is meant by the deflection sensitivity of a CRO ?

Ans. The deflection sensitivity of a CRO is defined as the vertical deflection of the beam on the screen per unit deflecting voltage.

Q. 9. What is meant by the deflection factor of a CRO ?

Ans. The deflection factor of a CRO is the reciprocal of the deflection sensitivity.

Q. 10. What is Astigmatism control ?

Ans. Astigmatism control is an adjustment that will provide sharp focus over the entire screen.

Q. 11. What is graticule ?

Ans. Graticule is a scale on transparent material that is fitted to the face of CRT for the purpose of measurement.

Q. 12. What is aquadag ?

Ans. Coating of a conducting material, known as aquadag, is provided over the interior surface of CRT in order to accelerate the electron beam after passing between the deflection plates and to collect the electrons produced by secondary emission when electron beam strikes the screen.

Q. 13. What is meant by retrace time ?

Ans. Retrace time is the time required by the electron beam to return to its original position on a CRT screen after being deflected to the right by a sawtooth waveform.

Q. 14. What is sweep time ?

Ans. Sweep time is time duration during which the beam is swept from left to right on the screen of a CRT by the linearly increasing sawtooth voltage.

Q. 15. For what an attenuator is used ?

Ans. An attenuator is used to reduce the signal level by a desired amount. The ordinary resistance voltage divider is one of the simplest type of attenuators.

Q. 16. What is a Lissajous pattern ?

Ans. Lissajous pattern is a pattern that results from applying periodic signals to the deflection plates of a CRO.

Q. 17. Enlist the applications of CRO.

Ans. The applications of CRO are given below :

1. Measurement of voltage and current.
2. Measurement of phase.
3. Measurement of frequency.
4. Comparison of two frequencies.

CRO is a versatile instrument that can be used for display, measurement, waveform analysis and other phenomenon in electrical and electronic circuits.

Q. 18. What are various focusing techniques ?

Ans. Two most commonly employed focusing techniques are :

1. Electrostatic focusing.
2. Electromagnetic focusing.

Q. 19. How is focusing achieved ?

Ans. Focusing is achieved by various sets of plates forming a sort of electronic lens. In general, there are three sets of plates. Voltages at anode 1 and 2 are kept fixed while it is variable on the third anode. By varying the voltage on the third anode, the spot may be focused.

Q. 20. What is a dual beam CRO ?

Ans. A dual beam CRO has got two electron beams that are controlled independently. The constructional features are almost the same as in the case of single beam CRO.

Q. 21. What is a dual trace oscilloscope ? **[PTU December 2004]**

Ans. Dual trace oscilloscope employs a single electron gun and provides two traces by switching the deflection plates from one input signal to another which implies that Y-channel is time shared by two signals. The display of the input is continuous and simultaneous to the eyes although it is sampled display.

Q. 22. What is the difference between a dual trace and dual beam CRO ?

Ans. Dual beam CRO uses two separate electron beams and deflection systems mounted in the same CRT to display two different electron beams simultaneously whereas the dual trace oscilloscope alternately displays two signals on one screen by time sharing one electron gun and deflection system.

Q. 23. How is dual trace oscilloscope superior to dual beam oscilloscope ?

Ans. Dual beam CROs are not as versatile as the dual trace CROs because the later can operate in a number of modes.

Q. 24. Why is it preferable not to use CRO in direct sunlight or in a brightly lighted room ?

Ans. Because this will allow a low intensity setting.

Q. 25. What is an oscilloscope probe ?

Ans. Probe is a means that connects the test circuit to the oscilloscope without altering, loading, or disturbing the test circuit.

Q. 26. Enumerate the passive probes.

Ans. The passive oscilloscope probes are : Direct probe, isolation probe and high impedance or 10 : 1 probe.

Q. 27. What is attenuation factor of probe?

Ans. The attenuation factor is the amount by which the probe reduces the amplitude of the signal.

Q. 28. What are the typical attenuation factors for probes?

Ans. The typical attenuation factors for probes are 1, 10 and 100.

Q. 29. What is meant by CMRR of a differential probe?

Ans. Common mode rejection ratio (CMRR) is a differential probe's ability to reject any signal that is common to both test points in a differential measurement. CMRR is a key figure of merit for differential probes and amplifiers, and is given as

$$\text{CMRR} = \frac{\text{Voltage gain for the differential mode signal, } A_d}{\text{Voltage gain for the common-mode signal, } A_{CM}}$$

Q. 30. Define the bandwidth (BW) of a CRO.

Ans. Bandwidth of a CRO means the 3 dB bandwidth of the vertical amplifier of the CRO. 20 MHz bandwidth means that a signal from dc to 20 MHz will get amplified without distortion.

PROBLEMS

1. Determine the velocity of the electron beam in an oscilloscope when the voltage applied to its accelerating anode is 2,000 V. **[Ans.** 26.5×10^6 m/s]
2. The deflection sensitivity of cathode ray tube is 0.06 mm/V and unknown voltage applied to the deflection plates shifts the spot by 4 mm towards the left in the horizontal direction. Determine the unknown applied voltage. **[Ans.** 66.67 V]
3. The X-deflecting plates in a cathode ray tube are 15 mm long and 6 mm apart. The centre of the plates is 20 cm from the screen. The accelerating voltage is 2,500 V. Determine deflection sensitivity and deflection factor of the cathode ray tube. **[Ans.** 0.1 mm/V; 10 V/mm]
4. A Lissajous pattern on an oscilloscope is stationary and has 5 vertical maximum values and 4 horizontal maximum values. The frequency of the horizontal input is 1,200 Hz. Determine the frequency of vertical input. **[Ans.** 1,500 Hz]
5. Draw the Lissajous pattern when the ratio of vertical frequency to horizontal frequency is (i) 1 : 2 (ii) 1 : 1.

Ans.

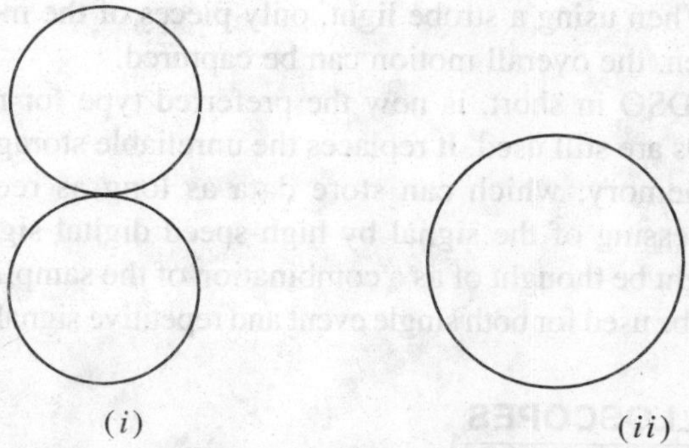

(*i*) (*ii*)

6. A complete sinusoidal wave cycle occupies 4 horizontal divisions and four vertical divisions from peak-to-peak. If the oscilloscope is set on the 20 ms/division and 500 mV/division. Determine (*i*) time period τ (*ii*) frequency *f* (*iii*) peak-to-peak, peak and rms values of applied voltage. **[Ans.** (*i*) 80 ms, (*ii*) 12.5 Hz, (*iii*) 2 V, 1 V, 0.707 V]

❑❑❑

9

Special Oscilloscopes

9.1 INTRODUCTION

In addition to the basic oscilloscope described in the preceding chapter, there are other special purpose oscilloscopes and features which extends some of the basic scope capabilities. The most important types will be briefly described here, and a few facts about their special features and functions will be mentioned. However, the operating details and applications of such oscilloscopes are too specialized and complex to be fully explored here.

Many laboratory quality oscilloscopes include a *delayed sweep feature*. This features enhances the versality of the instrument by making it possible to magnify a selected portion of the undelayed sweep, measure waveform jitter or rise time and check pulse-time modulation, as well as many other applications.

Sampling oscilloscopes achieve their large bandwidths by not taking the entire signal at a time. Instead only a sample of the signal is taken. The samples are then assembled to produce the waveform. This method can only work for repetitive signals, not transient events. The idea of sampling can be thought of as a stroboscopic technique. When using a strobe light, only pieces of the motion are seen, but when enough of these images are taken, the overall motion can be captured.

The *digital storage oscilloscope*, or DSO in short, is now the preferred type for most industrial applications, although simple analog CROs are still used. It replaces the unreliable storage method used in analog storage scopes with digital memory, which can store data as long as required without degradation. It also allows complex processing of the signal by high-speed digital signal processing circuits. A digital storage oscilloscope might be thought of as a combination of the sampling and storage oscilloscopes. The digital oscilloscopes can be used for both single event and repetitive signal investigations.

9.2 DELAYED TIME-BASE OSCILLOSCOPES

The delayed sweep feature allows small, selected portions of the oscilloscope trace to be displayed. The selected portion can be shown in an expanded version at whatever magnification is desired. The sweeper magnifier also allows parts of the display to be magnified up to 10 X. For example, assume that a digital signal such as the one depicted in Fig. 9.1(a) is to be displayed on the screen of an oscilloscope. If third pulse of this signal is required to be examined in detail, the *delayed sweep feature* could be used for display of the third pulse in expanded form.

In case, a faster time/division setting was tried for magnification of the trace in order to display an expanded version of this pulse, we would end up displaying only the first pulse (or part of the first pulse) of the signal. This would occur because a shorter time/division setting would expand only the part of the signal at the beginning of the sweep. The part of the sweep that contains the third part would be out of the area of the screen [Fig. 9.1(*b*)].

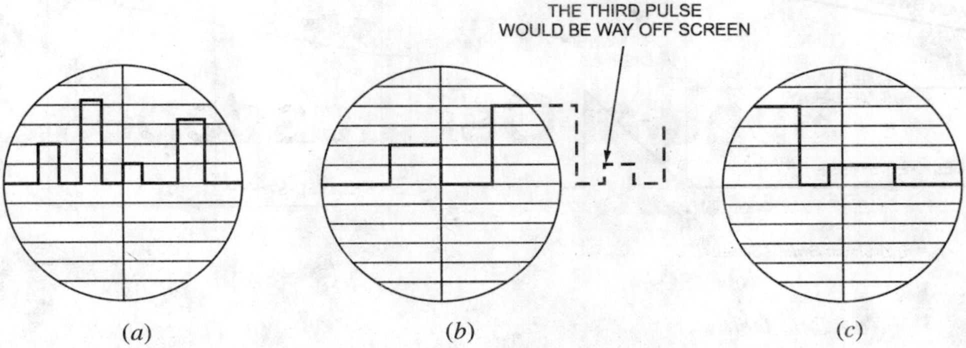

Fig. 9.1

The *delayed sweep feature* is made possible by the use of two time-base generators in the oscilloscope: a main time-base generator (A), also called the *delaying generator* and a secondary time-base generator (B), also called the *delayed generator*. The main time-base or delaying generator produces the same type of sweep waveform as in the basic oscilloscope. Time-base generator B also produces a sweep waveform, but the starting time of this waveform is controlled by the sweep signal of main time-base generator A. The amount of the time by which the start of sweep waveform B is delayed from the start of waveform A can also be varied.

The time-base ramp generated by the main time-base generator A is started by a trigger pulse at time t_0, it continues until attaining a comparator level set by the *division delay control*. At this time t_1, the main generator A stops, and the second time-base generator is ready for operation. Now, there is a possibility of two modes, depending upon the setting of the trigger for the delayed generator. If the delayed generator is in *auto*, it operates as soon as the delaying sweep attains the *division-delay comparator level*, as depicted in Fig. 9.2. This mode uses the main time-base generator A as a *time-delay generator*. If it is being triggered from an internal or external trigger pulse occurring at time t_b (after time instant t_1), then the delaying generator is being used as a *trigger-delay generator*. This trigger-delay mode is often called the *arming*, because it serves to make the delay generator to operate at the next trigger pulse. This is of great importance for a common pulse-train measurement situation where the pulses are subject to time jitter between t_0 and t_b.

Delayed sweep offers increased accuracy and resolution for many time-interval measurements. The *division delay control*, known as variously *delay-time multiplier*, *delay (division)*, or *delay (cm)* is the key control for such measurements. Normally, a 10-turn potentiometer with a readout of three significant numbers and a vernier scale is provided for the knob control.

Measurements may be carried out in different ways, with varying accuracy capability. For measurement of a single pulse for time delay from a reference t_0, the *delay knob* may be dialed until the pulse is centred on screen, and the pulse time delay is then, computed. If, for instance, the main sweep time is 10 μs/division, and the division delay setting is 6.215 divisions, the pulse time delay is 62.15 μs (± 3% for sweep-time accuracy) or 62.2 μs ± 1.9 μs.

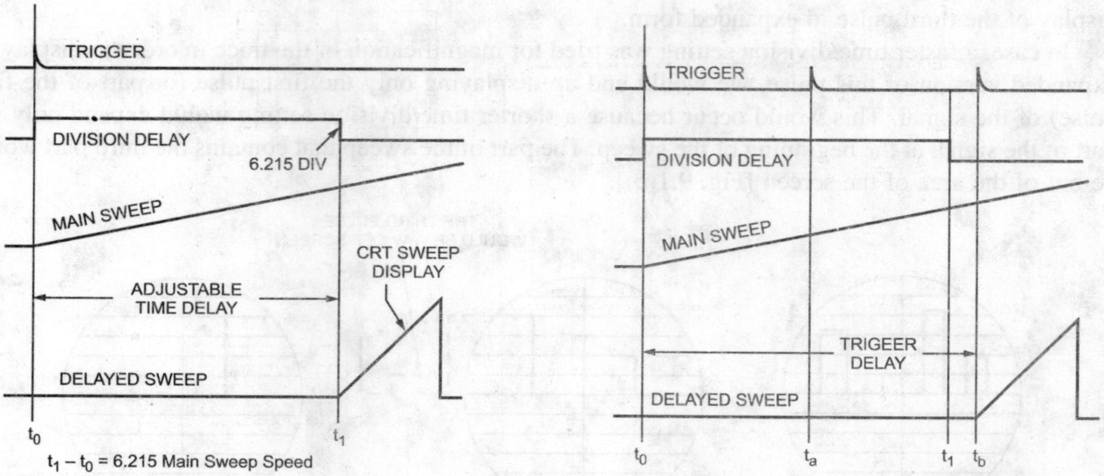

(a) Time Delay *(b) Trigger Delay*

Fig. 9.2 *Delay Mode Operation Illustrating The Time Sequence of Main (Delaying) Sweep and Delayed Sweep*

In case a reference time-mark pulse is used in conjunction with the pulse under measurement (with a dual-channel vertical oscilloscope), the accuracy may be increased substantially. Suppose, for instance, that a reference pulse occurs at t_0 + 50 μs. If the delaying sweep time is set to 2 μs/division, the CRT display should illustrate the two pulses separated by 6.1 divisions, which then provides a difference reading of 12.2 μs ± 3% for sweep-time accuracy and ± 3% for deflection nonlinearities and resolution. This results in an overall measurement of 62.2 μs ± 0.7 μs. Further improvement could be made if the division delay knob is used first to center the reference pulse, and then to centre the pulse of interest, for the deflection nonlinearity is cancelled. This provides a measurement of 62.2 μs ± 0.4 μs. Time difference between any two pulses can be handled in a similar way.

The other sweeps, in which two time-base generators can be used are mixed mode and dual sweep or switched mode displays [Fig. 9.3(*a*) & 9.3(*b*)].

The *mixed mode* provides a more convenient display pattern for many time-interval measurements, because both the reference signal and the signals of interest can be viewed in a wavetrain context with only the region of interest expanded for time-interval measurement. The mixed mode functions in almost the same way as the delay mode, except that the delaying sweep generator is displayed first. An external trigger is produced at time t_1, where upon the delayed sweep generator (at a faster sweep speed), is started. Once it attains the same ramp height as the delaying generator, it is displayed instead for the rest of the sweep cycle. This mode is especially useful in pulse-train studies, when the time interval of a specific set of pulses in a complex pulse-code train is of interest.

Dual sweep or *switched sweep mode* disp'ays two independently variable sweep speeds by electronically time sharing between the two time bas. s. Depending upon sweep speeds, the electronic switching is accomplished similarly to the *chop* and *alternate modes*. Some difficulty is encountered in practice with beam brightness (variations and flicker) if the two channels vary by much more than an order of magnitude in sweep times.

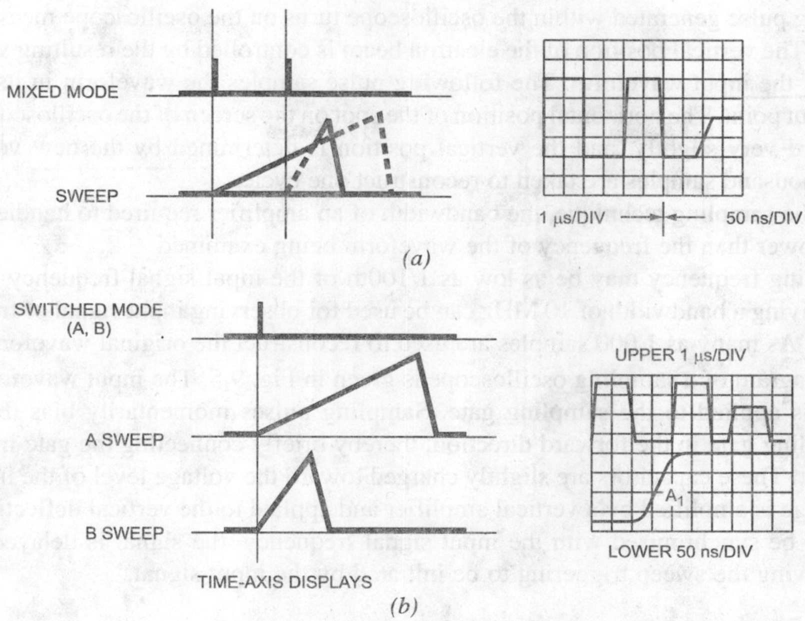

Fig. 9.3

9.3 SAMPLING OSCILLOSCOPES

Most ordinary oscilloscopes have an upper frequency limit in the range 20 MHz to 50 MHz. Higher input frequencies cause the electron beam to move so fast across the screen that only a very faint trace is produced.

To display repetitive signals whose frequencies are higher than the limits of high-frequency oscilloscopes, another technique, called *sampling,* is used. In this technique, the signal is reconstructed from sequential samples of its waveform. These samples are obtained at slightly different points along successive cycles of the signal waveform (Fig. 9.4). This means that the waveform being displayed must be repetitive one in order for sampling to be possible.

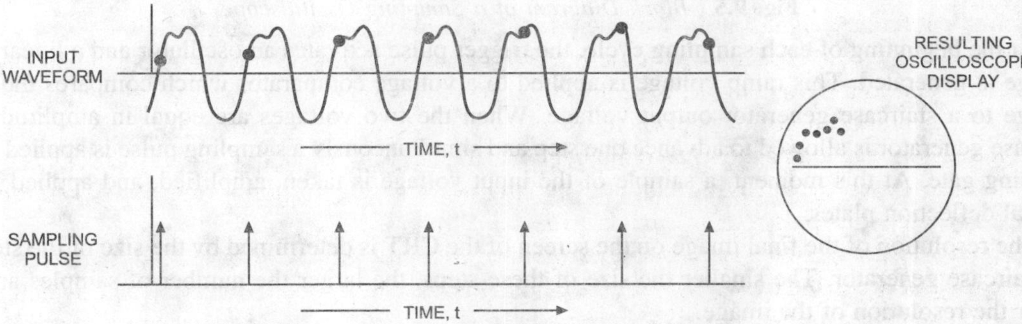

Fig. 9.4 *Relationship Between Input Waveform, Sampling Pulses, and Oscilloscope Display in Sampling CRO*

A sampling pulse generated within the oscilloscope turns on the oscilloscope measuring circuit for a brief instant. The vertical position of the electron beam is controlled by the resulting voltage observed at that point of the input waveform. The following pulse samples the waveform in its next cycle at a slightly different point. The horizontal position of the spot on the screen of the oscilloscope is meanwhile stepped forward very slightly, and the vertical position is determined by the new voltage value. As many as one thousand samples are taken to reconstruct one cycle.

By using the sampling technique, the bandwidth of an amplifier required to handle HF waveforms can be much lower than the frequency of the waveform being examined.

The sampling frequency may be as low as 1/100th of the input signal frequency *i.e.*, an ordinary oscilloscope having a bandwidth of 10 MHz can be used for observing input signal of frequency as high as 1,000 MHz. As many as 1,000 samples are used to reconstruct the original waveform.

A block diagram of a sampling oscilloscope is given in Fig. 9.5. The input waveform, which must be repetitive, is applied to the sampling gate. Sampling pulses momentarily bias the diodes of the balanced sampling gate in the forward direction, thereby briefly connecting the gate input capacitance to the test point. These capacitors are slightly charged toward the voltage level of the input circuit. The capacitor voltage is amplified by a vertical amplifier and applied to the vertical deflection plates. Since sampling must be synchronised with the input signal frequency, the signal is delayed in the vertical amplifier, allowing the sweep triggering to be initiated by the input signal.

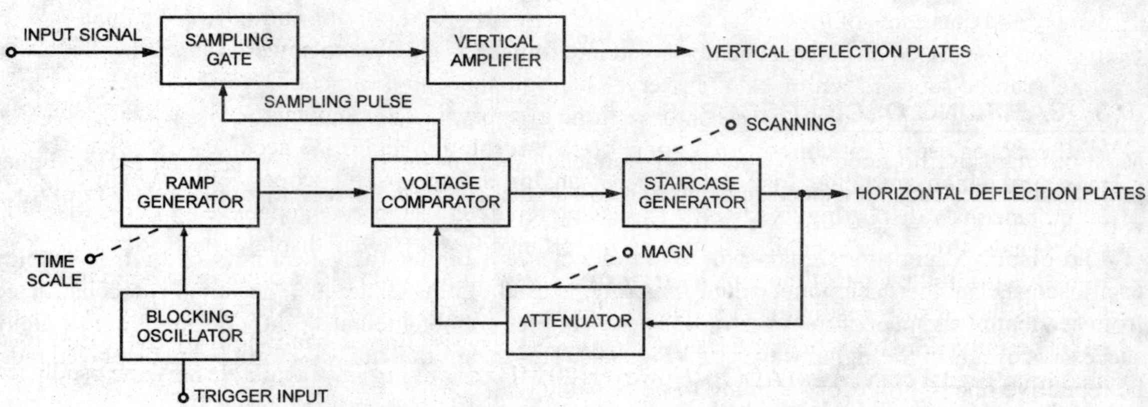

Fig. 9.5 *Block Diagram of a Sampling Oscilloscope*

At the beginning of each sampling cycle, the trigger pulse activates an oscillator and a linear ramp voltage is generated. This ramp voltage is applied to a voltage comparator which compares the ramp voltage to a staircase generator output voltage. When the two voltages are equal in amplitude, the staircase generator is allowed to advance one step and simultaneously a sampling pulse is applied to the sampling gate. At this moment, a sample of the input voltage is taken, amplified, and applied to the vertical deflection plates.

The resolution of the final image on the screen of the CRT is determined by the size of the steps of the staircase generator. The smaller the size of these steps, the larger the number of samples and the higher the resolution of the image.

The sampling oscilloscopes can be employed beyond 50 MHz into the UHF range around 500 MHz and beyond up to 10 GHz. However, sampling techniques cannot be used for the display of transient waveforms.

9.4 DIGITAL STORAGE OSCILLOSCOPES

The digital storage oscilloscope (DSO) is now the preferred type for most industrial applications, although simple analog CROs are still in use. It replaces the unreliable storage method used in analog storage oscilloscope with digit memory, which can store data as long as required without degradation. It allows complex processing of the signal by high-speed digital signal processing circuits. The basic structure of digital oscilloscope is that the input signal is converted into a digital form and stored in memory. In this state it is capable of being analyzed to generate a variety of different information. To view the display on the CRT the data from memory is reconstructed in analog form. Digitizing occurs by taking a sample of the input waveform at periodic intervals. In order to ensure that no information is lost, according to sampling theory, the sampling rate must be at least twice as far as the highest frequency in the input signal.

Whether repetitive sampling is used or not, a fast ADC is an important part of digital storage oscilloscope. Perhaps the fastest analog-to-digital conversion technology that can be applied to digitizing oscilloscopes is the *flash converter*.

Digital storage CROs use digital storage techniques while analog storage CROs use special CRTs called *storage tubes*. In this technique, the waveform to be stored is digitized, stored in a digital memory, and retrieved for display on the storage oscilloscope. The stored waveform is continually displayed by repeatedly scanning the stored waveform and, therefore, a conventional CRT can be employed for the display and thus some of the cost of the additional circuitry for digitizing and storing the input waveform is offset. The stored display can be displayed indefinitely as long as the power is applied to the memory, which can be supplied with a small battery. The digitized waveform can be further analyzed by either the oscilloscope or by loading the contents of the memory into a computer. Some of the digital storage oscilloscopes use 12 bit converters, giving 0.025% resolution and 0.1% accuracy on voltage and time readings, which are better than the 2–5% of analog storage oscilloscopes. Split screen capabilities (simultaneously displaying live analog traces and replayed stored ones) enable easy comparison of the two signals. Pretrigger capability is also an important advantage. The display of stored data is possible in both amplitude versus time and X–Y modes. In addition to the fast memory readout employed for CRT display, a slow readout is possible for developing hard copy with external plotters.

The only drawback of digital storage oscilloscopes is limited bandwidth by the speed of their analog-to-digital converters (ADCs). However, 20 MHz digitizing rates available on some oscilloscopes yield a bandwidth of 5 MHz, which is adequate for most of the applications.

An advantage of analog storage oscilloscope is that they can handle signals over a wide frequency range.

Figure 9.6 gives the block diagram of a digital storage oscilloscope (DSO). It uses both of digital-to-analog and analog-to-digital converters (DACs and ADCs) for digitizing, storing and displaying analog waveforms. The overall operation is controlled and synchronised by the control circuits, which usually have microprocessor executing a control programme stored in read-only memory (ROM). The data acquisition portion of the system contains a sample-and-hold (S/H) and an analog-to-digital converter (ADC) that repetitively samples and digitizes the input signal at a rate determined by the sample clock, and transmits the digitized data to memory for storage. The control circuits make sure that successive data points are stored in successive memory locations by continually updating the memory's ADDRESS COUNTER.

When memory is full, the next data point from the ADC is stored in the first memory location writing over the old data, and so on for successive data points. This data acquisition and the storage process continues until the control circuits receive a trigger signal from either the input waveform

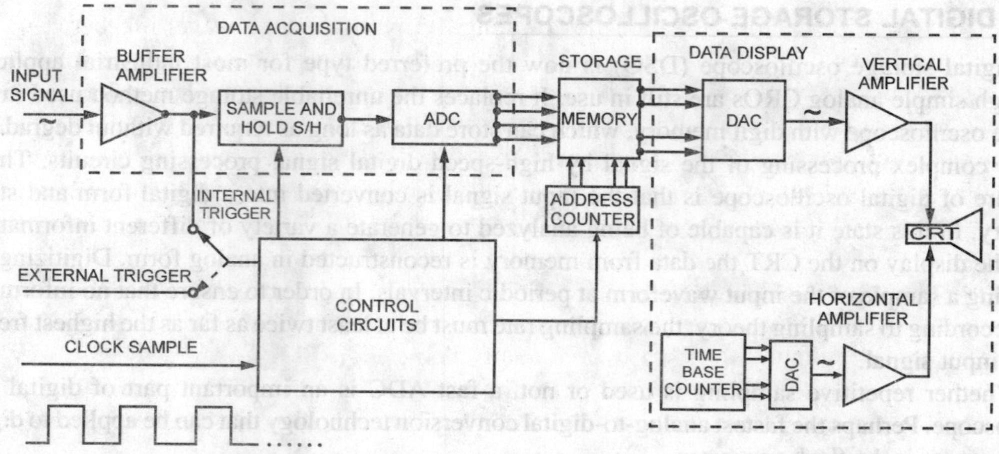

Fig. 9.6 *Block Diagram of a DSO*

(internal trigger) or an external trigger source. When the triggering occurs, the system stops acquiring data further and enters the display mode of operation, in which all or part of the memory data is repetitively displayed on the cathode ray tube (CRT).

In display operation two DACs are employed for providing the vertical and horizontal deflection voltages for the cathode ray tube. Data from memory produce the vertical deflection of the electron beam, while the time-base counter provides the horizontal deflection in the form of a staircase sweep signal. The control circuits synchronise the display operation by incrementing the memory ADDRESS COUNTER and the time-base counter at the same time so that each horizontal step of the electron beam is accompanied by a new data value from the memory to the vertical DAC. The counters are continuously recycled so that the stored data points are repetitively replotted on the screen of the CRT. The screen display consists of discrete dots representing the various data points but the number of dots is usually so large (typically 1,000 or more) that they tend to blend together and appear to be a smooth continuous waveform.

The display operation is terminated when the operator presses a front-panel button that commands the digital storage oscilloscope to begin a new data acquisition cycle.

9.5 DIGITAL STORAGE OSCILLOSCOPE APPLICATIONS

1. Multichannel Displays. Digital storage oscilloscopes, like their counterparts, analog oscilloscopes, have usually the capability of displaying waveform applied to two input channels. However, there are many DSOs which can also display four or more channels of new and stored waveforms. Some DSOs have four separate input channels. As in case of an analog oscilloscope, input waveforms can be displayed on the screen (sampled and interpolated) as they are received. For comparison purposes, stored waveforms as well as new input waveforms can also be displayed simultaneously.

2. Autoset. Like analog oscilloscopes, some DSOs have vertical (volts/division) and horizontal (time/division) controls. Others have the facilities of selecting the best amplitude and time settings for the displayed waveforms, and print the settings on the screen. In this case, manual selection of the quantities is also usually available.

3. Waveform Processing. DSOs normally have DVMs, digital frequency meters and time measuring circuits, which print the required measurements on the screen of the oscilloscope as the waveform is being displayed. Two *cursors* are used for defining the points on the waveform between which measurement is to be carried out. The cursors may be crossed lines, arrowheads, or dots, and they can be moved around by controls on the front panel of the oscilloscope. The voltage, time, and frequency are measured much more accurately than could be with an analog oscilloscope. Some digital storage oscilloscopes process the displayed waveforms without reference to cursors, measuring rms value of input voltage (V_{rms}), peak-to-peak value of input voltage ($V_{p\text{-to-}p}$), frequency (f), rise time (t_r), time period (τ), pulse width, duty cycle, and so on, and printing these quantities on the scope screen.

4. Pre-triggering and Post-triggering. A DSO can be employed for displaying portions of a waveform that occur before the normal trigger point or after the trigger point. Pre-triggering is possible because the input waveform is continually being digitized and stored in memory. "Pre-trigger view" means that the oscilloscope can display what happened before a trigger input is applied. This mode of operation is very useful when a failure occurs and is marked by the appearance of a signal. For determination of cause of failure it would be necessary to see various waveforms before the failure. The DSO will continuously record a selected waveform, and when the trigger waveform appears signifying a failure, the storage will stop and the waveform in the memory is available for viewing. Random or repetitive sampling type of storage oscilloscopes will not provide this function, as a failure is a one-time event and only the consecutive type of storage oscilloscope will provide this feature. The pre-trigger time may be up to 50% or 100% of the waveform time period, depending on the particular oscilloscope. In post-triggering, the display commencing point is delayed after the trigger point by a variable *hold-off time*, which may be several times the time period of the displayed waveform [Fig. 9.7(*b*)]. This is similar to the delay time introduced in delayed sweep analog oscilloscopes. Instead of delaying by a specified time, the display may be delayed by a selected number of waveform cycles. The combination of pre- and post-triggering allows any part of the input waveform to be investigated.

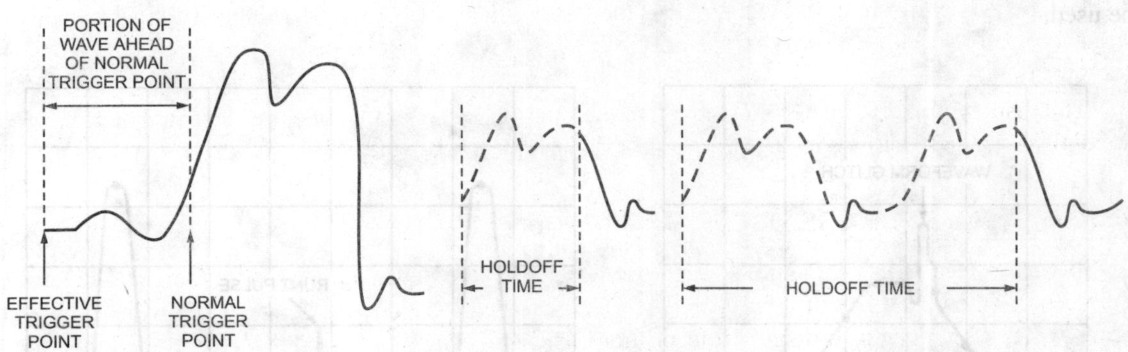

(*a*) *Pre-triggering*

(*b*) *Post-triggering*

Fig. 9.7

5. Zoom and Restart. Some of the DSOs are provided with *zoom* and *restart*, additional methods of time-delay selection. In the former case, the portion of the displayed waveform to be examined is first centred on the screen or marked by cursors. A zoom button is pushed to time-expand the waveform, or *zoom-in* on the portion of interest. This is shown in Fig. 9.8. The required time delay and time-base changes are introduced automatically. Restart selection also makes it possible to view part of a waveform in detail. However, instead of just time expanding, the waveform is resampled at a faster rate for maximum information.

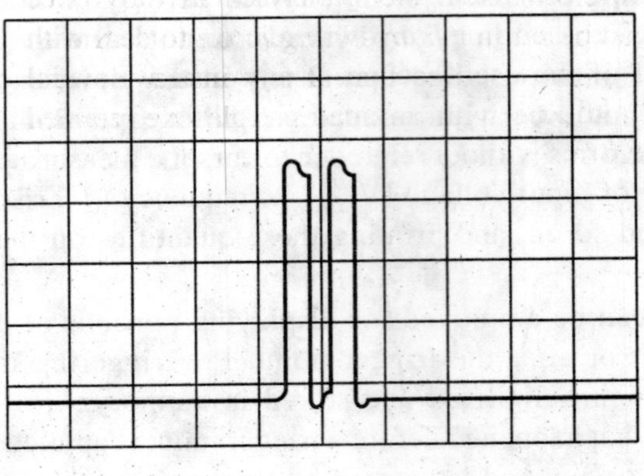

(a) *Waveform Centred on Screen*

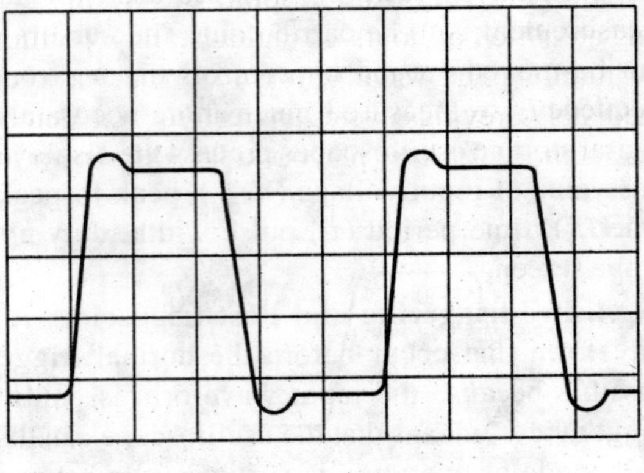

(b) *Zoom Facility Time-Expands The Waveform*

Fig. 9.8

6. Glitches and Runt Catching. Another problem with the sampling technique used in digital oscilloscopes is that it can miss short term transient, or 'glitches' which occur in between the sample points. Figure 9.9(*a*) indicates the fact that very fast spike-type changes in a displayed waveform (glitches) can be missed by a DSO if the sampling rate is not high enough. Occurrence of glitches during the sampling-time interval are invisible to the oscilloscope, therefore, where such waveform anomalies are significant, an oscilloscope with a very high sampling rate is to be used.

A *run pulse* is a special type of glitch that occurs in logic systems, as illustrated in Fig. 9.9(*b*). This is a pulse which is not quite large enough to produce circuit triggering. A *maximum/minimum* level detector circuit is used in some digital storage oscilloscopes to detect the pressure of a glitch. The oscilloscope may then be made to trigger on glitch, so that the waveform before and after the glitch can be used.

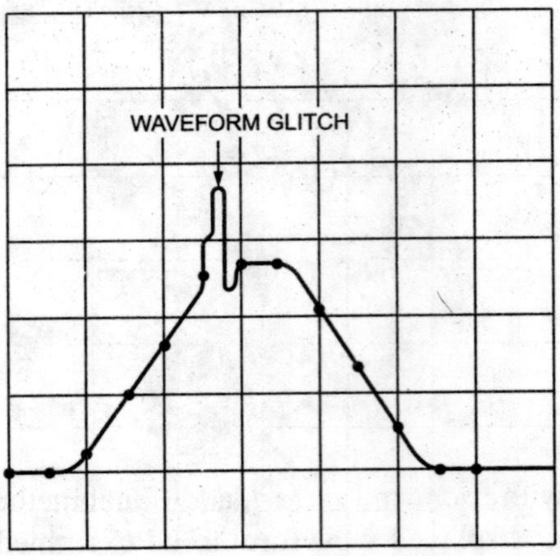

(a) *A Waveform Glitch That Can Be Missed By The Sampling Time Interval*

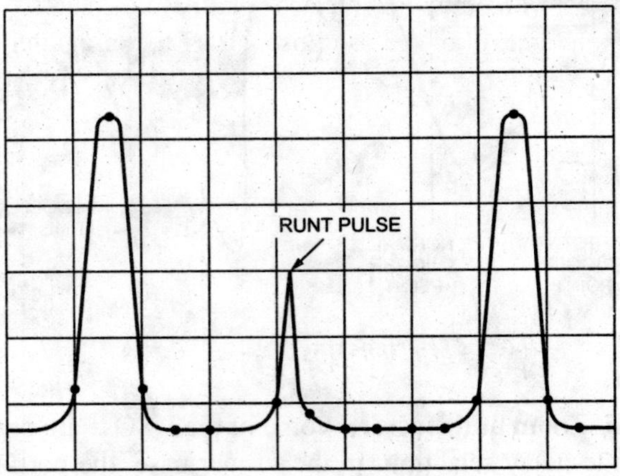

(b) *A Runt Pulse That Can Occur in Logic Circuits*

Fig. 9.9

7. Baby-Sitting Mode. Some waveform transients in electronic systems may occur only once during a period of hours. A digital storage oscilloscope can be set in a *baby-sitting mode* to deal with this situation. The waveform is sampled and recorded continuously, so that at any instant several immediately previous cycles are restored. On detection of any anomaly, the stored samples are retained in the memory for later playback. This is shown in Fig. 9.10.

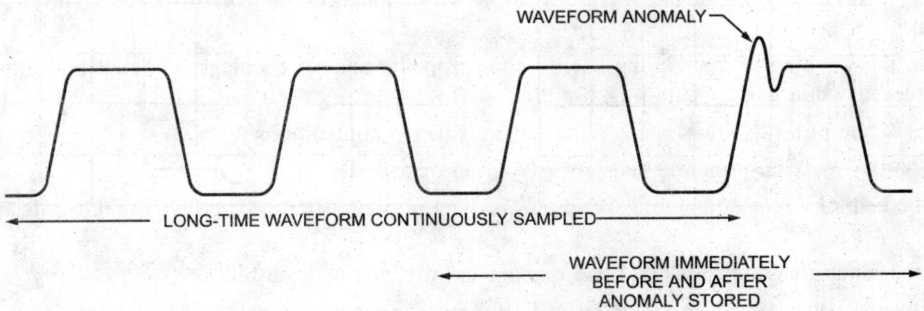

Fig. 9.10 *Baby-sitting Modes*

8. Roll Mode. Chart recording instruments are traditionally used to monitor the slowly varying quantities over a period of hours. A digital storage oscilloscope time-base, which is usually derived by frequency division of the output of a crystal oscillator, can be adjusted to provide an accurate sampling time interval of minutes or hours. Thus, the amplitude of the slowly varying quantity can be sampled and stored over a long period of time. The stored waveform can be displayed on the oscilloscope screen at a speeded-up rate, rolling from right to left as illustrated in Fig. 9.11. When operated in this way, the DSO is said to be in *roll mode*.

9. Documentation and Analysis. Most digital storage oscilloscopes can provide outputs to a plotter or computer type printer to generate hard copy of any waveform stored in memory. Although the waveform samples can be stored indefinitely, a printed copy is convenient for publication and/or further study. This allows the DSO memory to be employed for further storage of waveform. A facility for interfacing with computers is also usually made available for further analysis of stored waveforms, and for remote control of the oscilloscope by the computer.

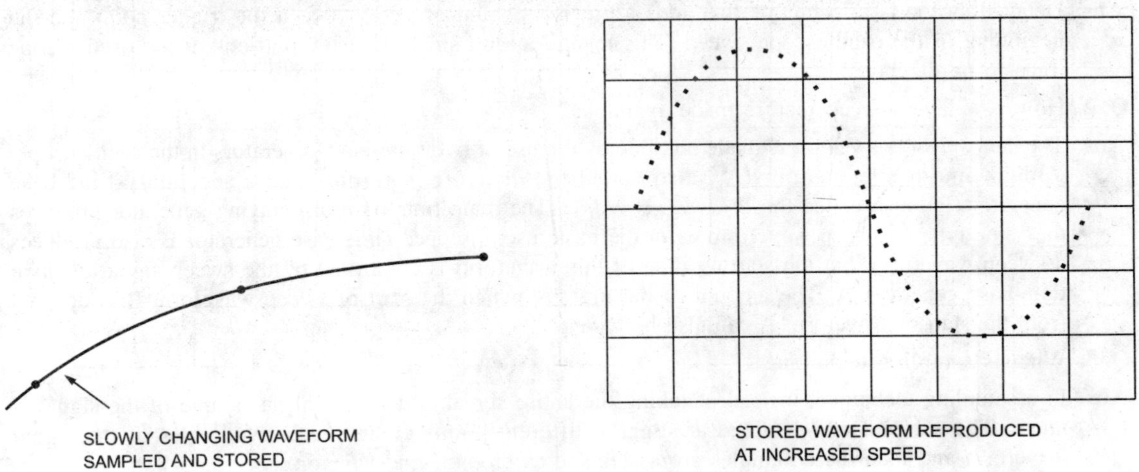

Fig. 9.11 *Roll Mode Operation*

EXERCISES

1. What is delayed sweep feature in an oscilloscope ? What is its need ? How is it provided in an oscilloscope ?
2. Explain how the portion of the waveform to be investigated is intensified by the Delayed Time Base system.
3. Explain the working of a sampling oscilloscope with the help of its diagram. What precautions should be observed while using sampling oscilloscope.
4. What is a sampling oscilloscope ? What are its various applications ?
5. Describe the working principle and block diagram of a DSO.
6. With the help of a block diagram explain the working of digital storage oscilloscope. Give its applications too.
7. Draw the block diagram of a DSO and explain the working of each block.
8. Describe the principle of operation of a digital storage oscilloscope. How does it differ from analog storage oscilloscope ?
9. For a digital storage oscilloscope discuss the following:
Multichannel displays, Autoset, Waveform processing, Pre-triggering and post-triggering.
10. Discuss zoom, and restart facilities for a DSO.
11. Explain glitches and runt catching, baby sitting mode and roll mode operation.
12. Write short notes on the following:
    (*i*) Delayed time-base oscilloscope. (*ii*) Sampling oscilloscope.
    (*iii*) Digital storage oscilloscope. (*iv*) DSO applications.

SHORT ANSWER TYPE QUESTIONS WITH ANSWERS

Q. 1. What is a delayed sweep feature ?

Ans. Delayed sweep is a technique that adds a precise amount of time between the trigger point and the beginning of the oscilloscope sweep. This feature allows small, selected portions of the oscilloscope trace to be displayed.

Q. 2. How is delayed sweep feature made available ?

Ans. The *delayed sweep feature* is made possible by the use of two time-base generators in the oscilloscope: a main time-base generator (A), also called the *delaying generator* and a secondary time-base generator (B), also called the *delayed generator*. The main time-base or delaying generator produces the same type of sweep waveform as in the basic oscilloscope. Time-base generator B also produces a sweep waveform, but the starting time of this waveform is controlled by the sweep signal of main time-base generator A. The amount of the time by which the start of sweep waveform B is delayed from the start of waveform A can also be varied.

Q. 3. What is sampling technique ?

Ans. In a sampling technique, instead of taking the entire signal at a time, only a sample of the signal is taken. The samples are obtained at slightly different points along successive cycles of the signal waveform and then these samples are assembled to produce the waveform.

Q. 4. What is drawback of sampling technique ?

Ans. One main drawback of sampling technique is that only purely repetitive waveform can be investigated. If the waveform changes over several cycles, the dot display will be in error. A transient waveform cannot be investigated by this technique.

Q. 5. What is the main drawback of DSO ?

Ans. The only drawback of DSOs is limited bandwidth by the speed of their ADCs.

Q. 6. Enumerate the applications of DSOs.

Ans. Applications of DSOs are enumerated below:

(*i*) Display and analysis of transient waveforms.

(*ii*) Display of low-frequency waveforms without flicker.

(*iii*) Pre-triggering viewing.

(*iv*) Comparison of stored and real time waveforms.

(*vi*) Interfacing to computer/printer etc.

❑❑❑

UNIT-V
Instrument Calibration and Recorders

CONTENTS IN THIS UNIT

Instrument Calibration: Comparison methods, Digital multimeters as standard instrument, Calibration instrument.

Recorders: X-Y recorders, Plotters.

CHAPTER 10: Instrument Calibration and Recorders.

10

Instrument Calibration and Recorders

10.1 INSTRUMENT CALIBRATION

Calibration is an essential process to be undertaken for each instrument periodically in order to ensure the accuracy of the instrument against the national and international standards. A reference standard at least four times more accurate than the instrument being tested is normally used. The calibration process is simple. It consists of reading the standard and test instruments simultaneously when the input quantity is held constant at several values over the range of the test instrument. The calibration is better carried out under the stipulated environmental conditions. All industrial grade instruments can be checked for accuracy in the laboratory by using the working standards. All modern electronic instruments that are likely to drift in sensitivity are provided with built-in calibration facility by using suitable circuits along with zener diodes. While calibrating, it is customary to take readings both in the ascending and descending order. The test instrument is calibrated under several environmental conditions in order to ensure that the grade of performance conforms to its stated specifications.

Bench type digital meters are usually sufficiently accurate for calibrating other analog and digital instruments. Highly accurate voltage and current sources, called the *calibrators*, are available for the sole purpose of calibration of instruments.

Before undertaking calibration of any instrument, it is necessary that the instrument be cleaned, and all required repairs (such as straightening of pointer, checking of zero position of pointer, and adjustment of mechanical zero control if required while tapping the instrument gently to relieve friction, be carried out.

The supply should be adjusted for increasing the voltage or current, progressively setting the pointer at each major scale division. Again, the instrument should be tapped

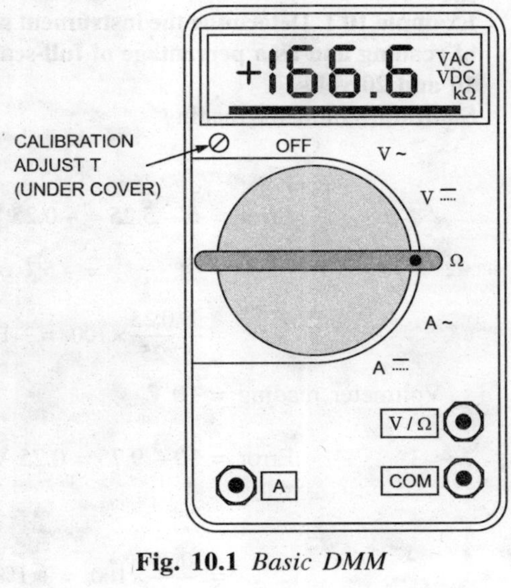

Fig. 10.1 *Basic DMM*

lightly to relieve friction at each setting. At each pointer position of the instrument under calibration, the precise quantity measured by the standard instrument is noted and recorded. The correction, difference between the indicated and precise level, is also tabulated, as shown.

While calibrating a digital instrument, if it is found inaccurate, a correction may be made by adjusting a variable resistor within the circuit (Fig. 10.1).

For instance, the dual-slope integrator DVM makes use of a constant-current source that determines the accuracy of the instrument. This can be adjusted while calibrating, to make the DVM indicate precisely the standard instrument voltage. Preparation of a calibration chart would be very useful. For both digital and analog instruments, to indicate whether the instrument is within its specified accuracy or not.

10.2 COMPARISON METHODS

10.2.1. DC Voltmeter Calibration. Voltmeter can be calibrated against other high accurate voltmeter, the test setup is illustrated in Fig. 10.2. In this calibration procedure, the power supply voltage should be about 10 percent greater than the maximum full-scale indication of the voltmeter under test. The value of standard resistor R should be 10 percent or less than the combined resistance of the voltmeters. The voltage is adjusted in steps over the desired range by varying the rheostat resistance and the two instrument readings are noted as illustrated in Table 10.1.

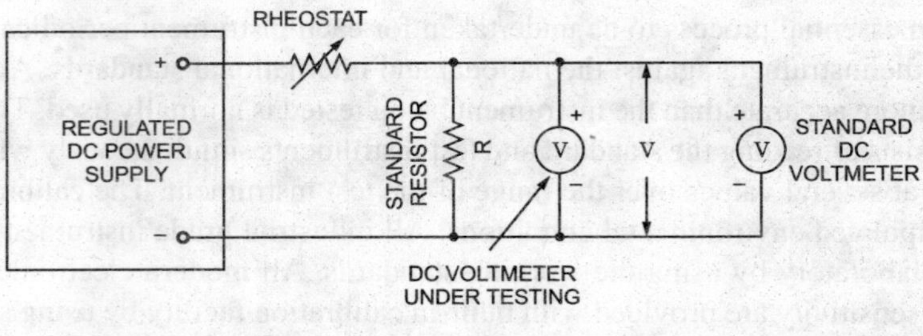

Fig. 10.2 *Calibration of DC Voltmeter*

Example 10.1. Determine the instrument accuracy as a percentage of reading and as a percentage of full-scale for the readings of 5, 10 and 20 volts.

Solution: Voltmeter reading $= 5$ V

$$\text{Error} = 5 - 5.25 = -0.25 \text{ V} = -\frac{0.25}{5}\times 100$$

$$= -5\% \text{ of reading} \quad \textbf{Ans.}$$

$$= -\frac{0.25}{25}\times 100 = -1\% \text{ of full-scale} \quad \textbf{Ans.}$$

Voltmeter reading $= 10$ V

$$\text{Error} = 10 - 9.75 = 0.25 \text{ V} = \frac{0.25}{10}\times 100$$

$$= +2.5\% \text{ of reading} \quad \textbf{Ans.}$$

$$= \frac{0.25}{25}\times 100 = +1\% \text{ of full-scale} \quad \textbf{Ans.}$$

TABLE 10.1
Calibration of DC Voltmeter

Readings in Volts		
Voltmeter Under Calibration	*Standard DC Voltmeter*	*Error in Volts*
0	0	0
5	5.25	– 0.25
10	9.75	+ 0.25
15	15.6	– 0.6
20	20.4	– 0.4
25	24.8	+ 0.2

Voltmeter reading = 20 V

$$\text{Error} = 20 - 20.4 = -0.4 \text{ V} = \frac{-0.4}{20} \times 100 = -2\% \text{ of reading} \quad \textbf{Ans.}$$

$$= \frac{-0.4}{25} \times 100 = -1.6\% \text{ of full-scale reading} \quad \textbf{Ans.}$$

10.2.2. DC Ammeter Calibration. Ammeters can be calibrated by using a test setup such as the one illustrated in Fig. 10.3. In this setup, a source of constant current is provided by a precision dc power supply and a rheostat. The value of the resistance R in the circuit should be at least 10 times the value of the combined resistance of the ammeters. The maximum current in the test circuit $\left(\frac{V}{R}\right)$ should be about 10 percent greater than the full-scale deflection of the meter under test. The current is adjusted in steps over the desired range by varying the rheostat resistance and the two instrument readings are noted, as illustrated in Table 10.2. No adjustment in normally made in the case of DMM used as an ammeter, because the DMM current measurement function uses the voltage measuring facility.

TABLE 10.2

Calibration of DC Ammeter

Reading in Amperes		
Ammeter Under Calibration	*Standard DC Ammeter*	*Error in Amperes*
0	0	
1	1.2	– 0.2
2	1.9	+ 0.1
3	3.24	– 0.24
4	3.85	+ 0.15
5	4.8	+ 0.2

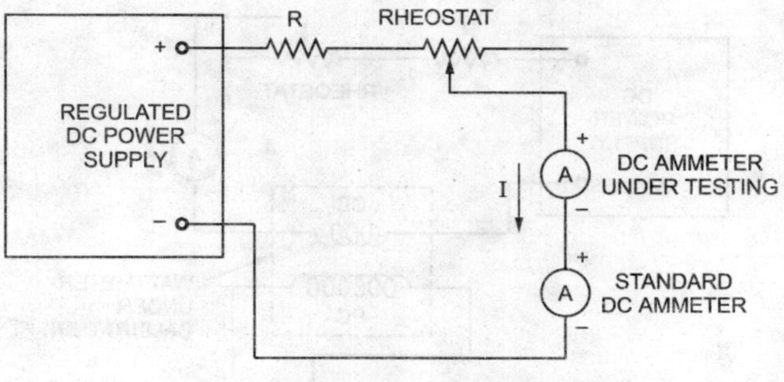

Fig. 10.3 *Calibration of DC Ammeter*

10.2.3. AC Instrument Calibration. AC voltmeters and ammeters can be tested exactly in the same way as the dc voltmeters and ammeters, except that an ac supply and ac standard instrument are used. Usually, the ac supply is derived from an auto-transformer connected to the 230 V, 50 Hz supply. The test setups are illustrated in Figs. 10.4(*a*) and 10.4(*b*).

For calibration of instruments at higher frequencies, the supply may be derived from an oscillator with a suitable output voltage level.

10.2.4. Ohmmeter Calibration. The ohmmeter is considered to be of moderate accuracy and low precision. It can be calibrated simply by measuring values of standard resistors with it and recording the reading obtained.

10.2.5. Wattmeter Calibration. As we know that a wattmeter is simply a combination of the ammeter and voltmeter. The test setup for calibration of dc wattmeter is depicted in Fig. 10.5. The current coil (CC) of wattmeter is connected, in series with a standard ammeter, rheostat and a current limiting resistor R, to one dc power supply. The potential coil (PC) of the wattmeter is connected across standard resistor R supplied from another DC power supply through rheostat. A standard voltmeter measures the voltage applied to the potential coil of the wattmeter. By measuring precise values of voltage and current for a given reading on the wattmeter, the exact power level is determined by the relation P = VI. Care must be taken to avoid the specified maximum voltage and maximum current in order to avoid any damage to the instrument.

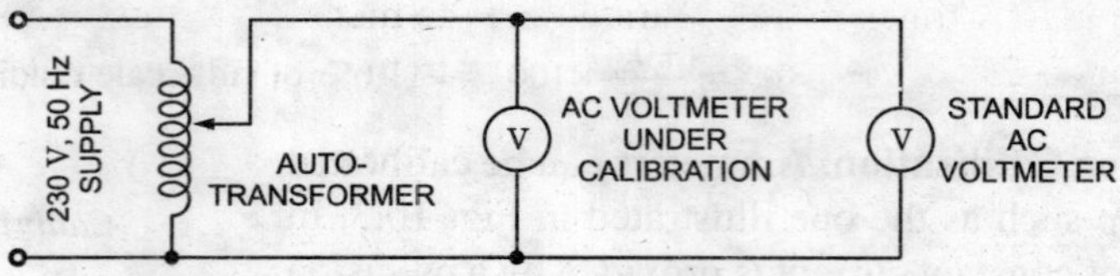

(a) Calibration of An AC Voltmeter

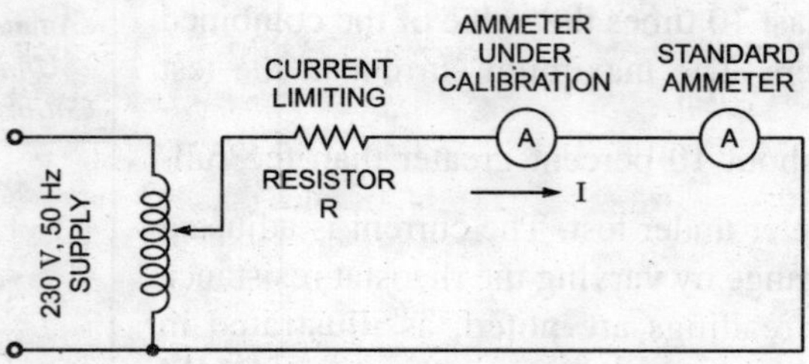

(b) Calibration of An AC Ammeter

Fig. 10.4

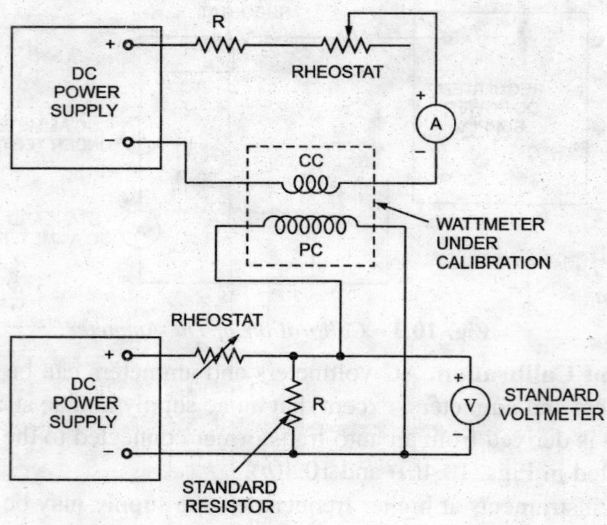

Fig. 10.5 *Calibration of a DC Wattmeter*

Example 10.2. An electrodynamic wattmeter being calibrated indicates full scale of 1,250 watts on its 250 V, 5 A range. The measured voltage and current by standard instruments are 4.8 A and 255 V, respectively. Determine the wattmeter error.

Solution: Power indicated, P_m = 1,250 watts

True value of power, P = V × I = 255 × 4.8 = 1,224 watts

Absolute error = P_m − P = 1,250 − 1,224 = + 26 watts **Ans.**

$$\text{Percentage error} = \frac{+26}{1,250} \times 100 = +2.08\%$$ **Ans.**

10.3 DIGITAL MULTIMETERS AS STANDARD INSTRUMENTS

The standard instrument used for calibration of another instrument by comparison method discussed in Art. 10.2 must be at least 4 times more accurate than the meter under test. The best accuracy of analog instruments is usually + 2% of full-scale while the accuracy of a typical hand-held DMM is ± (0.5% of reading + 1 division). Thus , for instance, a $3\frac{1}{2}$ digit DMM reading 100 V would have a maximum error of $\left(\frac{0.5\times100}{100}+0.1\right)$ or ± 0.6 V or ± 0.6%. So such a DMM can be used for calibration of an analog instrument, as shown in Fig. 10.6(*a*).

The accuracies of a high performance hand-held DMM and a bench type DMM are of the order of ± (0.1% of reading + 1 division) and ± (0.01% of reading + 1 division) respectively. Hence, a high performance hand-held DMM can be used as a standard instrument for calibration of either an analog instrument or a lower-accuracy DMM [Fig. 10.6(*b*)] and, a bench type digital instrument for calibration of any one of the three lower-accuracy instruments, as shown in Fig. 10.6(*c*).

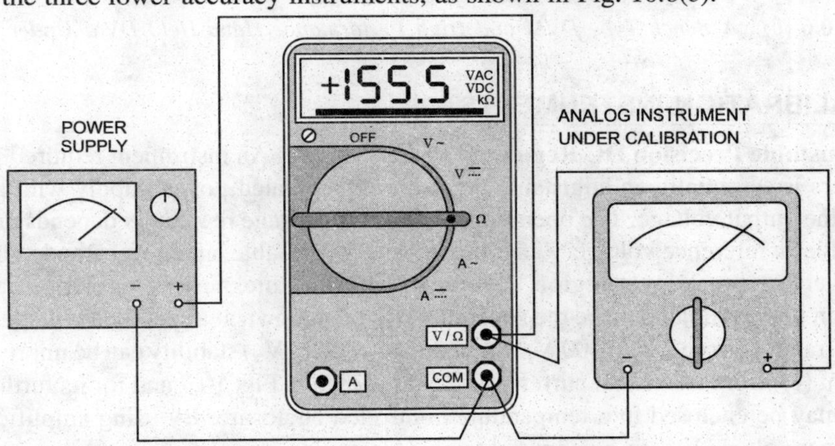

(a) An Hand Held Accurate DVM and an Analog Instrument Under Calibration

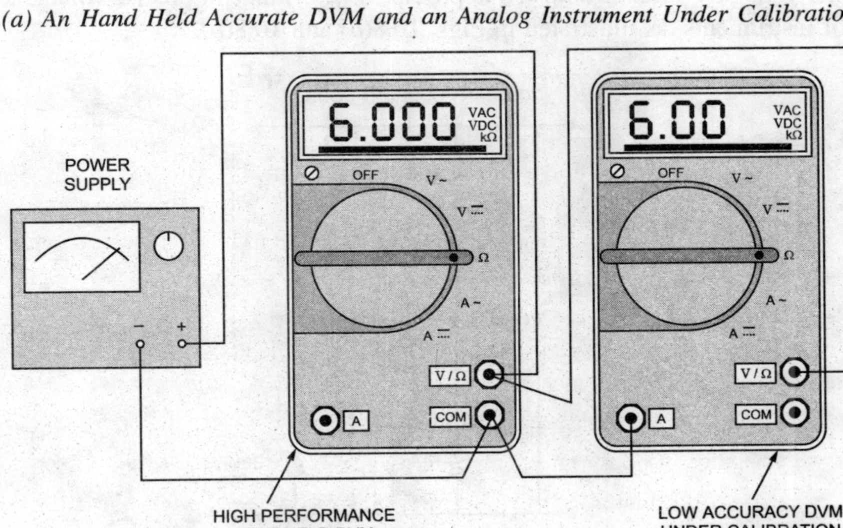

(b) An High Performance Hand-Held DVM and LOW Accuracy DVM Under Calibration

Fig. 10.6

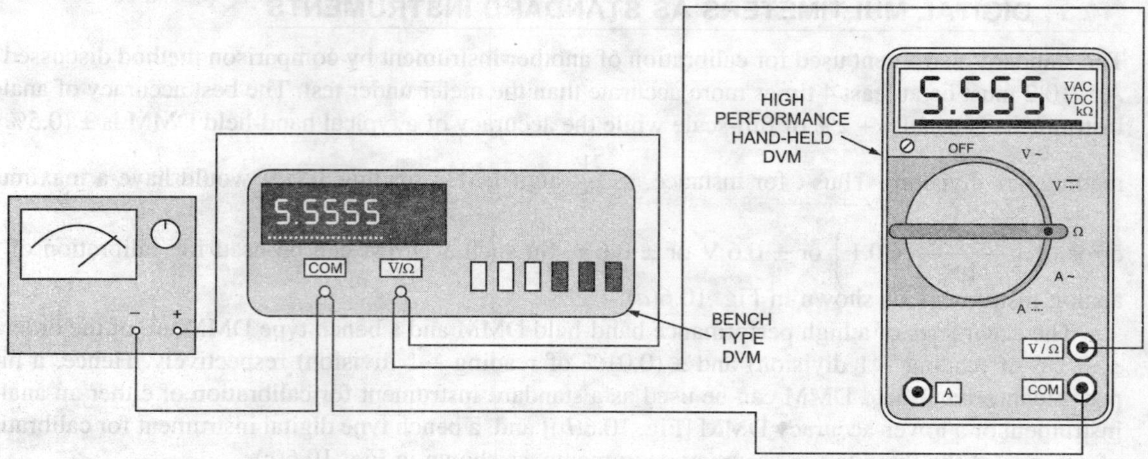

Fig. 10.6 *(c) A Bench Type DVM and High Performance Hand-Held DVM Under Calibration*

10.4 CALIBRATION INSTRUMENTS

10.4.1. Adjustable Precision DC Regulated Power Supply. An instrument required for calibration of dc voltmeters is essentially an adjustable precision dc regulated power supply with a high resolution display for the output voltage. The operation of all the dc voltage regulators depends on the zener diode which provides a reference voltage. Zener diode has a very stable anode-to-cathode voltage (V_Z) when operated over the breakdown region. By controlling the thickness of depletion layer and doping concentration it is possible to make the zener diode to breakdown at a specified voltage, called the zener voltage, V_Z (ranging from 2 V to 200 V). Breakdown voltage (V_Z) stability can be improved by supplying zener current (I_Z) from a constant current source, as shown in Fig. 10.7 and for its further improvement, the device may be enclosed in a temperature-controlled enclosure. By using amplifiers and precision resistors, zener voltage V_Z can be adjusted to provide a wide range of precise voltage level required for calibration of instruments, as illustrated in Figs. 10.8(*a*) and 10.8(*b*).

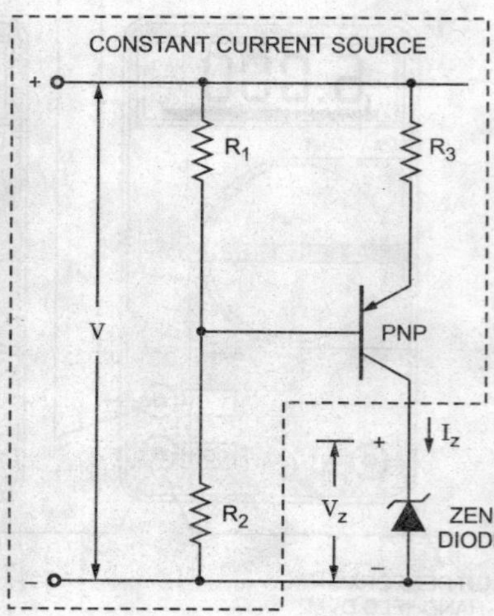

Fig. 10.7 *Zener Diode With a Constant Current Source*

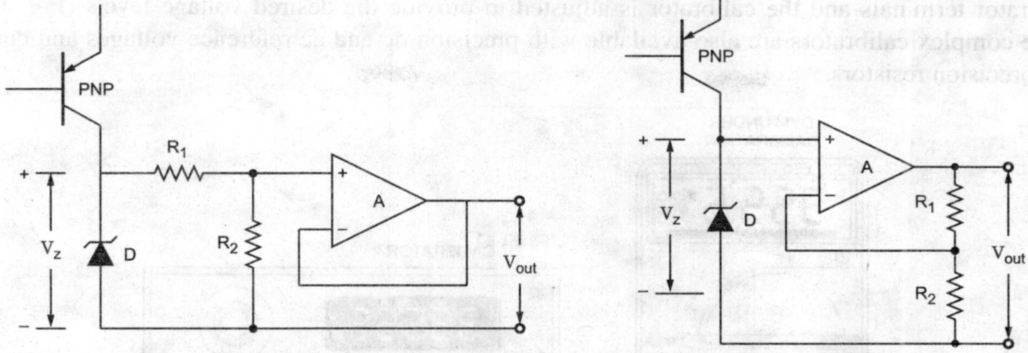

$$V_{out} = V_Z \frac{R_2}{R_1 + R_2}$$

(a)

$$\text{Output voltage } V_{out} = V_Z \frac{R_1 + R_2}{R_2}$$

(b)

Fig. 10.8

10.4.2. Voltage Calibrator. A voltage calibrator is essentially a precision voltage reference. The DATEL DVC 8500 may be used for calibration of voltmeters having accuracies of ± 0.03% or worse. The unit is a $4\frac{1}{2}$ digit voltage reference source with a full scale output range of + 19.999 Volts to – 19.999 Volts in 1 millivolt steps. An active buffered output amplifier provides very low output impedance and up to 25 milliamps output current at the rated accuracy of ± 25 ppm of setting, ± 500 μV. This short-circuit proof output is selected by panel digital lever switches. Voltage outputs may be continuously varied within ± 1.5 millivolts of selected readings by using the front panel vernier control. The DVC-8500A output is available from front panel banana jacks and a parallel near panel 36-pin gold-plated PC connector or lug terminals.

The worthnoting point is that the DVC-8500As are labeled DVC-8500 on the front panel and DVC-8500A on the rear panel. The letter designation describes the line voltage. With the exception of line voltage, the units are electrically identical.

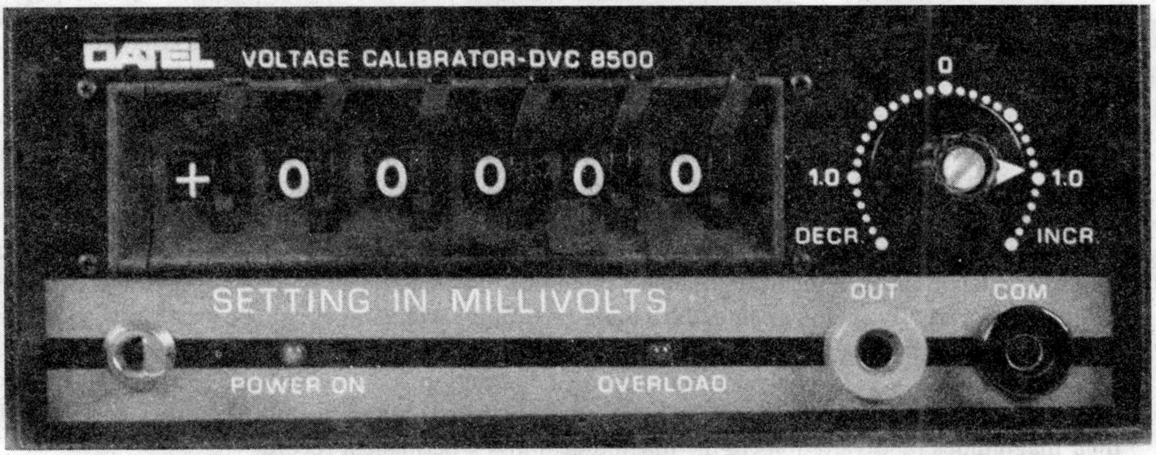

Fig. 10.9 *Voltage Calibrator [Courtesy DATEL, Inc.]*

For calibration of a dc voltmeter, the instrument under calibration is simply connected to the calibrator terminals and the calibrator is adjusted to provide the desired voltage levels (Fig. 10.10). More complex calibrators are also available with precision dc and ac reference voltages and currents, and precision resistors.

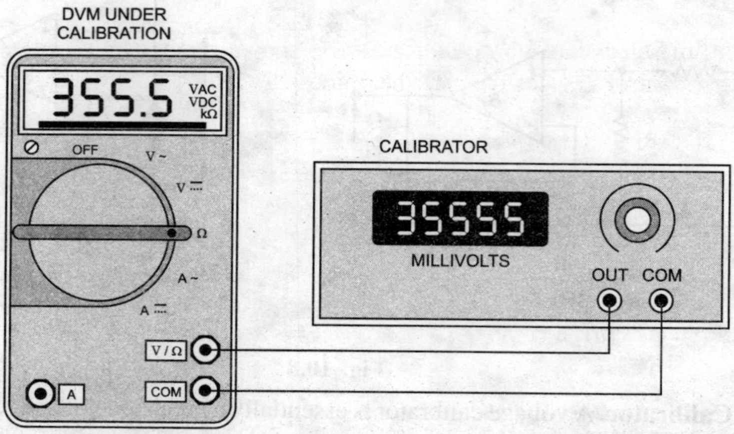

Fig. 10.10 *Voltmeter Calibration With DATEL DVC 8500*

10.5 RECORDERS

It is usually necessary that a permanent record of the state of phenomenon under investigation is maintained. In most of the industrial and research processes continuous monitoring of the condition, state, or value of the process variables (such as force, temperature, flow, humidity, voltage, current etc.) is necessary. All these needs can be met by electronic recording device, called the *recorder*.

A recorder is a measuring instrument that displays a time-varying signal in a form easy to examine and re-examine, perhaps long after the original signal has ceased to exist. In other words a recorder is a device whose function is to record the value of a quantity as it is being measured. Such devices may include graphic recording devices, computer printers, tape recorders or cathode ray tubes.

Recording preserves the experimental data in a managable and usable form. A recording system is very useful as (*i*) it preserves informations which could be obtained at any instant from indicating instruments and (*ii*) it gives informations about waveforms and transient behaviour or phase relationships in different parts of a circuit.

A recorder records electrical and non-electrical quantities as a function of time or relates two signals to each other. Electrical quantities, such as current, voltage etc., can be recorded directly while non-electrical quantities, such as pressure, temperature, speed, flow etc., are recorded indirectly by first converting them into the form of electrical signal with the help of sensors or transducers.

One of the important considerations in an instrumentation system is the method by which the data acquired is recorded. The recording method should be in consistence with the type of the system. For dealing with a wholly analog system, the *analog recording techniques* should be used while for dealing with a system having digital output, *digital recording devices* should be used.

All of these devices generally contain a built-in measuring apparatus as well as a recording mechanism. This gives recorders the capability of receiving electrical signals from detectors or sensors and converting their magnitudes (and variations in the magnitudes) into a permanent record.

The two basic laboratory chart recorders are the potentiometric strip-chart recorder and the X-Y recorder, which will be explained here.

10.6 POTENTIOMETRIC STRIP CHART RECORDER

The recorder of this type is basically a self-balancing potentiometer which apart from the indication is also capable of permanently recording the values of the voltages as they are being measured. This is accomplished by a marking device fitted on the pointer under which moves a motor driven chart for recording.

A block diagram of self-balancing potentiometer is shown in Fig. 10.11 which illustrates how a servo-motor and other control devices balance the potentiometer.

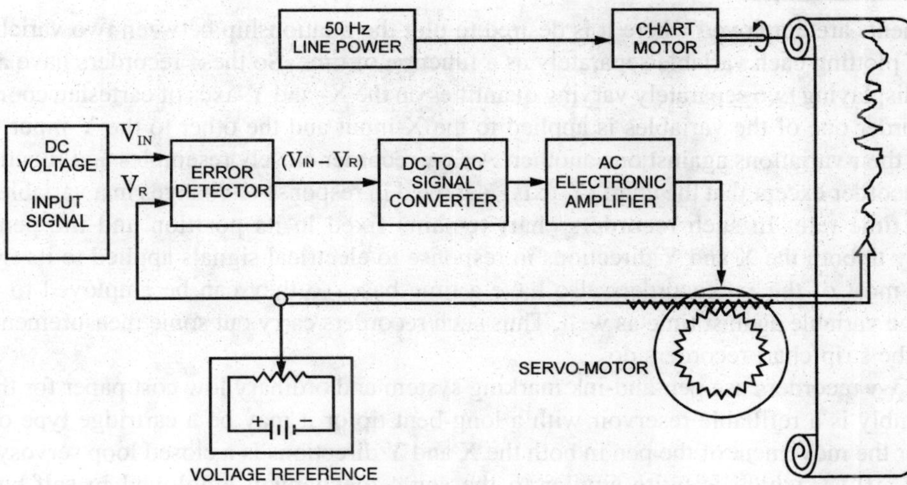

Fig. 10.11 *Block Diagram of Self-Balancing Potentiometer*

In this circuit, the voltage signal under measurement V_{IN}, is applied to one input terminal of a device, known as an *error detector,* and another adjustable reference voltage signal V_R is applied to the second input terminal of the error detector. Error detector compares the two signals and gives the result (i.e. the difference of two signals) at the output terminal.

This output signal from the error detector is called the *error signal* and it is employed to drive the servomotor. However, before the error signal is applied to the servomotor, it is first converted to an ac signal and then amplified up to a magnitude that is sufficiently large to activate the servomotor. This amplified error signal applied to the servomotor causes the motor shaft to rotate.

Servomotor shaft is attached mechanically to a potentiometer slider, which acts as a variable dc reference voltage source, V_R. As the shaft rotates it moves the slider of this potentiometer in such a direction that V_{IN} changes in value so that it approaches the value of input signal voltage, V_{IN} When the dc reference voltage, V_R, becomes equal to input signal voltage, V_{IN} the output of the error detector becomes zero and the servomotor stops rotating.

Servomotor shaft is also attached mechanically to a pointer which indicates the degree of rotation of the shaft. When the servomotor stops rotating, the reference voltage, V_R is equal to the input voltage, V_{IN}. Since the value of the reference voltage is known very accurately, the position of pointer controlled by motor shaft position can be calibrated to indicate the value of the input voltage, V_{IN} directly. If the pointer has a pen or any other type of marking device connected to it and a separate motor is employed to drive a ruled paper chart under the marking device, the movement of the pointer records permanently the voltage measured with respect to time on the chart.

Self-balancing potentiometer is employed for recording of input signal if it is in the form of a voltage. Such voltage quantities exist as the outputs of active transducers, such as thermocouples, piezoelectric crystals, photovoltaic cells etc.

Such recorders have a very high input impedance, infinity at balance conditions, and a high sensitivity of the order of 4 V/mm with an error of less than ± 0.25% with a bandwidth of 0.8 Hz. The most common application of these recorders is for recording and controlling of process temperatures. The chart for such a temperature recorder may be rectilinear or circular.

10.7 X-Y RECORDER

These instruments are employed where it is desired to plot the relationship between two variables, $y = f(x)$ instead of plotting each variable separately as a function of time. So these recorders have a special capability of displaying two separately varying quantities on the X- and Y-axes of cartesian coordinates. In such a recorder one of the variables is applied to the X-input and the other to the Y-input, and the recorder plots their variations against one another. An *x-y* recorder closely resembles and functions like a single pen recorder except that the chart (Y-axis) is moved in response to changes in a variable instead of a uniform time rate. In such recorders chart remains fixed in its position and the pen moves simultaneously in both the X and Y directions in response to electrical signals applied to its two input terminals. As most of the *x-y* recorders also have a time base, so they can be employed to plot the variation of one variable against time as well. Thus such recorders carry out some measurements in the same way as the strip-chart recorders do.

Normally *x-y* recorders use pen-and-ink marking system and ordinary low cost paper for the chart. The pen assembly is a refillable reservoir with a long-bent tip or it may be a cartridge type one. The mechanism for the movement of the pen in both the X and Y directions is a closed loop servosystem as shown in Fig. 10.12, which is quite similar to the servo-mechanism employed in self-balancing potentiometer. The rotation of the servo-motor moves the marking pen with the help of a string and pulley arrangement.

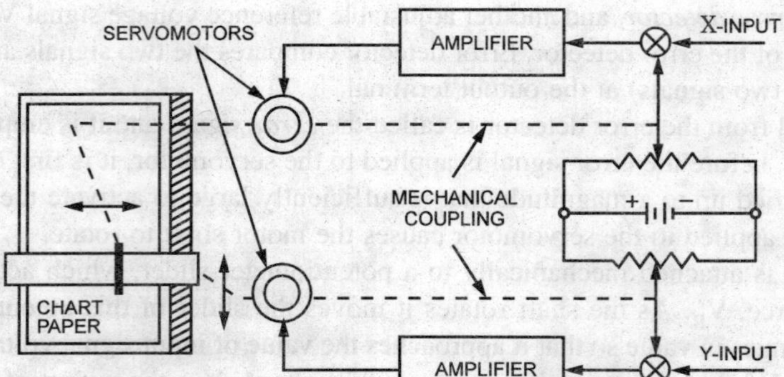

Fig. 10.12 *X-Y Recorder*

The advantages of *x-y* recorders are that they are rather economical to operate as ordinary low-cost paper can be used for the chart and they are easy and convenient to use. The disadvantages of *x-y* recorders are that in comparison to strip-chart recorders they are more expensive to buy, and cannot be employed for continuous recording. Because of electromechanical mechanism used for movement of pen the frequency response and speeds of *x-y* recorders are inherently limited.

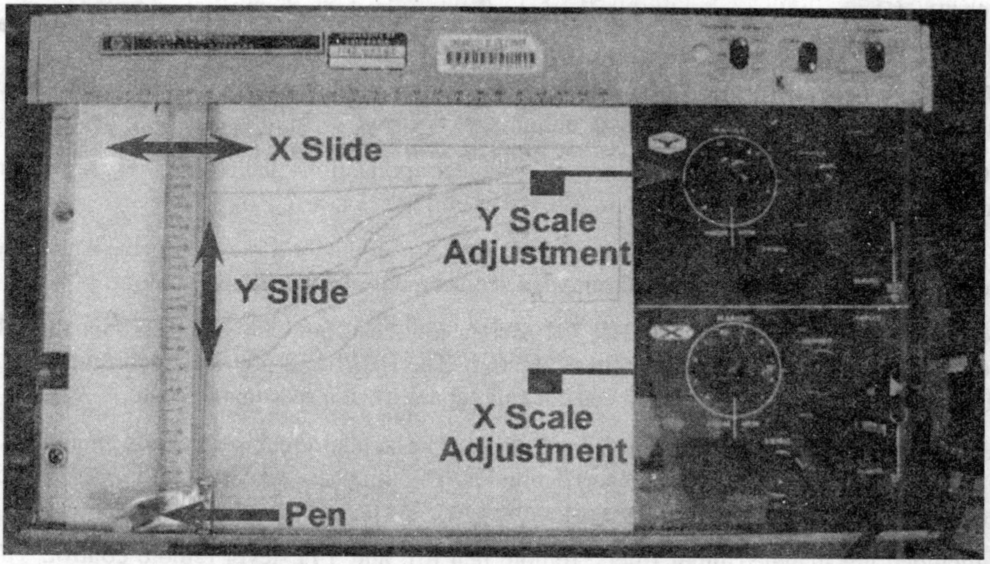

Fig. 10.13 *X-Y Recorder*

An *x-y* recorder may have the sensitivity of 10 μV/mm, slewing speed of 1.5 m/s, a frequency response of about 6 Hz in both the X and Y directions, and an accuracy of ± 0.2 per cent of full-scale. The paper size used in these recorders is 440 × 280 mm or 280 mm × 220 mm. Input impedance is from 100 kΩ to 50 MΩ or more depending upon the sensitivity range.

X-Y recorders have wide applications. These can be successfully employed for plotting current vs voltage curves of diodes and transistors, plotting B-H curves of magnetic materials and plotting speed-time curves for electric motors etc.

Low cost general purpose X-Y recorders, Time-Base Models 7015B, 7035B and 17018A manufactured by Hewlett-Packard are shown in Fig. 10.14.

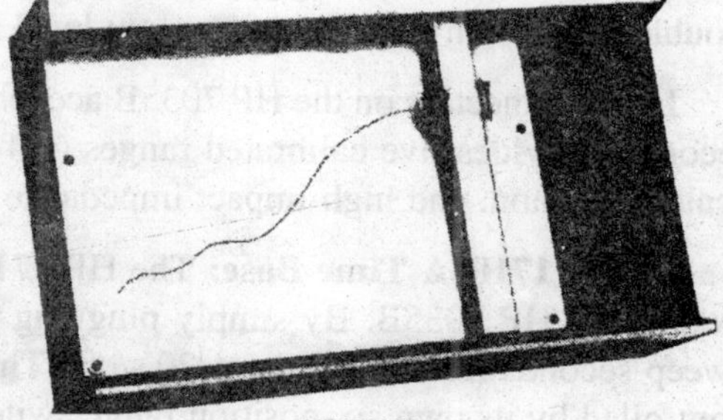

HP 7015B

HP 7035B

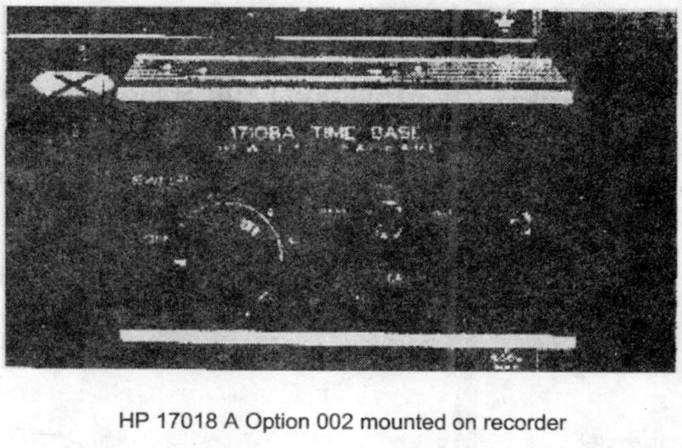

HP 17018 A Option 002 mounted on recorder

Fig. 10.14 *Low-Cost General Purpose X-Y Recorders [Courtesy Hewlett-Packard]*

HP 7015B: A low-cost recorder for applications that require high dynamic performance, medium sensitivity, and A4/A-size output.

HP 7035B: A general purpose laboratory recorder for applications that require low dynamic performance, medium sensitivity, and A-size output.

HP 7045B and 7046B: High performance recorders for applications that require very high dynamic performance, high sensitivity, and A4/A- or A3/B-size output.

All HP X-Y recorders have electrostatic hold-down for silent, trouble-free grip. And they use disposable pens for crisp, clean traces and one-step replacement of link, tip, and colour.

Model 7015B X-Y Recorder: The HP 7015B is a low-cost, one-pen X-Y recorder that allows charging on paper sizes up to ISO A/4 or 216 × 280 mm. The HP 7015B provides recording for a wide range of laboratory uses where there is a need for full capability at a reasonable cost.

The HP 7015B has a full complement of capabilities. The standard features include an internal time base with sweep selections from 5 seconds to 15 minutes. The time base provides automatic pen control and accepts remote triggering from sweep start and reset.

Also included are matched input filters, remote pen lift, and TTL-level remote control. The HP 7015B accepts TTL-level and low current (5 mA) contact closure for easy interface with external equipment.

Model 7035B X-Y Recorder: The HP 7035B is the ideal recorder for general lab use. Compact in design, the HP 7035B is well adapted to rack mounting with the addition of only two optional wing brackets. It features floating guarded inputs to help eliminate the common mode voltage effects that are troublesome when recording from low-level sources.

Input connectors on the HP 7035B accept both open wire and plug-type connectors. In addition, the recorder provides five calibrated ranges (0.4 mV/cm to 4 V/cm) for each axis; signal scaling for full-scale deflection, and high-impact impedance (1 megohm, except for the first two ranges).

Model 17108A Time Base: The HP 17108A is a self-contained time base that operates on either axis of the HP 7035B. By simply plugging in the HP 17108A, the HP 7035B is provided with five sweep seconds from 0.2 s/cm to 20 s/cm. This module, powered by a single self-contained battery, is controlled by its own six-position range switch and three-position mode switch.

HP 17108A Specifications

Sweep seconds: 0.2, 0.4, 2, 4, 20 s/cm

Accuracy: 5% of recorder full scale

Linearity: 0.5% of full scale (20ºC to 30ºC)

Output voltage: 0 to 1.5 V

Power: Replaceable mercury battery (100 h)

HP 7015B and 7035B Specifications

Performance Specifications

	HP 7015B	*HP 7035B*
Type of input	Floating binding posts or circuit board rear connector	Floating guarded signal pair; rear connector
Input ranges Input resistance	5, 50, 500 mV/cm 1 MΩ constant on all ranges	0.4, 4, 40, 400 mV/cm; 4 V/cm 100 kΩ on 4 mV/cm and potentiometric or essentially infinite at null on 0.4 mV/cm
Source resistance	10 kΩ maximum on all ranges	20 kΩ maximum on most sensitive range; no restrictions on other ranges
Accuracy	± 0.3% of full scale on most sensitive range (includes linearity and resettability) plus ± 0.2% of deflection when on other ranges; temp. coefficient ± 0.02% per °C	± 0.2% of full scale
Deadband	± 0.2% of full scale	
Common mode rejection	100 dB dc, 90 dB ac with 1 k unbalanced in HI terminal CMR decreases 20 dB per decade step in attenuation	130 dB dc, 100 dB ac with 1 kΩ between the negative input and guard connection CMR decreases 20 dB per decade step in attenuation
Normal mode rejection	Greater than 50 dB at 60 Hz (40 dB/decade rolloff above 60 Hz)	Greater than 30 dB at 60 Hz; then 18 dB/octave

Time Base Specifications

Time base	Six speeds: 0.1, 0.5, 1, 5, 10, 50 s/cm remote sweep start and reset via TTL level or contact closure	Optional external time base (HP 17108A)
Time Base Accuracy	1.5% ± 0.1 per °C	± 5% of recorder full scale

General Specifications

Power	Switch selectable for 100, 120, 220, 240 Vac + 5 – 10%; 47.5 to 440 Hz; 70 VA maximum	115 or 230 V ± 10%; 50 to 60 Hz; approx. 45 VA
Pen lift	Electric (remote via TTL level or contact closure	Electric pen lift capable of being remotely controlled
Writing area	18 × 25 cm	
Weight	Net 7.2 kg	Net 8 kg
Size	267 H × 432 W × 135 mm D	265 H × 445 W × 121 mm D

10.8 PLOTTERS

A plotter can be essentially the same as a potentiometric strip chart recorder except that it usually plots on a single sheet of paper like an X-Y recorder instead of a strip of paper. The sheet is rolled-fedback and forward as required, and the pen is moved horizontally to produce the drawing. Typically plotters are used for drafting technical drawings. A typical plotter is shown in Fig. 10.15.

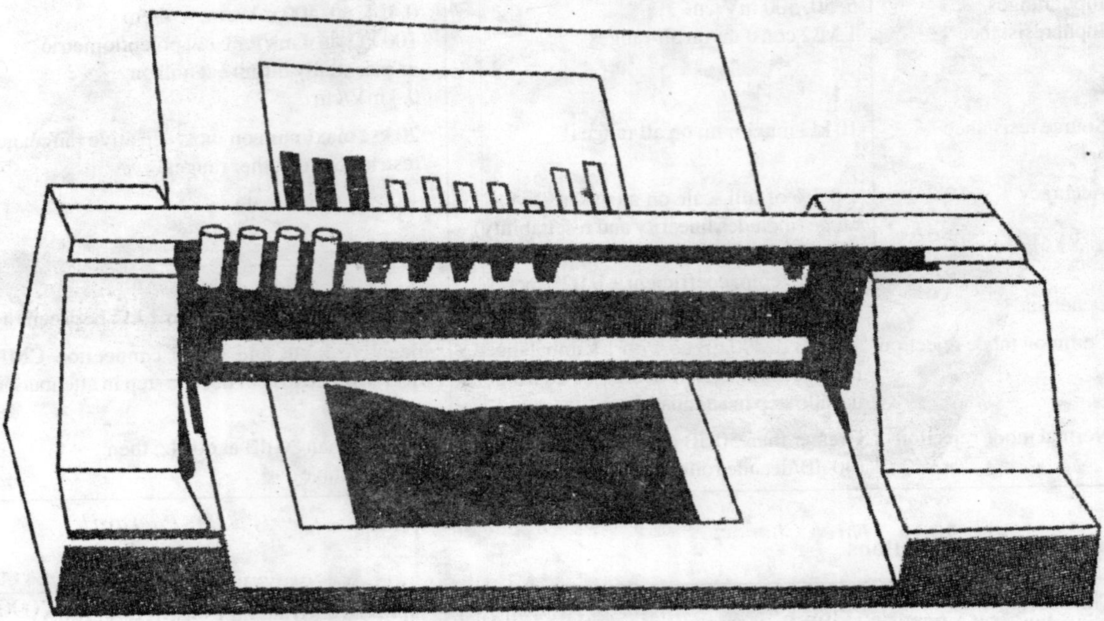

Fig. 10.15

The HP 7090A is designed for low-frequency (< 3 kHz) measurement, analysis, and documentation. The HP 7090A merges several technologies-waveform recording, digital plotting, analog recording, and automated measurement to provide a powerful solution to a broad range of measurement applications. It significantly increases the ability to measure and display low-frequency phenomena and substantially improves real-time recording and digital plotting . . . all in one low-cost system.

Signal Capture: Simultaneous sampling on each of three channels, 12-bit resolution, bandwidth of dc to 3 kHz (33.3 k samples/s maximum), and 1000 word memory per channel allow high resolution measurement, storage, and display.

Flexible Triggering: The HP 7090A has six trigger modes which allow virtually any signal change to initiate signal capture, even decaying repetitive signals such as faults in a power line voltage, or in a transducer's carrier. Combined with pre-trigger capability, these trigger modes make the HP 7090A ideal for turn-on/off characterization, fault monitoring and mechanical motion analysis.

A System Component: All panel functions are programmable via the HP-IB interface. Data can be transferred from the internal 1 k-buffers or streamed in real time from the analog-to-digital converters at up to 500 points/s. In addition, the menu-driven HP 17090B Measurement Graphics Software package is available for HP 9000 Series 200 computers (BASIC 3.0 only). The software allows easy data manipulation, storage and retrieval, and system integration.

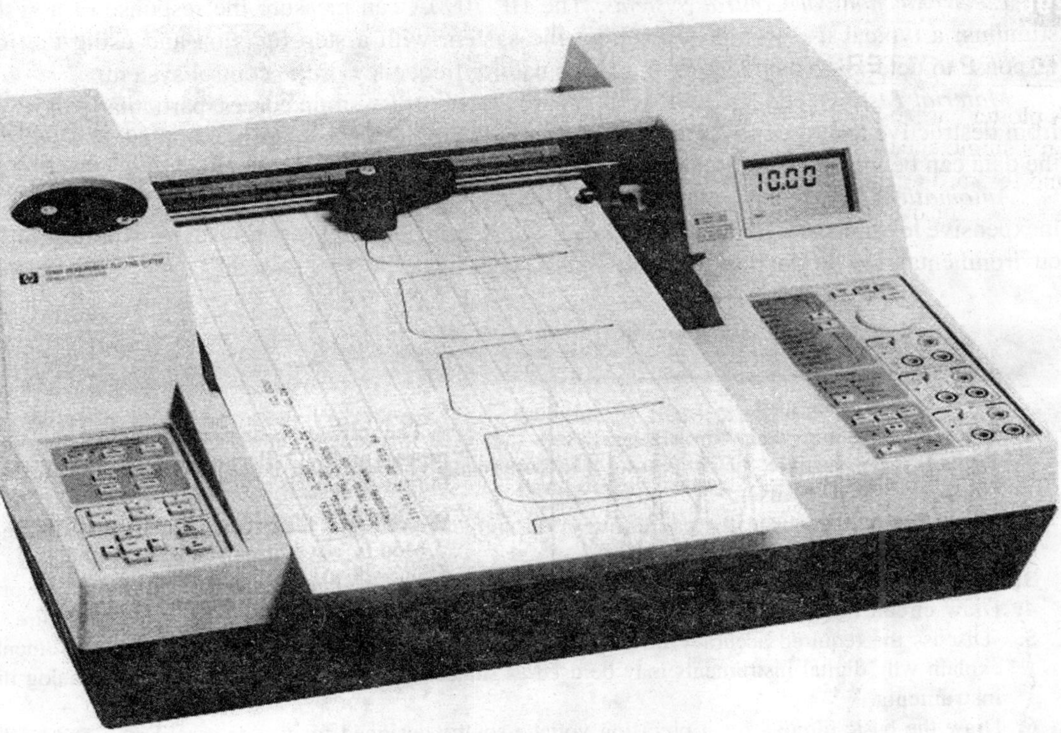

Fig. 10.16 *Three Channel Plotter With Digital Storage [Courtesy Hewlett-Packard]*

Versatile Capabilities: Uniquely, the HP 7090A is also a high performance analog/digital plotter. It is ideal for a graphics dump from a smart instrument or as part of an HP-IB system; The HP 7090 A can also be used to take an X-Y dump from an analog instrument. With the HP 7090A, hand annotation is unnecessary. The 7090A annotates setup conditions, date and time, selected data points from memory, and trigger information. It draws user-defined axes and grids, eliminating the need for pre-printed graph paper. The HP 7090A even lets you plot overhead transparencies for technical presentations using your current spread-sheet/graphics software.

Applications

Capturing Low Frequency Electrical Transients: General diagnostic monitoring (such as looking for relative timing sequences) and fault monitoring (capturing pre-trigger data for intermittent failure analysis) are natural applications for the HP 7090A.

Measuring Phase Relationships: The simultaneous sampling on all channels is ideal for measuring current/voltage phase relationships in power systems.

Analog Instrument/Digital System Link: The HP 7090A can integrate an analog instrument into an HP-IB system; the HP 7090A, with a controller can digitize output voltages from analog instruments for HP-IB system data entry.

Mechanical and Electromechanical Testing: Applications in which transducers convert velocity, acceleration, force, temperature or torque to voltage are a good fit for the HP 7090A. These applications have a maximum output frequency below 3 kHz. The HP 7090A's flexible trigger capabilities make it useful for one-shot electromechanical events such as clutch and mechanism engagements.

Electromechanical Control Systems: The HP 7090A can measure the response of a system to a stimulus; a typical use would be exciting the system with a step function and using the measured response to determine damping ratio and the natural frequency of the control system.

Material Testing: The HP 7090A can record classic stress-strain curves, particularly those obtained from destructive testing. The data is stored in a buffer, so even though the sample has been destroyed, the data can be viewed and rescaled in several different ways.

Automatic Test: When linked to an HP 9000 Series 200 computer, the HP 7090A is a good, inexpensive learning tool for small companies considering automatic test systems. Applications include environmental and production line testing and proof of performance records.

EXERCISES

1. What is calibration and why is it necessary that each instrument should be calibrated periodically. Explain the procedure for calibrating a dc voltmeter by comparing it with a standard voltmeter. Draw the circuit diagram also.
2. Draw a circuit diagram for calibrating a dc ammeter by comparing it with a standard ammeter and explain the calibration procedure.
3. Draw circuit diagrams for calibration of ac instruments (voltmeter and ammeter) and explain the procedure.
4. Draw circuit diagram for calibration of a dc wattmeter and explain the calibrating procedure.
5. Discuss the required accuracy for standard instruments used for calibration of other instruments. Also, explain why digital instruments may be used as standard instruments for calibration of analog indicating instruments.
6. Draw the basic circuits for a precision voltage source designed for use as a calibrator and explain their operation. Also discuss the necessity of a zener diode in such circuits.
7. What is recorder and what is the necessity of the recorder ?
   Explain the potentiometric strip-chart recorder with the help of its block diagram.
8. What is an X-Y recorder ? How is different from X-*t* or Y-*t* recorders ? Describe its operation with suitable diagram. Give its merits, demerits and applications.
9. Explain with the help of a functional block diagram, the principle of working of an X-Y recorder. **[U.P.S.C. I.E.S. Electrical Engineering-I, 1992]**
10. Write short notes on the following:
    (*i*) Calibration of a dc wattmeter by comparison method. (*ii*) Adjustable precision dc regulated power supply.
    (*iii*) Voltage calibrator. (*iv*) X-Y recorder.
    (*v*) Plotter.

SHORT ANSWER TYPE QUESTIONS WITH ANSWERS

Q. 1. Give the test setup for calibration of a dc voltmeter with the help of a standard voltmeter and a precision dc voltage source.

Ans. The required test setup is given in Fig. 1.

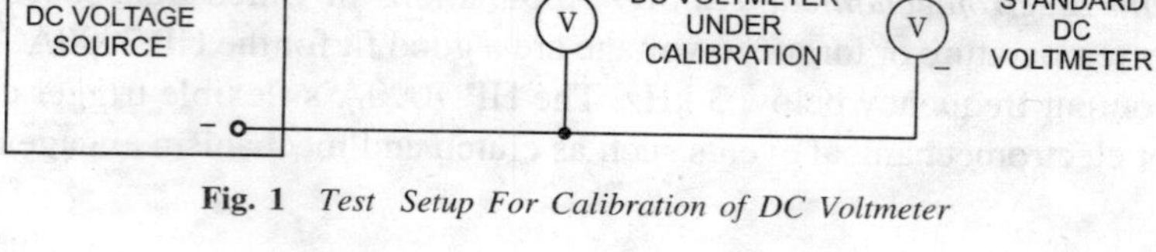

Fig. 1 *Test Setup For Calibration of DC Voltmeter*

Q. 2. Give the test setup for calibration of a dc ammeter with the help of a standard ammeter and a precision dc voltage source.

Ans. The required test setup is given in Fig. 2.

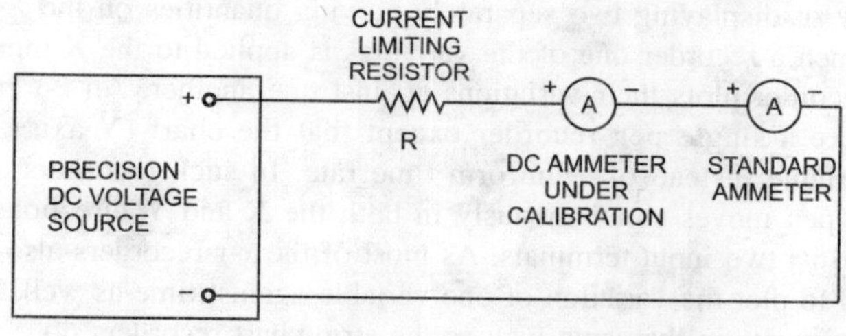

Fig. 2 *Test Setup For Calibration of DC Ammeter*

Q. 3. Give the test setup for calibration of a dc wattmeter with the help of a standard wattmeter and a precision dc voltage source.

Ans. The required test setup is given in Fig. 3.

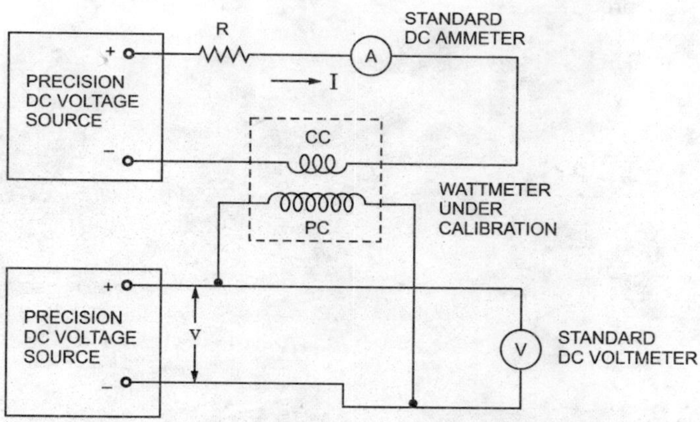

Fig. 3 *Test Setup For Calibration of DC Wattmeter*

Q. 4. What is a precision dc voltage source ?

Ans. A precision dc voltage source is essentially an adjustable precision dc regulated power supply which is capable to provide a desired precise and reliable level of voltage.

Q. 5. What is a recorder ?

Ans. A recorder is an instrument that displays a time-varying signal in a form of easy to re-examine, perhaps long after the original signal has ceased to exist.

Q. 6. What is the necessity of the recorders ? **[Pb. Technical Univ. December 2004]**

Ans. Recorders are required to keep a permanent record of the state of phenomenon under investigation and for continuous monitoring of the condition, state, or value of process variables such as force, pressure, temperature, flow, voltage, current etc.

Q. 7. What is the main recording requirement ?

Ans. The main requirement of the recording is that the recording method should be in consistence with the type of system. For dealing with a wholly analog system, the analog recording technique should be used while for dealing with a system having digital output, digital recording device should be used.

Q. 8. What are strip-chart recorders ? **[PTU December 2005]**

Ans. Strip chart recorders are recording instruments. It records one or more variables with respect to time. So it is also known as the *x-t* recorder.

Q. 9. What is *x-y* recorder and how is it different from *x-t* to *y-t* recorder ?

Ans. The *x-y* recoders are employed where it is desired to plot the relationship between two variables, $y = f(x)$ instead of plotting each variable separately as a function of time. So these recorders have a special capability of displaying two separately varying quantities on the X- and Y-axes of cartesian coordinates. In such a recorder one of the variables is applied to the X-input and the other to the Y-input, and the recorder plots their variations against one another. An *x-y* recorder closely resembles and functions like a single pen recorder except that the chart (Y-axis) is moved in response to changes in a variable instead of a uniform time rate. In such recorders chart remains fixed in its position and the pen moves simultaneously in both the X and Y directions in response to electrical signals applied to its two input terminals. As most of the *x-y* recorders also have a time base, so they can be employed to plot the variation of one variable against time as well. Thus such recorders carry out some measurements in the same way as the strip-chart recorders do.

❑❑❑

MODEL TEST PAPER I

ELECTRONIC INSTRUMENTATION AND MEASUREMENTS

Time : 3 Hours *Total Marks : 100*

Note: (*i*) Attempt all questions.
(*ii*) All questions carry equal marks.
(*iii*) Be precise in your answer.

1. Attempt any *Two* of the following:

(*a*) What is necessity of units in measurements ? What are three basic SI mechanical units. Define "fundamental units" and "derived units". Give examples of each.

(*b*) Explain the different types of errors that may occur in measurements.

A certain resistor has a voltage drop of 110.2 V and a current of 5.3 A. The uncertainties in the measurements are ± 0.2V and ± 0.6 A respectively. Calculate the power dissipated in the resistor and the uncertainty in power.

(*c*) Describe with the help of a labelled diagram, the constructional details of a PMMC meter and explain how 'control' and 'damping' torques are obtained.

2. Attempt any *Two* of the following:

(*a*) Draw and explain the circuit of an op-amp noninverting amplifier voltmeter.

An emitter-follower voltmeter circuit shown in Fig. 4.7 has emitter resistances of 3.3 kΩ each and $V_{CC} = \pm 12$ V

(*i*) Determine emitter currents when input voltage is zero volt.

(*ii*) Compute the meter circuit voltage when input dc voltage is (*i*) 2 V and (*ii*) 1 V.

(*b*) What are the different methods of digital measurement of voltage and current ? Explain one method with neat diagram.

(*c*) Write short note on multimeter probes.

3. Attempt any *Two* of the following:

(*a*) Explain the precision measurement of medium resistance with Wheatstone bridge. Also discuss the sensitivity of the bridge. Show when sensitivity becomes maximum and minimum.

(*b*) Describe how an unknown capacitance can be measured with the help of De Sauty's bridge. What are the limitations of this bridge and how are they overcome by using modified De Sauty's bridge ? Draw phasor diagram to illustrate your answer.

(*c*) Describe the working of Hay's bridge. It is said that this bridge is suitable for measuring high-Q inductors. Give reasons fore such a statement.

Four arms of Hay's ac bridge are arranged as follows:

AB-Coil of unknown impedance; BC-Non-reactive resistance of 1,000 Ω; CD-Non-reactive resistance of 833 Ω in series with a standard capacitor of 0.38 μF; DA is a non-reactive resistance of 16,800 Ω. If the supply frequency is 50 Hz, determine the inductance and resistance at the balance condition.

4. Attempt any Two of the following:

(*a*) What is the function of a time base circuit in a CRO ?

Draw the basic circuit of an oscilloscope time base generator and explain its operation.

(*b*) What is probe ? State the function and explain the working of 10 : 1 probe for a CRO.

Compute the minimum time division sensitivity for an oscilloscope that is to be used to investigate a 20 MHz waveform. Assume that the time base magnifier expands the horizontal display by a factor of 10.

(*c*) What is delayed sweep feature in an oscilloscope ? Explain how the portion of the waveform to be investigated is intensified by the delayed time base system.

5. Attempt any *Two* of the following:

(*a*) Draw a circuit diagram for calibrating a dc ammeter by comparing it with a standard ammeter and explain the calibration procedure.

(*b*) Draw the basic circuits for a precision voltage source designed for use as a calibrator and explain their operation. Also discuss the necessity of a zener diode for such circuits.

(*c*) What is an X-Y recorder ? How is it different front X-*t* or Y-*t* recorders ? Describe its operation with suitable diagram. Give its merits, demerits and applications.

❑❑❑

MODEL TEST PAPER II

ELECTRONIC INSTRUMENTATION AND MEASUREMENTS

Time : 3 Hours *Total Marks : 100*

Note: (*i*) Attempt all questions
(*ii*) All questions carry equal marks
(*iii*) Be precise in your answer.

1. Attempt any *Two* of the following:

(*a*) How does Engineering notation differ from Scientific notation and what is the advantage of engineering notation ?

What are the two SI temperature scales ? Explain them.

(*b*) Differentiate clearly between systematic and random errors in measurements.

A moving coil voltmeter has a uniform scale with 100 divisions and gives full-scale reading of 200 V. The instrument can read up to $\frac{1}{5}$th of a scale division with a fair degree of certainty. Determine the resolution of the instrument in volt.

(*c*) Give the circuit diagram of a multirange voltmeter and explain.

Explain voltmeter sensitivity, voltmeter loading effect and voltmeter swamping resistance.

2. Attempt any *Two* of the following:

(*a*) Draw and explain the operation of circuits of half-wave rectifier ac electronic voltmeters using (*i*) a voltage follower and (*ii*) precision rectifier.

(*b*) Draw the circuit of a digital frequency meter. Explain the difference between $3\frac{1}{2}$ digit and 4 digit displays.

(*c*) Explain with block diagram how a DMM can measure dc and ac signals, and various electrical parameters.

3. Attempt any *Two* of the following:

(*a*) Show how a voltmeter and ammeter can be used for measurement of resistance. Show the two possible connections and write the resistance equation for each. Discuss the errors involved.

(*b*) Explain Maxwell's bridge for measurement of unknown inductance. Determine condition for balance. Mention its applications.

The standard capacitance C_2 used in circuit shown in Fig. 7.3 is 0.2 μF, and R_2/R_1 can be set to any ratio between 10 : 1 and 1 : 10. Determine the range of measurements of unknown capacitance C_1.

(*c*) Describe the circuit and working of a Q meter. A coil was tested using a Q meter and following information was obtained.

At a frequency of 3 MHz, the value of capacitance to give maximum voltage across the capacitor was 251 pF while at 6 MHz frequency it was 50 pF. Calculate the self capacitance of the coil.

4. Attempt any *two* of the following:

(*a*) Explain the operation of the single-beam dual trace CRO. Also explain the use of this CRO in chop mode and in alternate mode. Sketch waveforms for illustration of the two modes of operation.

(*b*) Explain the following terms in reference to CRO.

(*i*) Sensitivity (*ii*) Accuracy (*iii*) Frequency response and (*iv*) Time base.

The sweep generator circuit shown in Fig. 8.19 has $R_{E1} = 5.6\ k\Omega$, $C_1 = 0.2\ \mu F$, $V_{B1} = 6.3\ V$, and the trigger levels for the Schmitt trigger are ± 2.5 V. Determine the peak-to-peak amplitude and time period of the ramp waveform.

(*c*) For a digital storage oscilloscope discuss the following:

(*i*) Multichannel display, (*ii*) Autoset (*iii*) Waveform processing (*iv*) Glitches and runt catching.

5. Attempt any *Two* of the following:

(*a*) Draw circuit diagram for calibration of a dc wattmeter and explain the calibration procedure.

(*b*) Discuss the required accuracy for standard instruments used for calibration of other instruments. Also, explain why digital instruments may be used as standard instruments for calibration of analog indicating instruments.

(*c*) Write a short note on 'plotter' or 'voltage calibrator'.

❑❑❑

B.TECH.

(SEM IV) EVEN SEMESTER THEORY EXAMINATION, 2009-1010

ELECTRONIC INSTRUMENTATION AND MEASUREMENTS

Time : 3 Hours *Total Marks : 100*

Note: Attempt all questions

1. Attempt any *four* parts of the following:

(*a*) What do you mean by term "Accuracy" in instruments? Differentiate it with term "Precision".

(*b*) The expected value of the voltage to be measured is 150 V and the measured value is 148 V. Calculate:

(*i*) Relative accuracy, (*ii*) Absolute error.

(*c*) Explain various causes for instrumental errors.

(*d*) Write merits and demerits of PMMC instruments.

(*e*) Explain the working of basic DC Ammeter with suitable diagram.

(*f*) Explain the operation of series OHM Meter with suitable diagram.

2. Attempt any *four* parts of the following:

(*a*) Explain the term "loading" in voltmeter and give the method to remove the adverse effect of loading.

(*b*) Explain digital voltmeter with suitable diagram. Also write its merits and demerits in comparison to analog voltmeter.

(*c*) Explain various specifications of digital multimeter (DMM) which are important while selecting for any application.

(*d*) Draw the block diagram of dual slope type Digital voltmeter and explain its working.

(*e*) Compare various techniques used in Digital voltmeter in tabular format with parameters such as circuit complexity stability, accuracy, noise effect, and operating speed.

(*f*) A $3\frac{1}{2}$ digit DVM has accuracy specification of ±5% of the reading ±1 digit. What is error in volts when reading is 5.00 V on its 10 V range?

3. Attempt any *two* parts of the following:

(*a*) Explain Wheatstone bridge and derive the expression for bridge sensitivity.

(*b*) Explain practical Q-meter with suitable diagram. Also mention various sources of errors in Q-meter.

(*c*) Explain the operation of Capacitance Bridge in general with suitable neat diagram.

4. Attempt any *four* parts of the following:

(*a*) Explain the three different modes of operation of Digital Storage Oscilloscope (DSO).

(*b*) In an experiment, the voltage across 10 kΩ resistor is applied to CRO. On the screen, the signal appears with total vertical and horizontal occupancy of 3 cm and 2 cm respectively. The front panel controls of V/div and Time/div are 2 V/div and 2 ms/div respectively. Calculate the RMS value of voltage across resistor and its frequency.

(*c*) Explain the working of sampling oscilloscope with suitable diagram.
(*d*) Define active and passive CRO probes along with their comparison in tabular format.
(*e*) Discuss the loading and measurement effects on CRO probes.
(*f*) Discuss the specification of CRO and probes for any particular laboratory application.

5. Attempt any *two* parts of the following:
(*a*) Write short notes on X-Y recorders.
(*b*) What is the importance of calibration in instrumentation? Also mention approximate duration of regular calibration of primary and secondary instruments along with reasons.
(*c*) Write short notes on various types of plotters.

ANSWERS/SOLUTION

1. (*a*) Refer to Art 2.4.
(*b*) True value, A = 150 V
Measured value, $A_m = 148$ V

(*ii*) Absolute error, $\varepsilon_0 = A_m - A = 148 - 150 = -2$ V **Ans.**

(*i*) Relative error, $\varepsilon_r = \frac{\varepsilon_0}{150} = \frac{-2}{150} = -0.0133$ or -1.33% **Ans.**

(*c*) Refer to Art 2.6(2) (*a*) page 31.
(*d*) Refer to Art 3.2.5 page 56-57.
(*e*) Refer to Art 3.4.
(*f*) Refer to Art 3.6.1.

2. (*a*) Refer to Art 3.5.4.
(*b*) Refer to Arts 5.1 and 5.7.
(*c*) Refer to Art 5.10.
(*d*) Refer to Art 5.7(2).
(*e*) Refer to Art 5.7.
(*f*) Number of full digits on display, $n = 3$

So resolution, $R = \frac{1}{10^n} = \frac{1}{10^3} = 0.001$

i.e. the meter cannot distinguish between values that differ from each other by less than 0.001 of full-scale.

For full-scale range of 10 V, the resolution is $10 \times 0.001 = 0.01$ V

The display for 5 V reading on 10 V scale of $3\frac{1}{2}$ digit meter would be 05.00.

The digit in the least significant digit has a value of full-scale × R i.e. $10 \times \frac{1}{10^3} = 0.01$ V

Total possible error $= \pm \frac{5}{100} \times$ reading ± value of digit in LSD $= \pm \frac{5}{100} \times 5 \pm 0.01 = \pm 0.26$ V **Ans.**

3. (*a*) Refer to Art 6.3.
   (*b*) Refer to Arts 7.5 and 7.5.3.
   (*c*) Refer to Art 7.3.

4. (*a*) Refer to Arts 9.4 and 9.5.
   (*b*) Peak-to-peak value of voltage across resistor,

$$V_{peak\text{-}to\text{-}peak} = \text{Vertical occupancy} \times \frac{V}{div} = 2 \times 3 = 6 \text{ V, assuming 1 div} = 1 \text{ cm}$$

RMS value of voltage across resistor, $V_{rms} = \dfrac{V_{peak\text{-}to\text{-}peak}}{2\sqrt{3}} = \dfrac{6}{2\sqrt{2}} = 2.12$ V **Ans.**

Time period of voltage waveform, $T = 2 \times 2 \text{ ms} = 4 \text{ ms} = 4 \times 10^{-3}$ s, assuming 1 div = 1 cm

Frequency of voltage waveform, $f = \dfrac{1}{T} = \dfrac{1}{4 \times 10^{-3}} = 250$ Hz **Ans.**

   (*c*) Refer to Art 9.3.
   (*d*) Refer to Art 8.9.
   (*e*) Refer to Art 8.9.
   (*f*) Refer to Art 8.10.

5. (*a*) Refer to Art 10.7.
   (*b*) Refer to Art 10.1.
   (*c*) Refer to Art 10.8.

❑❑❑

B.TECH.

(SEM IV) EVEN SEMESTER THEORY EXAMINATION, 2010-11

ELECTRONIC INSTRUMENTATION AND MEASUREMENTS

Time : 3 Hours *Total Marks : 100*

Note: Attempt **all** questions. All questions carry equal marks.

1. Attempt any *four* parts of the following: (5 × 4 = 20)

(*a*) Explain the terms accuracy, sensitivity and resolution as used for indicating instruments.

(*b*) Explain different types of errors that may occur in measurements.

(*c*) Describe the principle of operation and application of PMMC instruments.

(*d*) Describe the principle of operation and use of galvanometer.

(*e*) An ammeter reads 6.7 A and the true value of current is 6.5 A. Determine the error and the correction for this instrument.

(*f*) What is meant by arithmetic mean, average deviation and standard deviation ?

2. Attempt any *four* parts of the following: (5 × 4 = 20)

(*a*) Describe the working of binary weighted resistance D/A converter.

(*b*) Explain the difference between 3½ digit and 4 digit displays.

(*c*) Explain the working of integrating type DVM.

(*d*) Describe the principle of operation of electronic multimeters.

(*e*) Explain the principle of operation of electronic voltmeter.

(*f*) Explain how an electronic analog voltmeter can be used to measure alternating current.

3. Attempt any *two* parts of the following: (10 × 2 = 20)

(*a*) Define the Q-factor of a coil. Explain with a circuit diagram the construction and principle of operation of a basic Q-meter?

(*b*) State various methods of measurement of low resistance.
Why ammeter-voltmeter methods not suitable for the precise measurement of low resistance?

(*c*) Describe the operation of Kelvin double bridge and Schering bridge.

4. Attempt any *two* parts of the following: (10 × 2 = 20)

(*a*) Explain with neat sketches, the working principle of CRO.

(*b*) Draw the block diagram of D.S.O.P. and explain the function of each block.

(*c*) Describe CRO Probe and Sampling Oscilloscope.

5. Write short notes on any *two* of the following: (10 × 2 = 20)

(*a*) Strip chart recorder.

(*b*) Plotter.

(*c*) X-Y recorder.

ANSWERS/SOLUTION

1. (*a*) Refer to Arts 2.3 and 2.4.

(*b*) Refer to Art 2.6.

(*c*) Refer to Arts 3.1 and 3.2.1.

(*d*) Refer to Art 3.3.

(*e*) Measured value, $A_m = 6.7$ A

True value, A = 6.5 A

Absolute error or error, $\delta A = A_m - A = 6.7 - 6.5 = 0.2$ A **Ans.**

Correction = $-\delta A = -0.2$ A **Ans.**

(*f*) Refer to Art 2.8.

2. (*a*) Refer to Art 5.3.2.

(*b*) Refer to Art 5.9.

(*c*) Refer to Art 5.7(2).

(*d*) Refer to Art 4.6.

(*e*) Refer to Art 4.4.

(*f*) Refer to Art 4.5.

3. (*a*) Refer to Art 7.5.

(*b*) The methods used for measurement of low resistance are:

1. Ammeter-voltmeter method.
2. Potentiometer method.
3. Kelvin double bridge method.
4. Ohmmeter method.

Also, refer to Art 6.4.1.

(*c*) Refer to Arts 6.4.2 and 7.3.4.

4. (*a*) Refer to Arts 8.1, 8.2 and 8.5.

(*b*) Refer to Art 9.4.

(*c*) Refer to Arts 8.9 and 9.3.

5. (*a*) Refer to Art 10.6.

(*b*) Refer to Art 10.8.

(*c*) Refer to Art 10.7.

❑❑❑

INDEX

A

Absolute error, relative error and limiting or guarantee errors 22
AC bridge methods 176
AC bridge sensitivity 178
AC bridge theory 176
AC electronic voltmeters 93
  average reading 93
  peak reading 95
AC/DC current probes 105
Accuracy and precision 27
Active probes 259
ADC resolution and accuracy 123
Adjustable precision dc regulated power supply 288
Alternative method of determination of phase 253
Ammeter shunts 64
  advantages 65
  precautions 65
  theory 64
Analog electronic multimeter 101
Analog-to-digital (A/D) conversion 118
Anderson bridge 198
Astigmatism 238
Average reading ac voltmeters 93

B

Balance adjustment of ac bridges 178
Balanced bridge triode VTVM 87
Balanced bridge TVM 88
Balanced bridge voltmeters 87
Basic controls of CRO 235
Binary ladder or R-2R ladder DAC 115
Built-up Wheatstone bridges 163

C

Calibration circuit 238
Calibration instruments 288
Capacitance bridges 180
  De Sauty's bridge (or simple capacitance bridge) 180
  high voltage Schering bridge 188
  measurement of relative permittivity with Schering bridge 189
  parallel resistance-capacitance bridge 183
  Schering bridges 184
  series resistance-capacitance or modified De Sauty bridge 181
  Wien bridge 190
Cathode ray tube 225
Characteristics of digital meters 133
Characteristics of experimental data 40
Chopper type dc voltmeter 84
Comparison between CRT and TV picture tubes 228
Comparison methods 284
  ac instrument calibration 285
  dc ammeter calibration 285
  dc voltmeter calibration 284
  ohmmeter calibration 285
  wattmeter calibration 285
CRT 229
Current measurement with electronic instruments 100
Current probe 260

D

DAC accuracy and resolution 116
DC ammeter 63
  advantages 65
  ammeter shunts 64
  precautions 65
DC voltmeter 69
  chopper type 84
  direct coupled amplifier 82
  loading effect 72
De Sauty's bridge (or simple capacitance bridge) 180
Deflection plate assembly 226
Delayed time-base oscilloscopes 268
Determination of frequency 249
Determination of phase 252
Digital meters 111
Digital multimeters 135
Digital multimeters as standard instruments 287
Digital storage oscilloscope applications 274
Digital storage oscilloscopes 273
Digital voltmeters (DVMs) 126
Digital-to-analog (D/A) conversion 112
Digital-to-analog and analog-to-digital conversion 112
Dimensions 11
Dimensions of electrical and magnetic quantities 12
Direct coupled amplifier dc voltmeter 82
Direct probe 258
Dual slope A/D conversion 120

E

Electric current 6
Electron gun assembly 226
Electronic analog ohmmeter
  four-wire sensing 172
  two-wire sensing 172
Electrostatic deflection of moving electrons in CRO tubes 230
Electrostatic focusing 228
Emitter follower voltmeters 85
Emitter-follower type TVM 89
Errors common to all types of instruments
Errors in a Wheatstone bridge 162
Errors in PMMC instruments 55
Exponential sweep circuit 241

F

Fluke 104
Focus control 238
Focusing device 228
Frequency meter (counter) 141
Fundamental and derived quantities 3

G

Galvanometer 58
Gearwheel method 252
General purpose CRO 224
Glass body and base 228
Ground terminals and floating power supplies 90

H

Hay bridge 201
Heterodyne converter 143
High current probes 104
High voltage probe 103
High voltage Schering bridge 188
Horizontal deflection system 236
HP 17108A specifications 294
HP 7015B and 7035B specifications 295

I

Inductance bridges 194
  Anderson bridge 198
  Hay bridge 201
  inductance comparison or Maxwell's bridge 194
  Maxwell Wien bridge 195
  Own bridge 204
Instrument calibration 283
Intensity control 238
International and absolute units 17
Isolation probe 258

K

Kelvin double bridge method 169

L

Limitations of Wheatstone bridge 163
Limiting or guarantee errors 23
Lissajous pattern method 250
Low resistance measurement 168
  special features of 168
Low-resistance measuring instruments 172

M

Maxwell's bridge 194
Maxwell Wien bridge 195
Measurement error combination 32
Measurement errors 22
Measurement of
  high impedance components 209
  low impedance components 207
  low resistances;special features of 168
  relative permittivity with Schering bridge 189
  resistance by multimeter or VOM 102
  voltage and current 248
  voltage, frequency and phase 248
Measurement data specification 44
Merits and demerits of PMMC instruments 56
Multimeter probes 103
Multiple trace oscilloscope 244
Multi-range ammeters 66
Multi-range voltmeters 71

N

Nature of units 3
Normal law of error 41

O

Op-amp ac voltmeter circuits 97
Op-amp amplifier voltmeter 91
Op-amp voltage follower voltmeter 91
Op-amp voltage-to-current converter voltmeter circuit 92
Operation of Kelvin double bridge in practice 170
Operational amplifier voltmeter circuits 91
Oscilloscope probes 258

Oscilloscope specifications and performance 261
Oscilloscope time base 239
    exponential sweep circuit 241
Other unit systems 9
Own bridge 204

P

Parallel A/D conversion 122
Parallel resistance-capacitance bridge 183
Peak reading ac voltmeters 95
Permanent magnet moving coil (PMMC)
    instrument 50
    construction 51
    errors in PMMC instruments 55
    merits and demerits of 56
    principle of operation 50
    torque equation 54
Phosphor data chart 227
Plotters 296
PMMC as ohmmeter 73
Portable form of Wheatstone bridge test set-post office box 158
Position controls 238
Post deflection acceleration 233
Potentiometric strip chart recorder 291
Precision measurement of medium resistances with Wheatstone bridge 163
Prescaling 142
Probable error 43

Q

Q meter 206
    errors 210
    measurement of high impedance components 209
    measurement of low impedance components 207

R

Radio-frequency (RF) probe 106
Ramp A/D conversion 118
Range extension of frequency counter 142
Range selection 133
Recirculating remainder A/D conversion 122
Recorders 290
Relative error and percentage error 22
Resolution and sensitivity 26

S

Sampling oscilloscopes 271
Schering bridges 184
Scientific notation and metric prefixes 5
Screen for CRT 227
Sensitivity of Wheatstone bridge 159
Series ohmmeter 73
Series resistance-capacitance or modified De Sauty bridge 181
Seven segment displays 125
SI mechanical units 4
SI temperature scales 8
Significant figures 28
Slide-wire form of Wheatstone bridge 159
Special features of low resistance measurements 168
Special oscilloscopes 268
Specifications of oscilloscope (dual trace) 247
Spot wheel method of display 251
Standards 16
Standard of measurement 16
Statistical analysis 38
Successive-approximation A/D conversion 120

T

Time base specifications 295
Transistor voltmeter circuits 82
True rms reading voltmeters 96
Types of errors 29
    gross errors 29
    random (or accidental) errors 32
    systematic errors 30

V

Vertical deflection system 235
Voltage calibrator 289
Voltage-to-frequency A/D conversion 121
Voltmeter and ammeter methods 154
Voltmeter multipliers 69
Voltmeter sensitivity 72

W

Weighted-resistor DAC 113
Wheatstone bridge 163
    errors in a 162
    limitations of 163
    sensitivity of 159
Wheatstone bridge method 157
Wheatstone bridge test 158
Wien bridge 190

X

X-Y recorder 292

□□□